SEVENTH CANADIAN EDITION

ELECTRICAL WIRING:
COMMERCIAL

BASED ON THE 2015 CANADIAN ELECTRICAL CODE

RAY C. MULLIN • ROBERT J. FILICE • SAM MALTESE •
NICK PALAZZO • DENNIS J. MARCHETTI • PHIL SIMMONS

NELSON EDUCATION

Electrical Wiring: Commercial, Seventh Canadian Edition

by Ray C. Mullin, Robert J. Filice, Sam Maltese, Nick Palazzo, Dennis J. Marchetti, and Phil Simmons

Vice President, Editorial Higher Education:
Anne Williams

Publisher:
Jackie Wood

Marketing Manager:
Alexis Hood

Technical Reviewer:
Tony Poirier

Developmental Editor:
Katherine Goodes

Photo Researcher and Permissions Coordinator:
Julie Pratt

Production Project Manager:
Jennifer Hare

Production Service:
MPS Limited

Copy Editor:
Laurel Sparrow

Proofreader:
MPS Limited

Indexer:
Elizabeth Walker

Design Director:
Ken Phipps

Managing Designer:
Franca Amore

Cover Design:
Courtney Hellam

Cover and Chapter Opener Image:
idea for life/Shutterstock.com

Compositor:
MPS Limited

Library and Archives Canada Cataloguing in Publication Data

Mullin, Ray C., author
    Electrical wiring : commercial / Ray C. Mullin, Robert J. Filice, Sam Maltese, Nick Palazzo, Dennis J. Marchetti, Phil Simmons. — Seventh Canadian edition.

Includes index.
Revision of: Electrical wiring : commercial. 6th Canadian ed. Toronto : Nelson Education, [2012], ©2013
Based on the 2015 Canadian Electrical Code.
ISBN 978-0-17-657046-0 (paperback)

    1. Electric wiring, Interior.
2. Commercial buildings—Electric equipment. I. Title.

TK3284.M83 2015   621.319'24
C2015-903088-9

ISBN-13: 978-0-17-657046-0
ISBN-10: 0-17-657046-2

Brief Contents

Commercial Blueprints

Drawing A1	Site Plan; East Elevation; West Elevation; Index to Drawings
Drawing A2	Basement Floor Plan
Drawing A3	First Floor Plan
Drawing A4	Second Floor Plan
Drawing A5	North and South Elevations
Drawing A6	Transverse Building Section
Drawing A7	Longitudinal Building Section
Drawing E1	Site Plan; Legend; Index of Drawings
Drawing E2	Basement Electrical Floor Plan
Drawing E3	First Floor Electrical Floor Plan
Drawing E4	Second Floor Electrical Plan
Drawing E5	Power Riser Diagram
Drawing E6	Schedules
Drawing E7	Schedules and Details

Contents

UNIT 1

UNIT 2

UNIT 3

UNIT 4

UNIT 5

UNIT 6

UNIT 7

UNIT 8

UNIT 9

UNIT 10

UNIT 11

UNIT 12

UNIT 13

Commercial Blueprints

Introduction

The seventh Canadian edition of *Electrical Wiring: Commercial* is based on the 2015 edition of the *Canadian Electrical Code, Part I (CEC),* the safety standard for electrical installations. This Canadian edition thoroughly and clearly explains the *CEC* rules and changes that relate to commercial wiring.

Because *Electrical Wiring: Commercial* is both comprehensible and readable, it is suitable for colleges, technical institutes, and vocational technical schools. The *CEC* is the basic standard for the layout and construction of electrical systems in Canada; however, local and provincial codes may contain specific amendments that must be adhered to in electrical wiring installations within their jurisdictions.

To gain the greatest benefit from this text, the reader must refer to the *CEC* on a continuing basis. The authors encourage the reader to develop a detailed knowledge of the layout and content of the *CEC*.

This text takes the reader through the essential minimum requirements as set forth in the *CEC* for commercial installations, and often provides further information above and beyond the minimum *CEC* requirements.

The commercial electrician is required to work in three common situations: where the work is planned in advance; where there is no advance planning; and where repairs are needed. The first situation exists when the work is designed by a consulting engineer. In this case, the electrician must know the installation procedures, be able to read plans, and be able to understand and interpret specifications. The second situation occurs either during or after construction as a result of changes or remodelling. The third situation arises any time after a system is installed. Whenever a problem occurs with an installation, the electrician must understand the operation of all equipment included in the installation in order to solve the problem.

When the electrician is working on the initial installation or is modifying an existing installation, the circuit loads must be determined. In this book, thorough explanations and numerous examples of calculating these loads help prepare the reader for similar problems on the job. The text and assignments make frequent reference to the commercial building blueprint package included with the text.

Readers should be aware that many of the electrical loads and scenarios used in the building described in the text were contrived in order to create *CEC* problems. The authors' purpose is to demonstrate, and thus enhance, the learning of as many *CEC* problems as possible.

Preface

This text will provide a valuable resource to instructors and students alike. It includes all 2015 *Canadian Electrical Code, Part I (CEC)* references and wiring techniques.

Review questions are included at the end of each chapter to summarize the material being covered. The chapters are sequenced to introduce the student to the basic principles and wiring practices contained within a commercial building. Included in the text is information to introduce the reader to the Workplace Hazardous Materials Information System (WHMIS) and to fire safety in commercial applications.

The student requires a reasonable level of mechanical aptitude and skill in order to be successful in the practical application of the techniques discussed.

FEATURES

- New! The 2015 edition of the *Canadian Electrical Code, Part I (CEC),* the safety standard for electrical installation, is the basis for this text.

- New! We've included a full chapter on green technology and its use in the commercial sector.

- New! The four-colour design is fresh, engaging, and student-friendly.

- The references to the *CEC* are shown in *italics* to simplify cross-reference to the actual 2015 *CEC.*

- Easy-to-understand explanations set out the intent of the 2015 *CEC* sections and rules.

- A concise explanation of WHMIS legislation and valuable material relating to fire safety are provided.

- Straightforward, easy-to-follow calculations are illustrated with examples in the text.

- Clearly defined objectives are set out at the beginning of each unit, together with questions and problems dealing with unit objectives.

- Numerous diagrams and illustrations further emphasize unit material.

INSTRUCTOR RESOURCES

The **Nelson Education Teaching Advantage (NETA)** program delivers research-based instructor resources that promote student engagement and higher-order thinking to enable the success of Canadian students and educators. Be sure to visit Nelson Education's **Inspired Instruction** website at http://www.nelson.com/inspired/ to find out more about NETA.

Accessing Instructor Supplements from SSO Dashboard

1. Go to http://www.nelson.com/instructor and log in using the instructor e-mail address and password provided by your Nelson Learning Solutions Consultant.

2. Enter author, title, or ISBN in the **Add a title to your bookshelf** search.

3. Click **Add to my bookshelf** to add instructor resources.

4. At the Product page, click the **Instructor Companion site** link.

The following instructor resources have been created for *Electrical Wiring: Commercial*, Seventh Canadian Edition, and can be accessed from the SSO dashboard.

MindTap

Offering personalized paths of dynamic assignments and applications, **MindTap** is a digital learning solution that turns cookie-cutter into cutting-edge, apathy into engagement, and memorizers into higher-level thinkers. MindTap enables students to analyze and apply chapter concepts within relevant assignments, and allows instructors to measure skills and promote better outcomes with ease. A fully online learning solution, MindTap combines all student learning tools—readings, multimedia, activities, and assessments—into a single Learning Path that guides students through the curriculum. Instructors personalize the experience by customizing the presentation of these learning tools to their students, even seamlessly introducing their own content into the Learning Path.

NETA Test Bank

This resource was written by the text's co-author, Sam Maltese, Sheridan College. It includes over 515 multiple-choice questions written according to NETA guidelines for effective construction and development of higher-order questions. Also included are over 100 short-answer questions.

The NETA Test Bank is available in a new, cloud-based platform. **Nelson Testing Powered by Cognero®** is a secure online testing system that allows you to author, edit, and manage test bank content from any place you have Internet access. No special installations or downloads are needed, and the desktop-inspired interface, with its drop-down menus and familiar, intuitive tools, allows you to create and manage tests with ease. You can create multiple test versions in an instant, and import or export content into other systems. Tests can be delivered from your learning management system, your classroom, or wherever you want. Nelson Testing Powered by Cognero for *Electrical Wiring: Commercial*, Seventh Canadian Edition, can be accessed from the SSO front door.

NETA PowerPoint

Microsoft® PowerPoint® lecture slides for every chapter have been created by Sam Maltese, Sheridan College. There is an average of 28 slides per chapter, many featuring key figures, tables, and photographs from *Electrical Wiring: Commercial*, Seventh Canadian Edition. NETA principles of clear design and engaging content have been incorporated throughout, making it simple for instructors to customize the deck for their courses.

Instructor's Solutions Manual

The Instructor's Solutions Manual to accompany *Electrical Wiring: Commercial*, Seventh Canadian Edition, has been prepared by Sam Maltese, Sheridan College. To assist the instructor, questions from the end of each unit have been included with answers/solutions. These may be used as additional review or examination questions.

Image Library

This resource consists of digital copies of figures, short tables, and photographs used in the book. Instructors may use these jpegs to customize the NETA PowerPoint or create their own PowerPoint presentations.

Blueprints

Revised and updated by Nick Palazzo, Humber College, and technically checked by all text authors, a set of blueprints accompanies this textbook.

Day One Slides

Day One—Prof InClass is a PowerPoint presentation that instructors can customize to orient students to the class and their text at the beginning of the course.

MindTap

Stay organized and efficient with **MindTap**—a single destination with all the course material and study aids you need to succeed. Built-in apps leverage social media and the latest learning technology. For example:

- ReadSpeaker will read the text to you.
- Flashcards are pre-populated to provide you with a jump start for review—or you can create your own.
- You can highlight text and make notes in your MindTap Reader. Your notes will flow into Evernote, the electronic notebook app that you can access anywhere when it's time to study for the exam.
- Self-quizzing allows you to assess your understanding.

Visit http://www.nelson.com/student to start using **MindTap**. Enter the Online Access Code from the card included with your text. If a code card is *not* provided, you can purchase instant access at NELSONbrain.com.

Blueprints

Revised and updated by Nick Palazzo, Humber College, and technically checked by the text authors, a set of blueprints accompanies this textbook.

Acknowledgments

Every effort has been made to be technically correct and avoid typographical errors. The authors wish to acknowledge the valuable assistance of Bill Wright and Jeremy Groves in the preparation of the original architectural drawings for this book. Thank you to Tony Poirier, Durham College, for a thorough technical check of the text for adherence to the *CEC* 2015. Thank you as well to the people from Nelson Education Ltd. for their encouragement and professional advice.

The authors are also indebted to reviewers of each text in the *Electrical Wiring* series for their comments and suggestions:

Sam Agnew, Centennial College

Neil Blackadar, Nova Scotia Community College

Rob Brown, Fanshawe College

Michael Gillis, Nova Scotia Community College

Bruce Kellington, University College of the North

George Locke, Mohawk College

Shawn MacCaulay, Nova Scotia Community College

Ken Malczewski, Red River College

Mike Mankulich, Fanshawe College

Keith Mercer, Centennial College,

Tony Poirier, Durham College

Orest Staneckyj, Cambrian College

Violin Voynov, Durham College

We wish to thank the following companies for their contributions of data, illustrations, and technical information:

AFC/A Nortek Co.

Anchor Electric Division, Sola Basic Industries

Appleton Electric Co.

Arrow-Hart, Inc.

Brazos Technologies, Inc.

BRK Electronics, a Division of Pittway Corporation

Bussman Division, Cooper Industries

Carlon

Chromalox

Commander

Cummins Inc.

Eaton Corporation

Electri-Flex Co.

Erikson Electrical Equipment

Ferraz Shawmut

General Electric Co.

Halo Lighting Division, Cooper Industries

Heyco Molded Products, Inc.

Honeywell

Hubbell Incorporated, Wiring Devices Division

International Association of Electrical Inspectors

IPEX Juno
KEPTEL
The Kohler Co.
Leviton Manufacturing Co., Inc.
Lightolier Canada
Midwest Electric Products, Inc.
Milbank Manufacturing Co.
Moe Light Division, Thomas Industries
Noma Division, Danbel Industries Inc.
Nutone Inc.
Pass & Seymour, Inc.
Progress Lighting
Rheem Manufacturing Co.
Schneider Electric
Seatek Co., Inc.
Sierra Electric Division, Sola Basic Industries

SMART HOUSE, L.P.
Superior Electric Co.
Tekno Stamap
THERM-O-DISC
Thomas & Betts Corporation
Underwriters Laboratories of Canada (ULC
    is accredited by the Standards Council of
    Canada as a Certification Organization,
    a Testing Organization, a Registration
    Organization, and Standards Organization
    under the National Standards System
    of Canada.)
Winegard Company
Wiremold Co.
Wolberg Electrical Supply Co., Inc.
Woodhead Industries, Inc.

With the permission of the Canadian Standards Association (operating as CSA Group), material is reproduced from CSA Group code, C22.1-15 entitled *Canadian Electrical Code, Part I* (23rd edition), Safety Standard for Electrical Installations which is copyrighted by CSA Group, 178 Rexdale Blvd., Toronto, ON, M9W 1R3. This material is not the complete and official position of CSA Group on the referenced subject, which is represented solely by the code in its entirety. While use of the material has been authorized, CSA Group is not responsible for the manner in which the data is presented, nor for any interpretations thereof. For more information or to purchase standards from CSA Group, please visit http://shop.csa.ca/ or call 1-800-463-6727.

Logo on page 20 used courtesy of Canadian Standards Association
Logo on page 20 used courtesy of Standards Council of Canada
Logo on page 21 used courtesy of Underwriters Laboratories of Canada

For any further information or to give feedback on this text, you can contact the authors by email at

rob.filice@humber.ca
sam.maltese@sheridancollege.ca
nick.palazzo@humber.ca

Commercial Building Plans and Specifications

OBJECTIVES

After completing the study of this unit, the student will be able to

- define the job requirements from the contract documents
- explain the reasons for building plans and specifications
- locate specific information on the building plans
- obtain information from industry-related organizations

CONTRACTS

A contract is an agreement, enforceable by law, to supply goods or perform work at a stated price. A construction contract differs from most contracts in that the owner has the right to make changes to the contract as the project progresses. A construction contract is made up of working drawings and specifications, often referred to as the contract documents. A bound volume of specifications and a set of working drawings are provided to all contractors who undertake to bid a job.

SPECIFICATIONS

Specifications are written descriptions of materials, construction systems, and workmanship, and are generally laid out in the following manner:

- Notice to bidders
- Instruction to bidders
- Proposal (bid) form
- Owner–contractor agreement form
- Schedule of drawings
- General conditions
- Supplemental general conditions
- Alternates
- Technical sections

Notice to Bidders

This is the advertisement of the project. An electrical contractor may be a subcontractor (who signs a contract with the general contractor) or may be a prime contractor (who signs a contract with the owner or owner's representative, usually an architect or engineer).

The notice to bidders identifies the project, the project location, the closing date for bids, where the bidding documents may be obtained, and the amount of deposit required for the bidding documents.

Instruction to Bidders

The instruction to bidders provides a brief description of the project and its location. It includes information on the type of payment, with the most common types being lump sum, unit price, or cost plus fee. A lump sum agreement stipulates that the contractor will satisfactorily complete the job for a fixed sum, regardless of any difficulties the contractor may encounter. If the unit price agreement is used, all bidders base their bids on a unit price basis (per outlet, per pole, per luminaire). When the unit price method of payment is used, the actual quantities and classifications are determined by an engineer or quantity surveyor. Cost plus fee agreements are used where the job is extremely complex or construction must begin before complete drawings and specifications have been prepared.

In addition, bidders are told where and how the plans and specifications can be obtained prior to the preparation of the bid, how to make out the proposal form, where and when to deliver the proposal, the amount of any bid deposits required, and any performance bonds required.

The instruction to bidders normally requires that the bidder visit the site and stipulate the time and date of site visits.

The instruction to bidders often requires that the bidder provide the owner with information showing that the contractor has the ability to complete the contract. The information requested may include evidence that the bidder is licensed to do business in the jurisdiction where the project will be constructed, a list of work to be subcontracted, the names of the subcontractors, and what work the contractor currently has in progress.

Bonds

A surety bond is a guarantee made by the bonding company to meet the obligations of the contractor. If the contractor is unable to fulfill the terms of the contract, the bonding company must complete the contract (up to the amount of the bond) and then try to recover its money from the contractor. Owners require contractors to be bonded to ensure that they will complete the project and pay their bills. Common forms of bonds are bid bonds, performance bonds, and labour and material bonds.

Bid bonds ensure that, if a contractor is awarded a contract, the contractor will proceed with the contract within a stipulated time period. If the contractor does not accept the contract, the bond is forfeited to the owner. Bid bonds are normally 10% of the bidding price.

Performance bonds guarantee an owner that the contractor will complete the contract, and are normally 100% of the contract price. The premiums charged by the bonding company will vary according to the experience and financial stability of the contractor.

Labour and material bonds guarantee the payment of the contractor's bills to third parties that supply material or labour to the project.

Addenda

An addendum is a written addition to the contract documents issued before the bidding is closed and is used to:

- Clarify questions raised by bidders regarding conflicts, errors, omissions, and ambiguities
- Increase or decrease the scope of the work
- Provide additional information to bidders in the way of explanations
- Change the provisions of the contract

Proposals

A (bid) proposal form is included in a set of specifications. This allows the owner to evaluate all the bids on the same basis. A typical proposal form would include the following:

- Name of tendering company
- Name of owner
- Lump sum tender price
- Breakdown of lump sum price
- Schedule of unit prices
- Alternates
- Statutory declaration
- Agreement to bond
- Statement of experience
- Statement of senior supervisory staff

The proposal is a form that is filled out by the contractor and submitted at the proper time and place. The proposal is the contractor's bid on a project. The form is the legal instrument that binds the contractor to the owner provided that the contractor completes the proposal properly, the contractor does not forfeit the bid bond, and the owner accepts the proposal and signs the agreement. Unless the proposal is specifically marked as irrevocable, it may be revoked by the contractor without penalty at any time until the agreement has been signed by the owner.

If the owner fails to pay the contractor within the terms of the agreement, the contractor has a right to charge this as a lien against the property being upgraded. Under this "mechanic's lien," the owner may not deal with the property (e.g., increase the mortgage or sell it) until the lien has been satisfactorily discharged or the amount in dispute has been paid into court pending settlement.

The proposal may show that alternate bids were requested by the owner. In this case, the electrician on the job should study the proposal and consult with the contractor to learn which of the alternate bids has been accepted in order to determine the extent of the work to be completed.

On occasion, the proposal may include a specified time for the completion of the project. This information is important to the electrician on the job since the work must be scheduled to meet the completion date.

The Owner–Contractor Agreement. If the proposal is acceptable to the owner, then a legally binding agreement must be executed between the parties. This agreement may be included as part of the proposal document. The contractor and the owner sign the agreement and the result is a legal contract that, once signed, binds both parties by the terms and conditions given in the specification.

General Conditions. General conditions outline the parameters that apply to all projects. The following items are normally included under the General Conditions heading of the *General Clauses and Conditions* section.

- *General Note* specifies that the general conditions are part of the contract documents.
- *Definition*, as used in the contract documents, defines the owner, contractor, architect,

engineer, and other people and objects involved in the project.

- *Contract Documents* lists the documents involved in the contract, including plans, specifications, and agreement.

- *Insurance* specifies the insurance a contractor must carry on all employees and on the materials involved in the project.

- *Workmanship and Materials* specifies that the work must be done by skilled workers and that the materials must be new and of good quality.

- *Substitutions* stipulates that materials used must be as specified or equivalent materials must be shown to have the required properties.

- *Shop Drawings* identifies the drawings that must be submitted by the contractor to show how the specific pieces of equipment are to be installed.

- *Payments* specifies the method of paying the contractor during the construction.

- *Coordination of Work* specifies that each contractor on the job must cooperate with every other contractor to ensure that the final product is complete and functional.

- *Correction of Work* describes how work must be corrected, at no cost to the owner, if any part of the job is installed improperly by the contractor.

- *Guarantee* is the contractor's guarantee on the work for a certain length of time, usually one year.

- *Compliance With All Laws and Regulations* specifies that the contractor will perform all work in accordance with all required laws, ordinances, and codes, such as the *CEC* and local codes.

- *Others:* These sections are added as necessary by the owner, architect, and engineer when the complexity of the job and other circumstances require them. None of the items listed in the General Conditions has precedence over another item in terms of its effect on the contractor or the electrician on the job. The electrician must study each of the items before

taking a position and assuming responsibilities with respect to the job.

SUPPLEMENTARY GENERAL CONDITIONS

Supplementary General Conditions are more specific than General Conditions. While the General Conditions can be applied to any job or project in almost any location with little change, the Supplementary General Conditions are rewritten for each project. The following list covers the items normally specified by the Supplementary General Conditions.

- The contractor must instruct all crews to exercise caution while digging; any utilities damaged during the digging must be replaced by the contractor responsible.

- The contractor must verify the existing conditions and measurements.

- The contractor must employ qualified individuals to lay out the worksite accurately. A registered land surveyor or engineer may be part of the crew responsible for the layout work.

- Job offices are to be maintained as specified on the site by the contractor, and may include an area for the owner's representatives.

- The contractor may be required to provide telephones at the project site for use by the architect, engineer, subcontractor, or owner.

- Temporary toilet facilities and water are to be provided by the contractor for the construction personnel.

- The contractor must supply an electrical service of a specified size to provide temporary light and power at the site.

- It may be necessary for the contractor to supply a specified type of temporary heating to keep the temperature at the level specified for the structure.

- According to the terms of the guarantee, the contractor agrees to replace faulty equipment and correct construction errors for a period of one year.

The above list is by no means a complete catalogue of all of the items that can be included in Supplementary General Conditions.

Other titles may be applied to the Supplementary General Conditions section, including Special Conditions and Special Requirements. Regardless of the title used, these sections contain the same types of information. All sections of the specifications must be read and studied by all of the construction trades involved. In other words, the electrician must study the heating, plumbing, ventilating, air conditioning, and general construction specifications to determine if there is any equipment furnished by the other trades that the contract specifies is to be installed and wired by the electrical contractor. The electrician must also study the general construction specifications, since the roughing in of the electrical system will depend on the types of construction that will be encountered in the building.

This overview of the General Conditions and Supplementary General Conditions of a specification is intended to show the student that the construction worker on the job is affected by parts of the specifications other than the part designated for each particular trade.

Alternates

An alternate is a request by the owner for a price on alternate materials or methods of construction. Alternates are listed on the proposal form. The alternates may be "add price" or "subtract price," meaning the alternate will add to your base bid or subtract from it.

Technical Sections

Specifications are divided into a number of divisions that each cover some portion of the work. Technical sections contain descriptions of materials and instructions on methods of construction. Table 1-1 describes divisions commonly found in specifications.

Schedule of Drawings

This schedule is a list, by number and title, of all of the drawings related to the project. The contractor,

the estimator, and the electrician will each use this schedule prior to preparing the bid for the job. The contractor will determine if all the drawings required are at hand, the estimator will do a take-off from the drawings and formulate a bid, and the electrician will determine if all of the drawings necessary to do the installation are available.

The Drawing Set

A set of drawings is made up of numbered sheets. If there is a large quantity of sheets in a set, the set will be broken down into subsets of architectural, electrical, mechanical, and the like. The electrical drawings will be numbered E1, E2, E3, etc.

A typical set of construction drawings will include a site (plot) plan, floor plans, elevations, details and sections, legend of symbols, schedules, diagrams, and shop drawings.

Site (Plot) Plan.   The site plan will show:
- The dimensions of the lot
- A north-facing arrow
- The location of utilities and their connection points
- Any outdoor lighting
- Easements (rights to the property by someone other than the owner)
- Contour lines to show changes in elevation
- Accessory buildings
- Trees to be saved

Floor Plans.   Depending on the complexity of the building, a set of drawings may have only a single floor plan for each storey of the building. As the complexity of the building increases, there will be a number of floor plans for each storey, with one for each trade. A logical breakdown would be:
- Architectural
- Electrical
- Mechanical
- Plumbing

TABLE 1-1

Specification Divisions.

DIVISION	TITLE	INFORMATION RELATING TO THE ELECTRICAL TRADE
1	General Requirements	
2	Existing Conditions	
3	Concrete	
4	Masonry	
5	Metals	
6	Wood, Plastics, and Composites	
7	Thermal and Moisture Protection	Thermal Protection
		Fire and Smoke Protection
8	Openings	Type of Construction
9	Finishes	Acoustical Ceiling Suspension Assemblies
10	Specialties	Telephone Enclosures
		Signage
		Fire Protection
11	Equipment	Retail and Service Equipment
		Healthcare Equipment
		Office Equipment
		Foodservice Equipment
12	Furnishings	Portable Lamps
13	Special Construction	Controlled Environment Rooms
14	Conveying Equipment	
22	Plumbing	Facility Storm Drainage
		Plumbing Equipment
23	Heating, Ventilating, and Air Conditioning (HVAC)	Instrumentation and Control for HVAC
		Refrigerant Compressors
		Fuel-Fired Heaters
26	Electrical	Common Work Results for Electrical
		Schedules for Electrical
		Instrumentation and Control for Electrical Systems
		Low-Voltage Electrical Transmission
		Lighting
27	Communications	Communications
48	Electrical Power Generation	Fossil Fuel Plant Electrical Power Generation Equipment

- HVAC (Heating, Ventilating, and Air Conditioning)

- Structural

On very complex installations, different electrical systems such as power, lighting, voice/data, fire alarm, and security/access control may each have their own plan for each storey.

Use architectural floor plans to find room measurements, wall thickness, the sizes of doors and windows, and the swing of doors (see Figure 1-1). The floor plans of other trades will show the locations of the equipment associated with that trade.

Elevations.   Elevations show the vertical faces of buildings, structures, and equipment. Elevations

FIGURE 1-1 Partial architectural floor plan.

are used to show both interior and exterior vertical faces. Figure 1-2 shows the east exterior elevation of the building.

Details and Sections. Details are large-scale drawings that are used when sufficient detail cannot be provided on floor plans and elevations. If you look at Drawing A2 of the plan sheets that accompany the text, for example, find the symbols that look like a circle with the letters B-B in the top half of the circle and A2/A6 in the bottom half. There is also an arrow outside the circle pointing toward the west side of the building. Note that the two symbols (one at the top of the drawing and one at the bottom) are connected by a solid line. The

B-B is letting you know that this is Section B-B. The A2 notes that the section you are looking at is on Architectural Drawing A2. The A6 is telling you to look on Drawing A6 for the actual cutaway view as indicated by the arrow. The line connecting the two symbols is where the building would be split, and you are looking in the direction of the arrows, which in this case is the west side of the building. Sheet A7 shows details of structural elements of the building, such as the parapet in Figure 1-3.

Legend of Symbols. A legend of symbols provides the contractor or electrician with the information to interpret the symbols shown on the drawing.

EAST ELEVATION
SCALE = 1:100

FIGURE 1-2 East exterior elevation.

Refer to Figures 1-8 to 1-11 for examples of the various symbols normally found on construction drawings.

Schedules. Schedules are lists of equipment. Electrical schedules are made for equipment such as luminaires, panelboards, raceways and cables, motors, and mechanical equipment connections. Refer to Figure 1-4.

Diagrams. One-line and riser diagrams are provided to quickly show the flow of power throughout the building, as in Figure 1-5. Riser diagrams may be used to show power, telephone, or fire alarm systems.

Shop Drawings. Shop drawings are used by equipment manufacturers to show details of custom-made equipment for a specific job. Manufacturer's drawings of switchgear and MCCs are examples of shop drawings.

The Electrical Drawing Set

A set of electrical drawings will normally include:

- Legend of symbols
- Site plan
- One-line diagram
- Lighting layout
- Power layout
- Electrical details
- Schematic and wiring diagrams
- Schedules

Scale

A drawing of an object that is the same size as the object is said to be drawn *full scale* or at a scale of 1:1.

Construction drawings reduce the size of a building down to fit on a piece of paper by drawing

PRE-FABRICATED
CANAMOLD PARAPET

DRIP EDGE

DRIP EDGE
25 RIGID INSULATION
13 GYPSUM BOARD
PRE-FABRICATED 2 PIECE METAL FLASHING
FIBRE CANT

STEEL STUD TRACKING

140 CORE FILLED C.M.U. TYP.

EXTERIOR NON-LOAD BEARING WALL

64 STI. STUD C/C BATT INSUL.
140 C.M.U.
50 RIGID INSU.
25 AIR SPACE
100 BRICK

PROTECTED MEMBRANE ROOF
75 BALLAST
FILTER CLOTH
75 RIGID INSULATION
3 PLY ROOF MEMBRANE
SLOPED LW. CONC. TOPPING
64 CONC. ROOF SLAB W/ W.W.M.
355 HAMBRO JOISTS @ C1200 O.C.
C/W 38 METAL DECKING
SUSPENDED CEILING

D3 / A7 | A7 / **NON-LOAD BEARING WALL @ PARAPET DETAIL**
SCALE-1:10

FIGURE 1–3 Detail of parapet.

everything proportionally smaller than it actually is. This is called *reduced scale.* A drawing that uses a scale of 1:50 indicates that the building is 50 times larger than the drawing. A drawing with a scale of $\frac{1}{4}'' = 1$ ft has a scale of 1:48.

Enlarged scale would produce a drawing larger than the actual object. Printed circuit boards are drawn using enlarged scale since many of the components used in the manufacture of printed circuit boards are so small. An example of enlarged scale is 2:1, which indicates that the drawing of the component is twice as large as the component.

Types of Scales (Measuring Instruments)

Three types of scales are commonly used on construction drawings. When the drawing is in feet and inches, an architectural scale is used. Figure 1-6 shows a portion of an architect's triangular scale showing the $\frac{1}{8}''$ and $\frac{1}{4}''$ scales. For metric drawings, a metric scale is used (see Figure 1-7). When working with civil drawings (roads, dams, and bridges, etc.), an engineering scale will be used. Table 1-2 lists common construction drawing scales.

LUMINAIRE SCHEDULE							
STYLE	NOMINAL SIZE	LAMP	BALLAST	LENS-LOUVRE	MOUNTING	DESCRIPTION	VA	WATTS
A	450 mm × 1200 mm	Two F40/30U	Energy saving	Wraparound	Surface		87	75
B	600 mm × 600 mm	Two F40/30U	Energy saving	Flat opal	Surface		87	75
C	225 mm × 1275 mm	Two F40/30U	Energy saving	Clear acrylic	Surface	Enclosed gasketed	87	75
D	1200 mm strip	One F48CW	Std. Magnetic	Luminous ceiling	Surface		74	32
E	225 mm dia.	Two 26 W Quad T4	Electronic	Gold alzak reflector	Recessed		144	74
F	600 mm × 1200 mm	Four F032/35K	Electronic	24 cell lens-louvre	Recessed		132	106
G	300 mm × 2400 mm	Four F032/35K	Electronic	24 cell lens-louvre	Recessed		132	106
H	225 mm × 2400 mm	Two F40CWX	Energy saving	Translucent acrylic	Surface		87	75
I	300 mm × 1200 mm	Two F40CWX	Energy saving	Prismatic lens	Surface		87	75
J	350 mm dia.	One 150 W, A221	NA	Fresnel lens	Recessed	IC rated	150	150
K	400 mm square	One 70 W HPS	Std. Magnetic	Vandal resistant	Surface	With photo control	192	82
L	325 mm × 1200 mm	Two F40CWX	Energy saving	None	Surface or hung		87	75
M	175 mm dia	12 V 50 W NFL	Transformer	Coilex baffle	Recessed	Adjustable spot	50	50
N	500 mm long	One 60 W, A19	NA	None	Surface	Exposed lamp	60	60
O	1200 mm × 500 mm	Three F40CWX	Two ballasts	Low-brightness lens	Recessed		143	129
P	600 mm × 500 mm	Two F40CWX	One ballast	Low-brightness lens	Recessed		87	75
Q	1200 mm × 500 mm	Three F40CWX	Two ballasts	Small cell parabolic	Recessed		143	129

FIGURE 1-4 Luminaire schedule.

Working with the Drawings

In order to work successfully with the drawings, use the following thorough and systematic procedure.

- Check that the drawing set is complete.
- Review the floor plans and elevations to get a mental picture of the size and shape of the building.

FIGURE 1-5 Partial power riser diagram.

TABLE 1-2		
Common Construction Drawing Scales.		
METRIC	ARCHITECTURAL	ENGINEERING
1:20	$\frac{1}{2} = 1'0''$	1:1
1:50	$\frac{1}{4} = 1'0''$	1:10
1:100	$\frac{1}{8} = 1'0''$	1:50
1:200	$\frac{1}{16} = 1'0''$	1:100
1:500	$\frac{1}{32} = 1'0''$	1:1000

- Orient the building to the site using the plot plan. Be sure you know which sides are north, south, east, and west. Add this information to your floor plans.
- Check the scale of all drawings.
- Identify the type of construction (combustible, noncombustible), materials, and components shown in the drawings. Nonstandard items should be shown in a legend of symbols.
- Read all notes on the drawings carefully.

FIGURE 1-6   Architect's scale.

FIGURE 1-7   Metric scale.

- Relate details to larger views. The examples below show how details are identified.
- Note multiple or identical drawings. (*Typical* means uniform throughout the building.)

THE COMMERCIAL BUILDING PLANS

A blueprint package containing 14 drawings is included with the text and shows the general and electrical portions of the work specified.

- *Drawing A1—Site Plan, East Elevation, West Elevation, Index to Drawings:* The plot plan shows the location of the commercial building and gives needed elevations. The East Elevation is the street view of the building and the West Elevation is the back of the building. The index lists the content of all of the plan sheets.
- *Drawing A2—Architectural Floor Plan: Basement.*
- *Drawing A3—Architectural Floor Plan: First Floor.*
- *Drawing A4—Architectural Floor Plan: Second Floor.*

The architectural floor plans give the wall and partition details for the building. These drawings are dimensioned, so the electrician can find exact locations by referring to them. The electrician should also check the blueprints for the material used in the general construction, since this will affect when and how the system will be installed.

- *Drawing A5—Elevations; North and South:* The electrician must study the elevation dimensions, which are given in metres above sea level. For example, the finished second floor is shown at 66.547 m above sea level.
- *Drawings A6 and A7—Sections; Transverse, Longitudinal:* These drawings give details of the more important sections of the building. The locations of the sections are indicated on the floor plans. When looking at a section, imagine the building has been cut in two at the place indicated, and you are looking at it in the direction of the arrow. You should see the section exactly as you view the imaginary building from this point.

- *Drawing E2—Basement Electrical Plan.*
- *Drawing E3—First Floor Electrical Plan.*
- *Drawing E4—Second Floor Electrical Plan.*
- *Drawings E5, E6, and E7—Riser Diagrams, Schedules, and Details.*

Drawings E2 to E4 show the detailed electrical work on an outline of the building. Since dimensions usually are not shown on the electrical plans, the electrician must consult the other drawings for this information. It is recommended that the electrician refer frequently to the other blueprints to ensure that the electrical installation does not conflict with the work of the other construction trades.

- *Drawing E1—Site Plan, Legend, Index of Drawings.*

To assist the electrician in recognizing components used by other construction trades, the following illustrations are included: Figure 1-8, Architectural drafting symbols; Figure 1-9, Standard symbols for plumbing, piping, and valves; Figure 1-10, Sheet metal ductwork symbols; and Figure 1-11, Generic symbols for electrical plans. However, **the electrician should be aware that variations of these symbols may be used and the specifications and/or plans for a specific project must always be consulted**.

ESTIMATING

Estimating is the process of determining the cost of a project by counting, measuring, and listing items of material; assigning labour units to the material based on the conditions of the installation; and determining direct job expenses, overhead, and return on investment.

The main steps in estimating are:

- Preparing a material take-off
- Tabulating and pricing
- Summarizing

Preparing a Material Take-Off

A material take-off involves counting, measuring, and listing the quantities of materials and equipment

ELEVATION ⬇	PLAN ⬇	SECTION ⬇
Brick — With note telling kind of brick (Common, face, etc.) — Brick	Common / Face / Firebrick	Same as plan views
Concrete block	or	
Stone — Cut stone / Rubble	Cut stone / Rubble / Cast stone (Concrete)	Cut stone / Cast stone (Concrete) / Rubble or cut stone
Glass — or		Small scale / Large scale
Plaster	Stud, lath, and plaster / Solid plaster wall	
Wood — Siding / Panel	Floor areas are left blank; note indicates kind of wood used.	Ends of boards except trim / Trim
Insulation	Loose fill or batts / Board and quilt / Solid and cork	Same as plan views
Sheet metal flashing	Occasionally indicated by note	
Metals other than flashing — Indicated by note or drawn to scale	Same as elevation	Small scale / Steel / Cast iron / Aluminum / Bronze or brass / Large scale
Structural steel — Indicated by note or drawn to scale	or / or	• or —— ⊘ ------ Reinforcing bars / L I ⌐ I L-angles, I-beams, etc. / Small scale Large scale

FIGURE 1-8 Architectural drafting symbols.

© Cengage Learning®

ELEVATION VIEWS FROM OUTSIDE

Drip cap Meeting Rail Muntin Points to side of sash with hinges

Stile

Sill Outside casing

Mullion

OPENINGS IN A FRAME WALL

Mullion

Double hung window   Outswinging casement window   Inswinging casement window   Awning window   Door

ELEVATION VIEWS FROM OUTSIDE

OPENINGS IN 8 in. (203 mm) BRICK WALL

Double hung window   Outswinging casement window   Horizontal sliding window   Jalousie window   Door

Wall not plastered   Plaster line

OPENINGS IN BRICK VENEER WALL ELEVATIONS SIMILAR TO BRICK WALL

Double hung window   Inswinging casement window   DH window   Fixed glass   DH window   Door

ELEVATIONS SIMILAR TO BRICK WALL

Double hung window   Awning window   Door

Double hung window   Awning window   Door

OPENINGS IN A BRICK CAVITY WALL

OPENINGS IN AN SCR BRICK WALL

OPENINGS IN INTERIOR PARTITIONS

Interior door   Double-action door   Folding door (accordion)   Sliding door   Plastered arch

B

FIGURE 1-8 Architectural drafting symbols.

PIPING

Piping, in general	———————
(Lettered with name of material conveyed)	
Non-intersecting pipes	
Steam	
Condensate	
Cold water	
Hot water	
Air	
Vacuum	
Gas	
Refrigerant	
Oil	

PIPE FITTINGS

For welded or soldered fittings, use joint indication shown below	Screwed	Bell and spigot
Joint		
Elbow—90°		
Elbow—45°		
Elbow—turned up		
Elbow—turned down		
Elbow—long radius		
Side outlet elbow—outlet down		
Side outlet elbow—outlet up		
Base elbow		
Double-branch elbow		
Single-sweep tee		
Double-sweep tee		
Reducing elbow		
Tee		
Tee—outlet up		
Tee—outlet down		
Side outlet tee outlet up		
Side outlet tee outlet down		
Cross		
Reducer		
Eccentric reducer		

PIPE FITTINGS (continued)

For welded or soldered fittings, use joint indication shown below	Screwed	Bell and spigot
Lateral		
Expansion joint flanged		

VALVES

For welded or soldered fittings, use joint indication shown below	Screwed	Bell and spigot
Gate valve		
Globe valve		
Angle globe valve		
Angle gate valve		
Check valve		
Angle check valve		
Stop cock		
Safety valve		
Quick opening valve		
Float opening valve		
Motor operated gate valve		

PLUMBING

Corner bath	
Recessed bath	
Roll rim bath	
Sitz bath	SS
Foot bath	FB
Bidet	B
Shower stall	
Shower head	(Plan) (Elev)
Overhead gang shower	(Plan) (Elev)
Pedestal lavatory	PL
Wall lavatory	WL
Corner lavatory	Lav
Manicure lavatory Medical lavatory	ML
Dental lavatory	Dental lav

PLUMBING (continued)

Plain kitchen sink	S
Kitchen sink, R & L drain board	
Kitchen sink, L H drain board	
Combination sink & dishwasher	
Combination sink & laundry tray	S&T
Service sink	SS
Wash sink (Wall type)	
Wash sink	
Laundry tray	LT
Water closet (Low tank)	
Water closet (No tank)	
Urinal (Pedestal type)	
Urinal (Wall type)	
Urinal (Corner type)	
Urinal (Stall type)	
Urinal (Trough type)	TU
Drinking fountain (Pedestal type)	DF
Drinking fountain (Wall type)	DF
Drinking fountain (Trough type)	DF
Hot water tank	HWT
Water heater	WH
Meter	M
Hose rack	HR
Hose bibb	HB
Gas outlet	G
Vacuum outlet	
Drain	D
Grease separator	G
Oil separator	
Cleanout	
Garage drain	
Floor drain with backwater valve	
Roof sump	

Types of joints

Flanged	Screwed	Bell & spigot	Welded	Soldered

FIGURE 1-9 Standard symbols for plumbing, piping, and valves.

© Cengage Learning®

Blank Off. adjustable	Damper automatic	Damper deflecting	Damper deflecting up
Damper deflecting down	Damper volume PLAN ELEVATION	Duct flow direction	Duct inclined drop D
Duct inclined rise R	Duct section exhaust, return riser to 2nd floor riser to lst floor	Duct section supply riser to 2nd floor riser to lst floor	Duct section notation type exhaust K E place kitchen
Duct connection below joist	Fan flexible connection	Vanes	Louver & screen air intake
Ventilator, cowl PLAN ELEVATION	Ventilator, gooseneck PLAN ELEVATION	Ventilator, rainproof PLAN ELEVATION	Ventilator, standard PLAN ELEVATION

SINGLE LINE REPRESENTATION

| Supply
 S | Return
 R | Damper & retractor | Anchor
 PA. |
| Hanger
 H | Expansion joint | Louver opening
 L | Register or grille |

FIGURE 1-10 Sheet metal ductwork symbols.

GENERAL OUTLETS			PANELS, CIRCUITS, AND MISCELLANEOUS	
PREFERRED SYMBOL		**DESCRIPTION**	**PREFERRED SYMBOL**	**DESCRIPTION**

Ceiling wall

	Description
	Outlet
	Blanked outlet
	Drop cord
	Electrical outlet: for use only when circle used alone might be confused with columns, plumbing symbols, etc.
	Fan outlet
	Junction box
	Lampholder
	Lampholder with pull switch
	Pull switch
	Outlet for vapour discharge lamp
	Exit light outlet
	Clock outlet (specify voltage)

Panels, circuits, and miscellaneous descriptions:

- Lighting panel
- Power panel
- Branch circuit; concealed in ceiling or wall
- Branch circuit; concealed in floor
- Branch circuit; exposed
- Home run to panelboard; indicate number of circuits by numbers of arrows
- *Note:* Any circuit without further designation indicates a two-wire circuit. For a greater number of wires, indicate as follows: (3 wires), (4 wires), etc.
- Feeders. *Note:* Use heavy lines and designate by number corresponding to listing in feeder schedule.
- Underfloor duct and junction box. Triple system. *Note:* For single system use one line; for double systems use two lines. This symbol is equally adaptable to auxiliary system layouts.
- Generator
- Motor
- Instrument
- Power transformer (or draw to scale)
- Controller
- Isolating switch
- Overcurrent device (fuse, breaker, thermal overload)
- Switch and fuse

RECEPTACLE OUTLETS

Description
Single receptacle outlet
Duplex receptacle outlet
Triple receptacle outlet
Duplex receptacle outlet, split circuit
Duplex receptacle outlet with GFCI
Weatherproof receptacle outlet
Range receptacle outlet
Switch and receptacle outlet
Radio and receptacle outlet
Special purpose receptacle outlet
Floor receptacle outlet

SWITCH OUTLETS

Symbol	Description
S	Single-pole switch
S2	Double-pole switch
S3	Three-way switch
S4	Four-way switch
SD	Automatic door switch
SE	Electrolier switch
SK	Key operated switch
SP	Switch and pilot lamp
SCB	Circuit breaker
SWCB	Weatherproof circuit breaker
SMC	Momentary control switch
SRC	Remote control switch
SWP	Weatherproof switch
SF	Fused switch
SWF	Weatherproof fused switch

AUXILIARY SYSTEMS

Description
Push button
Buzzer
Bell
Annunciator
Outside telephone
Interconnecting telephone
Telephone switchboard
Bell ringing transformer
Electric door opener
Fire alarm bell
Fire alarm station
City fire alarm station
Fire alarm central station
Automatic fire alarm device
Watchman's station
Watchman's central station
Horn
Nurse's signal plug
Maid's signal plug
Radio outlet
Signal central station
Interconnection box
Battery
Auxiliary system circuits
Note: Any line without further designation indicates a 2-wire system. For a greater number of wires, designate with numerals in manner similar to 12–No. 1BW–3/4"–C, or designated by number corresponding to listing in schedule. Special auxiliary outlets
Subscript letters refer to notes on plans or detailed description in specifications.

SPECIAL OUTLETS

Symbol	Description
O a,b,c–etc.	Any standard symbol as given above with the addition of a lower case subscript letter may be used to designate some special variation of Standard Equipment of particular interest in a specific set of architectural plans.
a,b,c–etc.	
S a,b,c–etc.	When used they must be listed in the Key of Symbols on each drawing and if necessary further described in the specifications.

FIGURE 1-11   Generic symbols for electrical plans.

necessary to complete the installation. Material is normally taken off by systems and listed on the take-off sheets in a manner that will permit labour units to be easily assigned.

The exact method and forms used by a contractor will be determined by personal preference or the computerized estimating system used by the contractor. One breakdown of systems in the order they would be taken off might be:

- Luminaires
- Outlets
- Branch circuits
- Service, metering, and grounding
- Distribution equipment
- Feeders and busways
- Telephone
- Special systems

Tabulating and Pricing

Pricing is the process of assigning labour and prices to items of material that have been tabulated and listed on pricing sheets. Labour is assigned using standard labour units and adjusted to account for the conditions of the installation.

A standard labour unit is defined as the time required to handle and install an item under specified working conditions. Labour units are found in estimating manuals developed by such organizations as the Canadian Electrical Contractors Association and its provincial counterparts.

Summarizing

A summary bid sheet lists all of the projected costs of the project. The summary sheet includes such information as:

- Material costs
- Labour hours and costs
- Job factors (working height, multiple identical buildings, etc.)
- Direct job expenses (equipment, travel, warehousing)
- Overhead
- Profit

▶ JOB MANAGEMENT

It is the responsibility of the job manager (whoever is running the job) to ensure that:

- Tools and materials arrive at the job at the time they are required
- There are adequate facilities for storage and protection of materials, tools, and equipment
- There is adequate manpower to complete the work as scheduled
- Information is provided to the workforce to allow the work to proceed
- The paperwork associated with the job is handled properly
- Company policies are followed with respect to safety, employment regulations, and job requirements
- Local codes, laws, and regulations are enforced

The following steps will be helpful during the initial preparation for a project:

- Familiarize yourself with the project by studying the drawings and specifications.
- Prepare a list of materials required for the job. The estimator may provide you with the estimate sheets, or you may decide to do your own material take-off and prepare a bill of materials yourself. Material should arrive on site according to the activity that is being undertaken at the time. You will know how well you are doing by keeping records of all the material that you have ordered, what had to be returned, and any delays in construction that result from lack of material. Material arriving on site should be inspected, inventoried, and stored in a clean, dry, and secure location. Packing slips should be carefully checked and back-ordered items recorded.
- Using electrical estimate sheets and the general contractor's construction schedule, determine when electrical activities should occur. Table 1-3 provides a list of the main construction events and corresponding electrical activities.
- Prepare a schedule of electrical construction activities that corresponds to the general

TABLE 1-3

Main Construction Events and Electrical Activities.

CONSTRUCTION EVENT	ELECTRICAL ACTIVITY
Site preparation	Trenching and underground work
Footings and foundations	Embedded work
Superstructure and floors	Embedded work and hangers
Building enclosure	Feeders, equipment, and branch circuits
Interior walls, partitions, and ceilings	Panelboards, branch circuits, and pulling wire
Trimming and decorating	Lighting
Floor finishing	Finishing

contractor's construction schedule. The activity schedule will be the basis for determining man-hour requirements, when material should be delivered to the site, and what tooling will

be required. Figure 1-12 is an example of a construction activity schedule.

Change Orders

A change order is a change in the drawings or specifications that is made after the contract has been awarded. Change orders result from such things as:

• Requests for additional work from the owner

• Errors or omissions in the drawings and specifications

• Delays in construction

When an owner requests a change in the contract that will affect the contract price, an estimate of the costs involved in making the changes is prepared and a Request for Change Order is submitted to the owner. (This is a document that outlines the changes to be made, and authorizes the contractor to make the changes. It must be signed

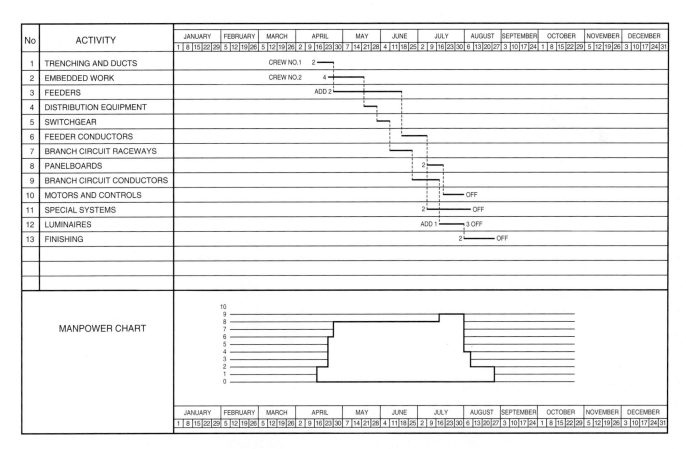

FIGURE 1-12   Construction activity schedule.

by the owner or owner's representative.) The most important thing about change orders is that they should be handled promptly and fairly, and the contractor should make sure change orders are signed by the owner's representative before starting work.

APPROVAL OF EQUIPMENT

All electrical equipment sold or installed in Canada is required to be approved for its intended use. Approved equipment has met specified safety standards of federal and provincial governments and will have an identifying label such as Canadian Standards Association (CSA) or Underwriters Laboratories of Canada (ULC) . For equipment that has not been approved, arrangements may be made for field approval by contacting your provincial inspection authority, which should also be able to provide you with a complete list of companies accredited by the Standards Council of Canada to approve electrical equipment in your province.

REGISTERED PROFESSIONAL ENGINEER (PEng)

Although the requirements may vary slightly in different jurisdictions, the general statement can be made that a registered professional engineer has demonstrated his or her competence by graduating from college or university and passing a difficult licensing examination. Following the successful completion of the examination, the engineer is authorized to practise engineering under the laws of the province.

It is usually required that a registered professional engineer must supervise the design of any building to be used by the public, and must indicate approval of the design by affixing a seal to the plans.

Information concerning the procedure for becoming a registered professional engineer and a definition of the duties of the professional engineer can be obtained by writing to the provincial government department that supervises licensing and registration.

CODES AND STANDARDS

Standards Council of Canada (SCC)

Standards Council of Canada
Conseil canadien des normes

Courtesy Standards Council of Canada

The Standards Council of Canada (SCC) is a federal Crown corporation whose mandate is to promote efficient and effective voluntary standardization as a means of advancing the national economy, and benefiting the health, safety, and well-being of the public.

SCC fulfills its mandate by accrediting organizations engaged in standards development, certification, and testing; by approving National Standards of Canada (NSC); by coordinating participation in international forums, such as the International Organization for Standardization (ISO) and the International Electrotechnical Commission (IEC); and by developing standardization solutions for government and industry.

Through these activities, SCC facilitates Canadian and international trade, protects consumers, and promotes international standards cooperation. For more information, please visit www.scc.ca.

CSA Group (CSA)

Courtesy of CSA Group

CSA Group is a not-for-profit, member-based organization formed in 1919 to develop standards that address the needs of a broad range of organizations and disciplines. These standards can be used for product testing and certification requirements and are often adopted by provincial and federal government authorities to form the basis for legislation and/or compliance requirements.

Thousands of manufacturers work directly with CSA Group to have their products tested and certified and the CSA mark (shown here) appears on billions of products around the world.

Courtesy of CSA Group

The Canadian Electrical Code is one of the many codes and standards that are developed by CSA Group. The *Canadian Electrical Code* has been adopted by provincial authorities to become law in the various provinces. For more information, visit www.csagroup.org.

Underwriters Laboratories of Canada (ULC)

Courtesy Underwriters Laboratories of Canada

Underwriters Laboratories of Canada (ULC) is an independent product safety testing, certification and inspection organization. ULC has tested products for public safety for over 90 years and is accredited by the Standards Council of Canada (SCC). ULC also publishes standards, specifications, and classification for products used in connection with fire, accident, and property hazards. For more information, see www.ulc.ca.

Canadian Electrical Code

The original *Canadian Electrical Code* was developed in 1920 and finalized in June 1927. At this time the revised draft was formally approved and printed as the *Canadian Electrical Code,* which is revised and updated every three years. The object of the *Code* is to establish safety standards for the installation and maintenance of electrical equipment. In its preparation, consideration has been given to the prevention of fire and shock hazards, as well as proper maintenance and operation, *Section 0.*

The *CEC* is the basic standard for the installation and construction of all electrical systems in Canada. However, some provincial and local codes contain specific amendments that must be adhered to in all electrical wiring installations under their jurisdiction. The authors encourage the reader to develop a detailed knowledge of the layout and content of the latest edition of the *CEC.* Also, refer to Table 1-4.

The *CEC* is divided into numbered sections, with each covering a major topic. These sections are further divided into numbered rules, subrules, items, and subitems; related rules are often grouped together under a subsection heading. All references in the text are to the *Section* or to the *Section and rule number.* For example, *Rule 8–200(1)(a)(i)* in the text refers to *Section 8, Rule 200, subrule 1, item a, subitem i.* Occasionally, a section reference will include a clause or item. See Table 1-4. It is hoped that this explanation will assist the student in locating *CEC* references in the text.

To help the user of this text, relevant *CEC* sections are paraphrased where appropriate. However,

TABLE 1-4		
Citing the *CEC*.		
DIVISION	**DESIGNATION**	**EXAMPLE**
Section	8	*Section 8*
Rule	Even numbers	*8–212*
Subrule	Numbered sequence	*8–106(1)*
Item	a–z	*8–108(1)(d)*
Subitem	Lower case/Roman numerals	*8–204(2)(a)(i)*
Clause/item	Capital A–Z	*8–204(2)(b)(ii)(B)*

the *CEC* must be consulted before any decision related to electrical installation is made.

Enforcement of the *CEC* is generally under the jurisdiction of the utility company providing electrical power in the particular province. In Ontario, for instance, electrical inspection is conducted by field inspectors employed by the Electrical Safety Authority (ESA).

Code Terms. The following terms are used throughout the *CEC*, and it is important to understand their meanings.

- *Approved* (as applied to electrical equipment): Equipment has been submitted for examination and testing to an accredited certification organization and the equipment conforms to appropriate CSA standards.
- *Shall:* Indicates a mandatory requirement.
- *Notwithstanding:* In spite of.
- *Practicable:* Feasible, possible.

Citing Code References. Every time an electrician makes a decision concerning the electrical wiring, the decision should be checked by reference to the *CEC*. Usually this is done from memory, but if there is any doubt in the electrician's mind, then the *CEC* should be referenced directly— just to make sure. It is a good idea to record the *CEC* location of the information—this is referred to as citing the reference. There is a very exact way in which the location of an item is to be cited, and the various levels of *CEC* referencing are shown in Table 1-4. Starting at the top of the table, each step becomes a more specific reference. A reference to *Section 2* includes all the information and requirements that are set forth in several pages. Citing a specific rule or exception narrows the reference to a few sentences or perhaps even a few words.

CEC Use of SI (Metric) Measurements

The 2015 *CEC* states measurements in the metric system, commonly known as the International System of Units (Système Internationale, SI). Figure 1-13 provides a list of SI units and symbols, as well as conversion factors for previously used units.

This system is international and many compromises were made to accommodate regional practice. In the United States, the practice is to use the period as the decimal marker and the comma to separate a string of numbers into groups of three for easier reading. In many countries, the comma is used as the decimal marker and spaces are left to separate the string of numbers into groups of three. The SI system, taking something from both, uses the period as the decimal marker and the space to separate a string of numbers into groups of three starting from the decimal point and counting in either direction (e.g., 12 345.789 99). The only exception to this occurs when there are four numbers on either side of the decimal point in the string. In this case, the third and fourth numbers from the decimal point are not separated (e.g., 2015.1415).

In the SI system, the units increase or decrease in multiples of 10, 100, 1000, and so on. For instance, one megawatt (1 000 000 watts) is 1000 times greater than one kilowatt (1000 watts).

The metric prefixes indicate the factor by which the basic unit is multiplied. For example, adding the prefix *kilo* to *watt* forms the new name *kilowatt*, meaning 1000 watts. Refer to Figure 1-14 for prefixes used in the SI system.

A few of the prefixes shown in Figure 1-14 have a preference in usage. These prefixes are *mega*, *kilo*, *centi*, *milli*, *micro*, and *nano*. Thus, commonly used multiples of a metre are a kilometre (1000 metres), a centimetre (0.01 metre), and a millimetre (0.001 metre).

The advantage of the SI system is that the prefixes allow measurements to be written with few zeros, which lessens the possibility of confusion. For example, a four-foot lamp is approximately 1200 millimetres, or 1.2 metres.

Some common measurements of length in the imperial system are shown with their SI equivalents in Figure 1-15.

The 2015 edition of the *CEC* indicates conduit sizes and box capacity in metric units. A conversion table in Figure 1-16 can be used to convert metric conduit sizes to traditional imperial sizes. Usable space and capacity of boxes are indicated in millilitres or cubic centimetres followed by the cubic

SYMBOL	SI UNIT	MULTIPLYING FACTOR FOR CONVERSION TO PREVIOUSLY USED UNIT	PREVIOUSLY USED UNIT
A	ampere(s)	1	ampere(s)
cm³	cubic centimetre(s)	0.061	cubic inch(es)
°(S)	degree(s) (angle)	1	degree(s) (angle)
°C rise	degree(s) Celsius	1.8	degree(s) Fahrenheit
°C temperature	degree(s) Celsius	1.8 plus 32	degree(s) Fahrenheit
h	hour(s)	1	hour(s) (time)
Hz	hertz	1	cycles per second
J	joule(s)	0.7376	foot-pound(s)
kg	kilogram(s)	2.205	pound(s)
kJ	kilojoule(s)	737.6	foot-pound(s)
km	kilometre(s)	0.621	mile(s)
kPa	kilopascal(s)	0.295	inch(es) of mercury
		0.334	feet of water
		0.145	pound(s) per square inch (psi)
kW	kilowatt(s)	3415.179	BTU/h
lx	lux	0.093	foot-candle(s)
L	litre(s)	0.220	gallon(s)
m	metre(s)	3.281	feet
m²	square metre(s)	10.764	square feet
m³	cubic metre(s)	35.315	cubic feet
MHz	megahertz	1	megacycles per second
min	minute(s)	1	minute(s)
mL	millilitre(s)	0.061	cubic inch(es)
mm	millimetre(s)	0.03937	inch(es)
mm²	square millimetre(s)	0.00155	square inch(es)
Ω	ohm(s)	1	ohm(s)
Pa	pascal(s)	0.000295	inch(es) of mercury
		0.000334	feet of water
		0.000145	pounds per square inch (psi)
V	volt(s)	1	volt(s)
W	watt(s)	1	watt(s)
μF	microfarad(s)	1	microfarad(s)

FIGURE 1-13   Symbols for SI units.

Courtesy of CSA Group

mega	1 000 000	(one million)
kilo	1 000	(one thousand)
hecto	100	(one hundred)
deca	10	(ten)
	1	(one)
deci	0.1	(one-tenth) (1/10)
centi	0.01	(one-hundredth) (1/100)
milli	0.001	(one-thousandth) (1/1 000)
micro	0.000 001	(one-millionth) (1/1 000 000)
nano	0.000 000 001	(one-billionth) (1/1 000 000 000)

FIGURE 1-14   SI prefixes and their values.

inch value in parentheses. It is important to note that one millilitre is equal to one cubic centimetre.

A metric (SI) measurement is not shown for box size, wire size, horsepower designation for motors, and other "trade sizes" that are nominal rather than exact measurements.

ELECTRICAL INSPECTION

The *CEC* requires that a permit be obtained for the inspection of all work with respect to the installation, alteration, or repair of electrical

one inch	=	2.54	centimetres
	=	25.4	millimetres
	=	0.0254	metre
one foot	=	12	inches
	=	0.3048	metre
	=	30.48	centimetres
	=	304.8	millimetres
one yard	=	3	feet
	=	36	inches
	=	0.9144	metre
	=	914.4	millimetres
one metre	=	100	centimetres
	=	1 000	millimetres
	=	1.093	yards
	=	3.281	feet
	=	39.370	inches

FIGURE 1–15   Conversion factors for some common measurements of length.

TRADE SIZE DESIGNATOR	
ENGLISH(INCH)	Metric(mm)
3/8	12
1/2	16
3/4	21
1	27
1-1/4	35
1-1/2	41
2	53
2-1/2	63
3	78
3-1/2	91
4	103
5	129
6	155

FIGURE 1-16   Conduit size conversion.

equipment. It is the responsibility of the electrical contractor or person doing the work to obtain the permit. When the installation has passed inspection, the inspection authority will issue a connection permit (sometimes referred to as authorization for connection or current permit) to the supply authority. The supply authority will then make connection to the installation.

REVIEW

Note: Refer to the *CEC* or the blueprint package as necessary.

1. What section of the specification contains a list of contract documents?

   _____

   _____

   _____

   _____

2. The requirement for temporary light and power at the job site will be found in what portion of the specification?

   _____

   _____

   _____

   _____

   _____

3. The electrician uses the Schedule of Drawings for what purpose?

_____

_____

_____

_____

_____

_____

Complete the following items by indicating the letter(s) designating the correct source(s) of information for:

4. Room width   _____   a. Site plan

5. Grading elevations   _____   b. Architectural floor plan

6. Ceiling height   _____   c. Elevations

7. Panelboard schedules _____   d. Details

8. Exterior wall finishes _____   e. Electrical layout drawings

9. View of interior wall _____   f. Specification

10. Electrical outlet location _____ g. Sections

11. Electrical receptacle style _____ h. Electrical symbol schedule

12. Swing of door   _____

Match the initialism on the left with the phrase or word that best relates to that organization, document, or person.

13. SCC   _____   a. Accrediting organizations

14. *CEC*   _____   b. Seal

15. ULC   _____   c. Manufacturers' standards

16. PEng   _____   d. Listing service

17. CSA   _____   e. Electrical code

Write the appropriate letters (a, b, c, or d) to indicate the proper interpretation of the *CEC*.

18. Must be done   _____   a. Shall

19. May be done   _____   b. Special permission

20. Up to the electrician _____   c. Not allowed

21. Is required   _____   d. Allowed

22. Is allowed   _____

23. List the drawings that are normally included in an electrical drawing set.

_____   _____

_____   _____

_____   _____

_____   _____

24. List the steps to be followed when working with a set of drawings.

_____   _____

_____   _____

_____   _____

_____   _____

25. Measure the length of each line using the scale indicated.   Answer

a. 1 : 100   _____   _____

b. 1 : 50   _____   _____

c. 1 : 25   _____   _____

d. 1 : 75   _____   _____

e. 1 : 50   _____   _____

f. 1 : 125   _____   _____

g. 1 : 100   _____   _____

h. 1 : 25   _____   _____

i. 1 : 50   _____   _____

j. 1 : 25   _____   _____

k. $\frac{1}{8}'' = 1$ ft   _____   _____

l. $\frac{1}{4}'' = 1$ ft   _____   _____

m. $\frac{1}{2}'' = 1$ ft   _____   _____

n. $1\frac{1}{2}'' = 1$ ft   _____   _____

o. $\frac{3}{8}'' = 1$ ft   _____   _____

p. $\frac{3}{4}'' = 1$ ft   _____   _____

q. $\frac{1}{4}'' = 1$ ft   _____   _____

r. $\frac{1}{8}'' = 1$ ft   _____   _____

s. $\frac{1}{4}'' = 1$ ft   _____   _____

t. $\frac{1}{2}'' = 1$ ft   _____   _____

The Electric Service

OBJECTIVES

After completing the study of this unit, the student will be able to

- install power transformers to meet the *CEC* requirements
- draw the basic transformer connection diagrams
- recognize different service types
- connect metering equipment
- apply ground-fault requirements to an installation
- install a grounding system

The installation of the electric service to a building requires the cooperation of the electrician, the local utility company, and the electrical inspector. The availability of high voltage and the power company requirements determine the type of service to be installed. This unit will investigate several common variations in electrical service installations. The consumer's service for this building begins with a 300 KVA, 120/208 V 3PH 4 Wire pad-mounted transformer (Figure 2-1), mounted on the transformer pad shown in Figure 2-2. (See Site Plan.)

FIGURE 2-1   300 KVA, 120/208 V 3PH 4 W pad-mounted transformer.

Courtesy of Natalie Barrington

FIGURE 2-2   300 KVA transformer pad dimensions.

Primary service conductors run in conduit down the pole and underground to the pad-mounted enclosure

Transformer in weatherproof enclosure

First floor

Seals required
Rule 6–300

Basement

Main switch

Enclosure must be kept locked if accessible to the public
Rule 26–240(2)

Service lateral conductors run in conduit underground to the main switch
Rule 12–012

© Cengage Learning®

FIGURE 2-3   Pad-mounted enclosure supplying underground service entrance.

Utility company may feed
primary supply to transformers
overhead or underground

Building wall

Cable connections
between transformers
and busway

Busway (bus duct)
service entrance

Secondary
switches

Current
transformer

Main
switch

FIGURE 2-4 Distribution transformers supplying bus duct service entrance.

Pad-Mounted Transformers

Liquid-insulated transformers as well as dry-type transformers are used for this type of installation. These transformers may be fed by an underground or overhead service, Figure 2-3. The secondary normally enters the building through a bus duct or large cable (Figure 2-4).

Underground Vault

This type of service is used when available space is an important factor and an attractive site is desired. The metering may be at the utility pole or in the building, Figure 2-5.

Unit Substation

For unit substation installations (Figure 2-6), the primary runs directly to the unit substation where all of the necessary equipment is located, Figure 2-7.

The utility company requires the building owner to buy the equipment for a unit substation installation.

Rules 26–240(2) and (3)

Building wall

Seals required
Rule 6–312

Primary
conductors
in conduit

Service conductors
run in conduit
underground
to vault

Metal grating
(bonded)

Main switch

Transformer in vault.
May be one 3-phase
transformer or three
single-phase transformers
properly connected

FIGURE 2-5 Underground vault supplying underground service entrance.

Higher Voltage
Section

Transformer
Section

Distribution
Section

© Cengage Learning®

FIGURE 2-6   A unit substation.

LIQUID-FILLED TRANSFORMERS

Modern liquid-filled transformers are filled with mineral oil or synthetic cooling liquids. The liquid performs several important functions: (1) it is part of the required insulation dielectric, and (2) it acts as a

coolant by conducting heat from the core and the winding of the transformer to the surface of the enclosing tank, which then is cooled by radiation or fans. Many older transformers used askarel as a coolant since it is not flammable. *Askarel* is a trade name for oil containing PCBs, and this type of oil is no longer permitted in applications such as transformers or ballasts. If it is found in an existing installation, approved steps regulated by the Ministry of the Environment must be taken to ensure containment of the area or the safe disposal of the oil.

DRY-TYPE TRANSFORMERS

Dry-type transformers are widely used because they are lighter than comparable liquid-filled transformers. Installation is simpler because there is no need to take precautions against liquid leaks.

Dry-type transformers are constructed so that the core and coil are open to allow for cooling by

FIGURE 2-7   High-voltage service entrance.

© Cengage Learning®

the free movement of air. Fans may be installed to increase the cooling, in which case the transformer can be used at a greater load level. A typical dry-type transformer installation is shown in Figure 2-6. An installation of this type is known as a *unit substation* and consists of three main components: (1) the high-voltage switch, (2) the dry-type transformer, and (3) the secondary distribution section.

TRANSFORMER OVERCURRENT PROTECTION

Rule 26–240 covers transformer installations in general; *Rules 26–252* and *26–254* group transformers into two voltage levels. Requirements for the overcurrent protection of transformers rated over 750 volts are covered in *Rule 26–252* and *Table 50*.

Figure 2-8 illustrates four of the more common situations found in commercial building transformer installations. The overcurrent devices as specified in *Rule 26–252* protect the transformer only. The conductors supplying or leaving the transformer may require additional overcurrent protection. (See Unit 17 of this text for information concerning fuses and circuit breakers.)

Figure 2-9 shows one method of installing a dry-type transformer in a commercial or industrial building. Table 2-1 provides full load current ratings for both three-phase and single-phase transformers, based on incoming line voltage. *Rule 26–254* details the overcurrent protection for power and distribution, other than dry-type transformers rated at 750 volts or less.

TRANSFORMER CONNECTIONS

A transformer is used in a commercial building primarily to change the transmission-line high voltage to the value specified for the building, such as 347/600Y or 120/208Y volts. A number of connection methods can be used to accomplish the changing of the voltage. The connection used depends upon the requirements of the building. The following paragraphs describe several of the more commonly used secondary connection methods.

TABLE 2-1

Transformer Full Load Current Table.

RATED LINE VOLTAGE

KVA	120	240	480	600
6	16.6	14.4	7.2	5.8
9	25	21.6	10.8	8.7
10	28	24	12	9.6
15	41.6	36.1	18.1	14.4
30	83.3	72.3	36.1	28.9
45	125	108	54	43.2
75	208	181	90.5	72.1
112.5	312	272	136	108
150	416	360	180	144
225	624	541	270	216
300	832	722	361	288
450	1250	1080	542	433
500	1388	1203	601	481
600	1666	1445	722	578

Three-Phase

$$Amp = \frac{KVA \times 1000}{Volts \times 1.732}$$

Single-Phase

$$Amp = \frac{KVA \times 1000}{Volts}$$

Transformer Load Current per line in amps—three-phase.

RATED LINE VOLTAGE

KVA	120	240	480	600
1	8.3	4.2	2.1	1.6
2	16.7	8.4	4.2	3.2
3	25	12.5	6.3	5
5	41.7	20.8	10.4	8.3
7.5	62.5	31.2	15.6	12.5
10	83.4	41.6	20.8	16.7
15	125	62.5	31.2	25
25	208	104	52.1	41.6
37.5	312	156	78.2	62.5
50	416	208	104	83.3
75	625	312	156	125
100	833	416	208	167
150	1250	625	312	250
167	1391	696	348	278
250	2083	1041	521	417
333	2775	1387	694	555
500	4166	2083	1042	833

Transformer Full Load Current in amps—single-phase

A

Transformers over 750 volts with primary and secondary protection. *Rule 26–252(4), Table 50.*

Transformer rated impedance	Primary over 750 volts**	Secondary over 750 volts	Secondary 750 volts or less
Not over 7.5%	• Max. fuse 300% • Max. breaker 600%	• Max. fuse 150% • Max. breaker 300%	• Max. fuse 250% • Max. breaker 250%
Over 7.5% but not over 10%	• Max. fuse 200% • Max. breaker 400%	• Max. fuse 125% • Max. breaker 250%	• Max. fuse 250% • Max. breaker 250%

B

Transformers over 750 volts with primary protection only. *Rule 26–252(1) & (2).*

Primary over 750 volts**	Secondary
• Maximum fuse—150%* • Maximum breaker—300%	• No overcurrent protection required

* Where this percentage does not correspond to a standard fuse rating, the next higher standard rating or setting may be used.

C

Transformers 750 volts and less with primary protection only. *Rules 26–254(1) and 26–256(1).*

Primary 750 volts or less	Secondary 750 volts or less
Other than dry type Maximum fuse size not to exceed 150% of the transformer's rated primary current, *Rule 26–254(l).* If the rated primary current is 9 amperes or more and 150% of this current does not correspond to a standard fuse or breaker size, then the next larger standard size may be used. For transformers having rated primary current of less than 9 amperes a fuse or nonadjustable-trip circuit breaker not to exceed 167% may be used, *Rule 26–254(2).* **Dry type** An individual overcurrent device is required. Maximum fuse or breaker size is not to exceed 125% of the transformer's rated primary current. Where 125% of the rated primary current does not correspond to a standard overcurrent device rating, the next higher standard rating may be used, *Rule 26–256.*	**Other than dry type** Transformer secondary protection is not required when the primary fuse does not exceed 150% of the rated primary current, *Rule 26–254(1).* **Dry type** Transformer secondary protection is not required when the primary fuse does not exceed 125% of the rated primary current, *Rule 26–256(1).* Transformer secondary protection is not required when the primary fuse does not exceed 125% of the rated primary current and the secondary conductors are rated at not less than 125% of the secondary current, *Rule 26–256(l).*

D

Transformers 750 volts and less with primary and secondary protection. *Rule 26–254(4) and 26–256(2).*

Primary 750 volts or less	Secondary 750 volts or less
Other than dry type Individual overcurrent protection is not required if the feeder overcurrent protection is not greater than 300% of the rated primary current and the secondary overcurrent protection is not greater than 125% of the rated secondary current, *Rule 26–256(2).* If the transformer is equipped with coordinated overload protection furnished by the manufacturer, individual primary overcurrent protection is not required if the primary feeder has overcurrent protection: • set not over six times primary current for transformers with not more than 7.5% impedance. • set not over four times primary current for transformers with more than 7.5% but not over 10% impedance. This may allow more than one transformer to be connected to one feeder. **Dry type** Individual overcurrent protection is not required if the feeder overcurrent protection is not greater than 300% of the rated primary current and the secondary overcurrent protection is not greater than 125% of the rated secondary current.	**Other than dry type** Maximum fuse or breaker is not to exceed 125% of the transformer's rated secondary current when the primary does not have individual overcurrent protection, *Rule 26–254(4).* For transformers having rated secondary current of 9 amperes or more, if the 125% sizing does not correspond to a standard size, then the next higher standard-size fuse or nonadjustable breaker may be used. For transformers having rated secondary current of less than 9 amperes, if the 125% sizing does not correspond to a standard size, then a fuse or breaker not to exceed 167% may be used. **Dry type** Maximum fuse or breaker is not to exceed 125% of the transformer's rated secondary current when the primary does not have individual overcurrent protection, *Rule 26–256(2).*

** Individual primary overcurrent protection is not required if the primary feeder overcurrent device provides the required protection, *Rules 26–252(3) and 26–254(3).* This may allow more than one transformer to be connected to the same feeder.

FIGURE 2-8 Typical transformer overcurrent protection requirements.

Be sure that proper overcurrent protection
is provided for each of the following:
  • Transformer, *Rule 25–256*
  • Conductors, *Rule 14–104*
  • Paneboard, *Rule 14–606*

Recommended
fuse size 60 amperes
in 60-ampere,
600-v switch

Neutral
pad

Maximum
circuit breaker
150 amperes

Lighting
panel

45kvA
Primary 600v
Secondary 208Y/120v

Note: In this example
the system is grounded
at the first point of
distribution.

Grounding electrode
conductor, *Rule 10–206*

$$\text{Primary} = \frac{kVA \times 1{,}000}{E \times 1.73} = \frac{45 \times 1{,}000}{600 \times 1.73} = 43.35 \text{ amperes}$$

Suggested size primary fuse: $43.35 \times 1.25 = 54.2$ amperes.
Install next standard size 60-ampere, dual-element, time-delay fuses in a 60-ampere switch.

$$\text{Secondary} = \frac{kVA \times 1{,}000}{E \times 1.73} = \frac{45 \times 1{,}000}{208 \times 1.73} = 125 \text{ amperes}$$

Suggested size secondary fuse: $125 \times 1.25 = 156.25$. Install 150-ampere dual-element,
time-delay fuses.

FIGURE 2-9 Diagram of how a dry-type transformer might be installed in a commercial or industrial
building served by 600 volts, three-phase, where it is necessary to step down the voltage to 120/208 volts,
three-phase, four-wire to supply lighting panels.

▶ SINGLE-PHASE SYSTEM

Single-phase systems usually provide 120 and/
or 240 volts with a two- or three-wire connection,
Figure 2-10. The centre tap of the transformer secondary shall be grounded in accordance with *Section 10*,
as will be discussed later. Grounding is a vital safety
measure and should be installed with great care.

Open Delta System

This connection scheme has the advantage of being
able to provide either three-phase or three-phase and
single-phase power using only two transformers. It
is usually installed where there is a strong probability that the power requirement will increase, at
which time a third transformer can be added. The
open delta connection is illustrated in Figure 2-11.

© Cengage Learning®

FIGURE 2-10 Single-phase transformer connection.

In an open delta transformer bank, 86.6% of the capacity of the transformers is available. For example, if each transformer in Figure 2-11 has a 100-kVA rating, then the capacity of the bank is:

$$(100 \text{ kVA} + 100 \text{ kVA}) \times 86.6\% = 173 \text{ kVA}$$

Another way of determining the capacity of an open delta bank is to use 57.7% of the capacity of a full delta bank. The capacity of three 100-kVA transformers connected in full delta is 300 kVA. Thus, two 100-kVA transformers connected in open delta have a capacity of

$$300 \text{ kVA} \times 57.7\% = 173 \text{ kVA}$$

When an open delta transformer bank is to serve three-phase power loads only, the centre tap is not connected.

Four-Wire Delta System

This connection, also illustrated in Figure 2-11, has the advantage of providing both three-phase and single-phase power from either an open (two transformers) or a closed (three transformers) delta system. A centre tap is brought out of one of the transformers, which is grounded and becomes the neutral conductor to a single-phase, three-wire power system. The voltage to ground from phases B and C, which are connected to the transformer that has been tapped, will be equal. These voltages are additive when measured phase to phase, for example, 120 volts to ground and 240 volts between phases. The voltage measured between the grounded tap and the A phase will be higher than 120 volts and lower than 240. This phase is called the *high leg* and cannot be used for lighting purposes. This high

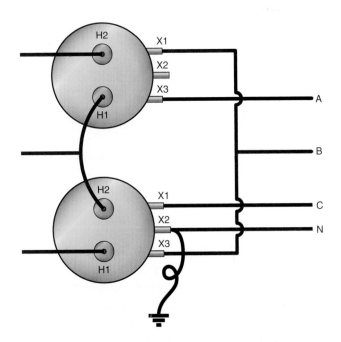

FIGURE 2-11 Three-phase open delta connection for light and power.

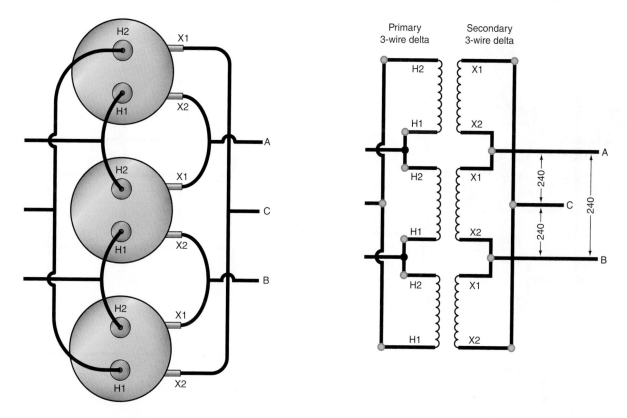

FIGURE 2-12   Three-phase, delta-delta transformer connection.

leg must be identified by using a red coloured conductor or by clearly marking it as the A phase conductor whenever it is present in a box or cabinet with the neutral of the system. See *Rule 4–038(4,5)*.

Three-Wire Delta System

This connection, as illustrated in Figure 2-12, provides only three-phase power. One phase of the system may be grounded, in which case it is often referred to as corner grounded delta. The power delivered and the voltages measured between phases remain unchanged. Overcurrent devices are not to be installed in the grounded phase. See *Rule 14–010*.

Three-Phase, Four-Wire Wye System

The most commonly used system for modern commercial buildings is the three-phase, four-wire wye (Y), Figure 2-13. This system has the advantage of being able to provide three-phase power, and also permits lighting to be connected between any of the three phases and the neutral. Typical voltages available with this type of system are 120/208, 240/416, 277/480, and 347/600. In each case, the transformer connections are the same.

Notes to All Connection Diagrams

All connection diagrams are shown with additive polarity and standard angular displacement. All three-phase connections are shown with the primary connected in delta. For other connections, it is recommended that qualified engineering assistance be obtained.

▶ THE SERVICE ENTRANCE

The regulations governing the method of bringing the electric power into a building are established by the local utility company, and vary considerably among utility companies. Several of the more common methods of installing the service entrance are shown in Figures 2-3, 2-4, 2-5, and 2-7.

FIGURE 2-13   Three-phase, delta-wye transformer connection commonly used to supply three-phase power and lighting.

METERING

The electrician working on commercial installations seldom makes metering connections. However, the electrician should be familiar with the following two basic methods of metering.

High-Voltage Metering

When a commercial building is occupied by a single tenant, the utility may elect to meter the high-voltage side of the transformer. To accomplish this, a potential transformer and two current transformers are installed on the high-voltage lines and the leads are brought to the meter, as shown in Figure 2-14.

The left-hand meter socket in the illustration is connected to receive a standard socket-type watt-hour meter; the right-hand meter socket will receive a varhour meter (volt-ampere reactive meter). The two meters are provided with 15-minute demand attachments that register kilowatt (kW) and kilovolt-ampere reactive (kVAR) values respectively. These demand attachments will indicate the maximum usage of electrical energy for a 15-minute period during the interval between the readings made by the utility company. The rates charged by the utility company for electrical energy are based on the maximum demand and the power factor as determined from the two meters. A high demand or a low power factor will result in higher rates.

Low-Voltage Metering

Low-voltage metering of loads greater than 200 amperes is accomplished in the same manner as high-voltage metering. In other words, potential and current transformers are used. For loads of 200 amperes or less, the feed wires from the primary supply are run directly to the meter socket, Figure 2-15. For multiple occupancy buildings, such as the commercial building discussed in this text, the meters are usually installed as a part of service-entrance equipment called a *switchboard*.

Notes:
1. When meter Ⓑ reads one-half of meter Ⓐ, power factor is 90%.
2. When meter Ⓑ reads less than one-half of meter Ⓐ, power factor is above 90%.
3. When meter Ⓑ reads more than one-half of meter Ⓐ, power factor is below 90% and penalty rate is charged by the power company.

FIGURE 2-14 Connections for high-voltage watt-hour meter and varhour meter. Demand attachments and timers are not shown.

Courtesy of Milbank Manufacturing Company

FIGURE 2-15 Meter socket.

SERVICE-ENTRANCE EQUIPMENT

When the transformer is installed at a location far from the building, the service-entrance equipment consists of the service-entrance conductors, the main switch or switches, the metering equipment, and the secondary distribution switches, Figure 2-16. The commercial building shown in the blueprints is equipped in this manner and will be used as an example for the following paragraphs. Table 2-2 provides the main service loading schedule.

The Service

The service for the commercial building is similar to the service shown in Figure 2-3. A pad-mounted, three-phase transformer is located outside the building and three 78 mm, rigid PVC conduits running underground serve as the service entrance. The conductors—phase A, phase B,

FIGURE 2-16 Commercial building service-entrance equipment, one line diagram.

phase C, and the neutral—are installed in each conduit.

According to *Rule 4–004(1)(d) Appendix B*, the conductors must be installed in accordance with configurations described in *Diagrams B4-1 to B4-4* located in *Appendix B:* in an underground run, directly buried, or in a raceway as specified in *Tables D8A* through *D15B*. According to *Table D14A* column *3/Phase Detail 3*, our service calculation value of 342.89 A will require 500 kcmil conductors per phase and neutral. (*Diagram B4-4* is reproduced in Table 2-3; *Table D14A* is reproduced in Table 2-4.)

▶ SERVICE DISCONNECTING MEANS

The *CEC* permits as many as four consumer services, *Rule 6–104*, unless special permission is given for more, *Rule 2–030*. The sum of the overcurrent

protection installed in the switches may exceed the ampacity of the service-entrance conductors.

The service laterals are tied into a 1200-ampere switch with 1000-ampere fuses. The load side of this switch feeds the secondary distribution switches and the 800-ampere switch used for the boiler. These switches are shown in Figures 2-17, 2-18A, and 2-18B.

The metering for the service entrance of this commercial installation is located in the third section of the main service-entrance equipment. The meters for six of the occupancies in the building are installed directly in the line. The seventh meter and current transformers are installed to measure the load to the boiler. The service equipment can be purchased in a single enclosure, such as the enclosure illustrated for the commercial building in this text. The contractor and electrician should consult the *CEC* and local regulations, and obtain the supply authority's approval of location and installation requirements before installing any service, *Rule 6–206*.

TABLE 2-2

Main Service Loading Schedule.

TENANT AREA	DEMAND ON SERVICE CONDUCTOR
Owner's area	11 240 W
Bakery	44 470 W
Drugstore	21 849 W
Insurance office	32 515 W
Beauty salon	21 825 W
Doctor's office	14 234 W
Subtotal	146 133 W
Boiler	150 000 W *Rule 62–116(4b)*
Total	296 133 W

$$I = \frac{P}{E \times 1.73} = \frac{296\ 133}{208 \times 1.73} = 822.95 \text{ A.}$$

Using 75°C rated equipment, *Rule 4–006*, and conductor ampacities based on *Table 2, CEC,* the minimum ampacity of the service conductors would be $822.95 \times 1.25 = 1028.69$ A.

Using three conductors per phase: $\frac{1028.69}{3} = 342.89$ A

The minimum conductor size from the 75°C column within *Table 2* would be 500 kcmil R90-XLPE.

Note: The conductors supplying this building will be based on the requirements for the secondary of the transformer supplying the building and will therefore be larger than required by this service calculation.

TABLE 2-3

Diagram B4-4: Installation configuration—raceway.

▶ GROUNDING

Why is grounding necessary? Besides saving lives and protecting equipment, electrical systems and their associated conductors are grounded to keep

TABLE 2-4

CEC Table D14A.

TABLE D14A
ALLOWABLE COPPER CONDUCTOR AMPACITIES FOR THE INSTALLATION CONFIGURATION OF DIAGRAM B4-4
(SEE APPENDIX B NOTE TO RULE 4–004.)

Size, AWG or kcmil	1/Phase Detail 1	2/Phase Detail 2	3/Phase Detail 3	4/Phase Detail 4	5/Phase Detail 5	6/Phase Detail 6	8/Phase Detail 8
1/0	180	164	152	141	131	125	116
2/0	206	187	172	160	149	142	131
3/0	235	213	196	181	168	160	148
4/0	269	242	223	205	190	181	168
250	298	267	244	225	208	198	183
350	361	321	293	268	248	235	217
500	437	386	350	319	294	279	258
600	480	423	383	349	321	303	281
750	538	471	425	386	355	335	310
1000	620	540	485	439	400	380	353
1250	676	583	521	472	433	408	378
1500	724	623	555	502	459	434	402
1750	756	548	576	521	476	449	416
2000	785	671	596	538	491	463	429

① Owners   ⑤ Beauty salon
② Doctor   ⑥ Bakery
③ Drug store   ⑦ Insurance office
④ Spare

FIGURE 2-17 Commercial building service-entrance equipment, pictorial view.

voltage spikes to a minimum should lightning strike or other line disturbances occur. Grounding stabilizes normal voltage to ground. By bonding equipment grounding conductors to the grounding conductor of the system, the low impedance path for fault current will facilitate the operation of overcurrent devices under ground-fault conditions. (To "facilitate" means to "make easier.")

Electrical metallic conduits and other electrical equipment are grounded to keep the voltage to ground to a minimum, thus reducing the electrical shock hazard.

It is very important that the ground connections and the grounding system be properly installed. To achieve the best possible grounding system, the electrician must use the recommended procedures when installing the proper equipment. Figure 2-19 illustrates some of the terminology used in *CEC, Section 10, Grounding and Bonding.*

Properly grounded electrical systems and enclosures provide a good path for ground currents if a ground fault occurs—for example, if insulation fails. The reason for this is that the lower the impedance (AC resistance to the current) of the grounding path, the greater the ground-fault current. This increased ground-fault current causes the overcurrent device protecting the circuit to respond faster. This is called

FIGURE 2-18A Bolted pressure contact switch for use with Class J, Class L, or Class T high-capacity fuses. The service for the commercial building in the plans has two of these switches as the main disconnecting means and a boiler disconnect.

FIGURE 2-18B These switches can be purchased with GFP, shunt tripping, auxiliary contacts, phase failure relays, and other optional attachments. Switch illustrated has GFP, phase failure relay, shunt tripping, and antisingle-phasing blown fuse indicator.

All sections of switchboard must be bonded to ground, *Rule 10–402*

Switchboard frame must be grounded, *Rule 10–400*

MAIN SERVICE EQUIPMENT

Service neutral conductor (grounded conductor)

Service raceway *Rule 10–300*

Neutral conductor *Rule 10–210*

Bonding bushing, *Rule 10–606(1)(d)*

Neutral bus

Bonding jumper, *Rule 10–614 and Table 41*

Grounding bus

System grounding conductor, *Rules 10–204(2), 10–806, 10–812*

Bonding jumper around meter *Rule 10–902 (App. B)*

• Equipment bonding conductor (Bonding jumpers) Rules *10–614 and 10–902 (App. B)*

M

Street side water supply to building

Hot water pipe

Ground clamp

Grounding electrode (metal water pipe) *Rules 10–700 and 10–902*

Gas pipe, *Rule 10–406(4)*

© Cengage Learning®

FIGURE 2-19 Terminology of service grounding and bonding, *Section 10.*

inverse time, which means that the higher the current, the less time it will take to operate the overcurrent device.

For example, with 208 volts at 0.1 Ω impedance:

$$I = E/Z = 208/0.1 = 2080 \text{ amperes}$$

but with 208 volts at 0.01 Ω impedance:

$$I = E/Z = 208/0.01 = 20\ 800 \text{ amperes}$$

When an overcurrent device opens the circuit faster, less equipment damage results.

The amount of ground-fault damage to electrical equipment is related to (1) the response time of the overcurrent device and (2) the amount of current. One common term used to relate the time and current to the ground-fault damage is *ampere-squared seconds* (I^2t):

$$I^2t = \text{amperes} \times \text{amperes} \times \text{time in seconds}$$

It can be seen in the expression for I^2t that when the current (I) and time (t) are kept to a minimum, a low value of I^2t results. Lower values of I^2t mean that less ground-fault damage will occur. Units 17,

18, and 19 provide detailed coverage of overcurrent protective devices, fuses, and circuit breakers.

System Grounding

System grounding calls for grounding of the entire system rather than a single item. By "system" it is meant, for example, the service neutral conductor, hot and cold water pipes, gas pipes, service-entrance equipment, and jumpers around water meters. If any of these system parts become disconnected or open in any way, the integrity of the system is still maintained through the other paths. In other words, everything is tied together. The concept of system grounding is shown in Figure 2-20.

Figure 2-20 and the following steps illustrate what can happen if an entire system is *not* grounded.

1. Bonding jumper A is not installed originally.

2. A live wire contacts the gas pipe. The gas pipe now has a voltage of 120 volts.

3. The insulating joint in the gas pipe results in a poor path to ground; assume the resistance is 8 ohms.

FIGURE 2-20   System grounding.

4. The 20-ampere fuse does not blow because:

$$I = E/R = 120/8 = 15 \text{ amperes}$$

5. If a person touches the "hot" (live) gas pipe and the water pipe at the same time, current flows through the person's body. If the body resistance is 12 000 ohms, then the current is:

$$I = E/R = 120/12\ 000 = 0.01 \text{ ampere}$$

This value of current passing through a human body can cause death.

6. The fuse is now "seeing" 15 amperes + 0.01 ampere = 15.01 amperes, so it still does not operate.

7. If the system grounding concept had been used, bonding jumper A would have kept the voltage difference between the water pipe and the gas pipe at zero, *Rule 10–406(4)*. Thus, the fuse would operate. If 3.05 m (10 ft) of No. 4 copper wire were used as the jumper, then the resistance of the jumper would be 0.003 08 ohms. The current is:

$$I = E/R = 120/0.003\ 08 = 38\ 961 \text{ amperes}$$

(In an actual system, the impedance of all of the parts of the circuit would be much higher. Thus, a much lower current would result, but it would be enough to cause the fuse to operate.)

Advantages of System Grounding

The advantages of system grounding are:

- To protect life from the danger of electric shock
- To limit the voltage upon a circuit
- To facilitate the operation of electrical apparatus and systems
- To limit voltage on a circuit when exposed to lightning
- To limit AC circuit voltages to ground to 150 volts or less on circuits supplying interior wiring systems
- To keep the potential voltage differentials between the different parts of the system at a minimum, thereby reducing shock hazard; and
- To keep impedance of a ground path at a minimum, which results in higher current flow in the event of a ground fault. The lower the impedance, the greater the current flow, and the faster the overcurrent device opens.

In Figures 2-16 and 2-19, the main service equipment, the service raceways, the neutral bus, the grounding bus, and the hot and cold water pipes have been bonded together to form a *grounding electrode system.*

In the commercial building, at least 6 m of No. 3/0 bare copper conductor is installed in the footing to serve as an additional supplemental electrode.

This conductor, buried in the concrete footing, is supplemental to the water pipe ground. Many interpretations of the *CEC* are possible concerning the grounding electrode systems concept. Therefore, the local inspection authority should be consulted. For instance, some electrical inspectors may not require the bonding jumper between the hot and cold water pipes shown in Figure 2-19. They may determine that an adequate bond is provided through either type of water heater (electric or gas). Other electrical inspectors may require that the hot and cold water pipes be bonded together because some water heaters contain insulating fittings that are intended to reduce corrosion inside the tank caused by electrolysis. According to their reasoning, even though the water heater originally installed contains no fittings of this type, such fittings may be included in a future replacement heater, thereby necessitating a bond between the cold and hot water pipes. The local electrical inspector should be consulted for the proper requirements. However, bonding the pipes together, as shown in Figure 2-19, is the recommended procedure.

Grounding Terms (*Section 0*)

Section 0 provides a definition for the various terms associated with grounding and bonding electrical systems.

- *Bonding:* a low-impedance path obtained by permanently joining all noncurrent-carrying metal parts to ensure electrical continuity. Any current likely to be imposed must be conducted safely.
- *Bonding Conductor:* a conductor that connects the noncurrent-carrying parts of electrical equipment to the service equipment or grounding conductor.
- *Grounding:* a permanent and continuous path to earth with sufficient ampacity to carry any fault current liable to be imposed on it.
- *Grounding Conductor:* the conductor used to connect the service equipment to the grounding electrode.
- *Grounding Electrode:* a buried metal water piping system, or metal rod, plate, or other means buried or driven into the ground so that it is in contact electrically with the ground.

Grounding Rules (*Section 10*)

When grounding service-entrance equipment, the following *CEC* rules must be observed.

- The system must be grounded when maximum voltage to ground does not exceed 150 volts, *Rule 10–106(1)(a)*.
- The system must be grounded if a neutral is used as a circuit conductor (e.g., a 347/600-volt, wye-connected, three-phase, four-wire system), *Rule 10–106(1)(b)*. This is similar to the connection of the system shown in Figure 2-13.
- The identified conductor must be grounded where the midpoint of one transformer is used to establish the grounded neutral circuit conductor, as on a 120/240-volt, three-phase, four-wire delta system, or a single-phase three-wire system, *Rule 10–210(1)*. Only one phase shall be grounded, *Rule 10–210(2)*. Refer to Figure 2-11.
- All grounding schemes shall be installed so that no objectionable currents will flow over the grounding conductors and other grounding paths, *Rule 10–200(1)*.
- The grounding electrode conductor must be connected to the supply side of the service disconnecting means, *Rule 10–204(1)(b)*. It must not be connected to any grounded circuit conductor on the load side of the service disconnect, *Rule 10–204(1)(c)*.
- The identified neutral conductor is the conductor that must be grounded, *Rule 10–210(1)*.
- Tie (bond) exposed noncurrent-carrying parts of the system together, *Rules 10–400, 10–402, and 10–406*.
- The grounding conductor used to connect the grounded neutral conductor to the grounding electrode must not be spliced, *Rule 10–806(1)*. There are exceptions for installations where the grounding electrode conductor may be spliced by either exothermic welding or irreversible compression-type connectors that are listed for that purpose.
- The grounding conductor is to be sized according to *Rule 10–812*.

- The minimum conductor size for concrete-encased electrodes is found in *Table 43, Rule 10–700(3)(a)*. This conductor may alternatively be direct earth buried, *Rule 10–700(3)(b)*.

- The metal water supply system shall be bonded to the grounding conductor of the electrical system, *Rule 10–406(2), Appendix B*.

- The grounding conductor may be connected to a metal water pipe that is run underground. This is also referred to as an in-situ grounding electrode, or also a manufactured grounding electrode or a concrete-encased grounding electrode, *Rule 10–700(2) & (4)*.

- Conductors used for grounding shall be either protected or resistant to any corrosive condition existing. Conductor can be insulated or bare and made of either copper, aluminum, or other acceptable material, *Rule 10–802*.

- A grounding conductor that is free from exposure to mechanical damage shall be permitted to be run along the surface of the building construction without metal covering or protection, if it is rigidly stapled to the construction; otherwise, it shall be in conduit, electrical metallic tubing, or cable armour. See *Rule 10–806(2)*.

- A No. 8 AWG or smaller grounding conductor shall be in conduit, electrical metallic tubing, or cable armour. *Rule 10–806(3)*

- The grounding conductor may be uninsulated in most installations. See *Rule 10–802*.

- Bonding shall be provided around all insulating joints or sections of the metal piping system that may be disconnected, *Rules 10–614(1)(d), 10–902,* and *10–906*.

- The connection to the grounding electrode must be accessible, *Rule 10–902(2)*.

- The grounding conductor must be connected tightly using the proper lugs, connectors, clamps, or other approved means, *Rule 10–906*.

Figure 2-21 illustrates the most common types of manufactured grounding electrodes. These are connected to the neutral in the same manner as a system grounding conductor from the public water main.

1) Concrete-encased electrode

Must be bare copper minimum of 6 m long and within 50 mm of bottom of concrete foundation footing or buried directly in earth. Conductor sized from *Rule 10–700*

For 200 A service minimum No. 3 AWG bare copper, as per *Table 43*.

2) Rod electrode

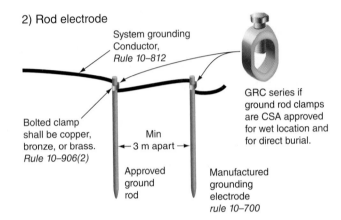

System grounding Conductor, *Rule 10–812*

GRC series if ground rod clamps are CSA approved for wet location and for direct burial.

Bolted clamp shall be copper, bronze, or brass. *Rule 10–906(2)*

Min 3 m apart

Approved ground rod

Manufactured grounding electrode *rule 10–700*

3) Plate electrode

System grounding Conductor, *Rule 10–812*

317 mm

317 mm

Approved ground plate

Non-corrosive (brass or copper) lug (SLU 70) and brass bolt

FIGURE 2-21 Common types of manufactured grounding electrodes.

BONDING

Rule 10–604 lists the parts of the service-entrance equipment that must be bonded. *Rule 10–606* lists the methods approved for bonding this service equipment. Grounding bushings, Figure 2-22, and bonding jumpers are installed on service-entrance equipment to ensure a low-impedance path to ground if a fault occurs on any of the service-entrance conductors, *Rule 10–606(1)*. Service-entrance conductors are not fused at the service head. Thus, the short-circuit current on

FIGURE 2-22 Insulated ground bushing with bonding lug.

FIGURE 2-23 Insulated plastic bushings.

these conductors is limited only by the following: (1) the capacity of the transformer or transformers supplying the service equipment, and (2) the distance between the service equipment and the transformers. The short-circuit current can easily reach 40 000 to 50 000 amperes or more in commercial buildings. These installations are usually served by a large-capacity transformer located close to the service-entrance equipment and the metering. This extremely high fault current produces severe arcing, which is a fire hazard. The use of proper bonding reduces this hazard to some extent.

Rule 10–614(1)(b) states that the main bonding jumpers must have an ampacity not less than the grounding electrode conductor. The grounding lugs on grounding bushings are sized by the trade size of the bushing. The lugs become larger as the size of the bushing increases.

The conductor may be sized according to *Table 41* if used in conjunction with two locknuts and a grounding bushing. *Rule 12–906(2)* states that if No. 8 AWG or larger conductors are installed in a raceway, an insulating bushing or equivalent must be used, Figure 2-23. This bushing protects the wire from shorting to ground as it passes through the metal bushing. Combination bushings can be used. These bushings are metallic (for mechanical strength) and have plastic insulation. When conductors are installed in electrical metallic tubing, they can be protected at the

fittings by the use of connectors with insulated throats.

If the conduit bushing is made of insulating material only, as in Figure 2-23, then two locknuts must be used, Figure 2-24, *Rule 10–610(b)*.

FIGURE 2-24 The use of locknuts. See *Rule 10–610(b)*.

© Thomas & Betts Corp. (www.tnb.com) Used with permission.

FIGURE 2-25 Typical pipe ground clamps.

© Thomas & Betts Corp. (www.tnb.com) Used with permission.

FIGURE 2-26 Grounding conductor connection to water pipe.

— Copper bonding jumper for water meter

Water pipe

FIGURE 2-27 Ground clamp for bonding jumper connection at water meter.

Various types of ground clamps are shown in Figures 2-25 through 2-28. These clamps and their connection to the grounding electrode must conform to *Rule 10–906*.

© Thomas & Betts Corp. (www.tnb.com) Used with permission.

FIGURE 2-28 Ground clamp used to attach grounding conductor to well casing.

Sizing the System Grounding Conductors

The system grounding conductor connects the grounding electrode to the equipment bonding conductor and to the grounded service conductor (neutral) on the supply side of the service disconnecting means. In the commercial building, this means that the system grounding conductor connects the main water pipe (grounding electrode) to the neutral bus, which is bonded to the ground bus, **only** on the line side of the main service (Figure 2-29).

Rule 10–812 is referred to when selecting grounding conductors for services of an AC system. This rule is also used to size the ground conductor when establishing a new service on the secondary side of a wye-connected transformer.

The main bonding jumper (see Figures 2-30 and 2-31) connects the ground bushings of the service conduit to the grounding bus bar in the service disconnect. The jumper is sized according to *Table 41*. The service to the commercial building consists of three 500-kcmil Type RW90 XLPE conductors in parallel. They have a total current-carrying

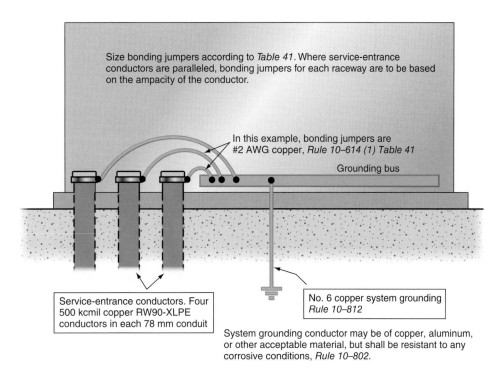

Size bonding jumpers according to *Table 41*. Where service-entrance conductors are paralleled, bonding jumpers for each raceway are to be based on the ampacity of the conductor.

In this example, bonding jumpers are #2 AWG copper, *Rule 10–614 (1) Table 41*

Grounding bus

Service-entrance conductors. Four 500 kcmil copper RW90-XLPE conductors in each 78 mm conduit

No. 6 copper system grounding *Rule 10–812*

System grounding conductor may be of copper, aluminum, or other acceptable material, but shall be resistant to any corrosive conditions, *Rule 10–802*.

FIGURE 2-29 Typical service entrance (grounding and bonding).

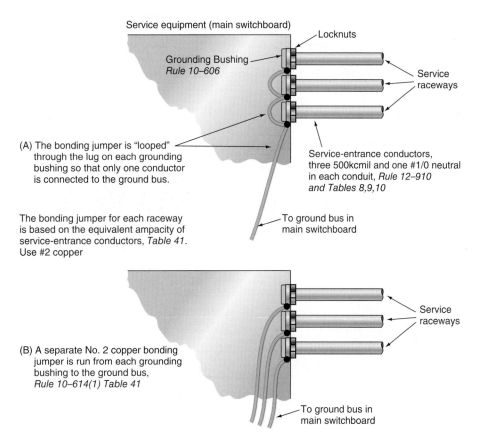

Service equipment (main switchboard)

Locknuts

Grounding Bushing *Rule 10–606*

Service raceways

(A) The bonding jumper is "looped" through the lug on each grounding bushing so that only one conductor is connected to the ground bus.

Service-entrance conductors, three 500kcmil and one #1/0 neutral in each conduit, *Rule 12–910 and Tables 8,9,10*

The bonding jumper for each raceway is based on the equivalent ampacity of service-entrance conductors, *Table 41*. Use #2 copper

To ground bus in main switchboard

Service raceways

(B) A separate No. 2 copper bonding jumper is run from each grounding bushing to the ground bus, *Rule 10–614(1) Table 41*

To ground bus in main switchboard

FIGURE 2-30 Sizing of main bonding jumper for the service on the commercial building, *Rule 10–614* and *Table 41*.

A On a service, standard locknuts and bushings are not adequate, *Rule 10–606(1)(d)*.

Service raceway

Bonding jumper for
service raceways sized
according to *Table 41*

Ground
bus

Service equipment

Solder lugs on connections are not permitted,
Rule 10–904(2)

Main bonding jumper, *Rule 10–614*

Grounding electrode conductor sized
according to *Rule 10–812*

Neutral bus bar

* When No.8 or larger
system conductors are used,
insulating bushing or other
substantial insulating material
securely fastened in place
shall be installed, *Rule 12–906(2)*

Grounding electrode conductor to be installed in one continuous
length without splices or joints,

The enclosure for grounding electrode conductor must be
electrically continuous, *Rule 10–806(4)*

Bonding jumpers sized according to *Rule 10–614* and *Table 41*

Exterior
wall

Grounding
clamps

To interior
water piping

M

Bonding is provided to ensure that the grounding will conduct safely any
fault current likely to be imposed on the system, *Rule 10–002*

B

Eccentric knockouts

Concentric knockouts

FIGURE 2-31 Bonding and grounding of service equipment, *Section 10.*

© Cengage Learning®

capacity of 1290 amperes. In *Table 41*, this requires a No. 3/0 bonding jumper.

According to *Rule 4–024(1)*, the minimum size of a neutral conductor is based on the maximum unbalanced current or the size of the grounding conductor required for the system. This will ensure that the neutral will be able to carry any fault current that might be imposed on it.

When parallel conductors are used, *Rule 10–204(3)* requires a neutral to be run in each raceway that contains the conductors. Based on a computed neutral load of 333 amps, the required neutral conductor for three runs in parallel would be 333/3 = 111 amps. If an RW90 conductor is used, *Table 2* indicates that No. 3 AWG would be the minimum size neutral permitted in each of the three raceways, *Rule 4–022(2)*.

Rule 12–108 requires conductors that are run in parallel to be a minimum of 1/0 AWG.

However, for service conductors in metal raceways, *Rule 10–204(3)* permits neutral (grounded) conductors smaller than 1/0 to be run in parallel as long as they meet the requirements of *Rule 4–024(3)*.

Refer to Figures 2-29 through 2-32 for the proper procedures to be used in bonding electrical equipment.

Table 16 is referred to when selecting equipment bonding conductors when there is overcurrent protection ahead of the conductor supplying the equipment. *Table 16* is based on the ampacity of the circuit conductor.

The electric boiler branch circuit in the commercial building consists of two 350-kcmil conductors per phase protected by 700-ampere fuses in the main switchboard. The use of flexible metal conduit at the boiler means that a bonding wire must be run through each conduit, as shown in Figure 2-33. A

Nonconductive paint, emamel, or similar coating must be removed at contact points when sections of electrical equipment are bolted together to ensure that the sections are effectively bonded. Tightly driven metal bushings and locknuts generally will bite through paints and enamels, thus making the removal of the paint unnecessary, *Rule 10–600*.

FIGURE 2-32 Ensuring bonding when electrical equipment is bolted together.

detailed discussion on flexible connections is given in Unit 11. An illustration of a typical application for the electric boiler feeder in the commercial building is shown in Figure 2-31.

▶ GROUND-FAULT PROTECTION (*RULE 14–102*)

The *CEC* requires the use of ground-fault protection (GFP) devices on services that meet the conditions outlined in *Rule 14–102*. Thus, GFP devices are installed

- On solidly grounded wye services above 150 volts to ground, less than 750 volts between phases, that are rated at 1000 amperes or more (e.g., on 347/600-volt systems)

- On solidly grounded wye services below 150 volts to ground rated at 2000 amperes or more (for example, on 120/208-volt systems)

- To operate at 1200 amperes or less

- So that the maximum time of opening the service switch or circuit breaker does not exceed one (1) second for ground-fault currents of 3000 amperes or more

- To limit damage to equipment and conductors on the *load side* of the service disconnecting means (GFP will *not* protect against damage caused by faults occurring on the *line side* of the service disconnect.)

ELECTRIC BOILER

Lug for No. 3 copper equipment grounding conductors

Flexible metal conduit *Rule 12–1000*

MAIN SWITCHBOARD

Equipment grounding conductors,
- must be run in each conduit, *Rule 10–618(3)*
- are sized according to *Table 16*

Overcurrent protective device

700-Amp Fuses

Equipment grounding bus

Conductors in parallel. *Rules 12–108* and *10–814(3)*

© Cengage Learning®

Each conduit contains three No. 500kcmil "hot" conductors and one No. 3 equipment bonding conductor solidly connected to the grounding bus in the switchboard and to the lugs in the terminal box on the boiler. The No. 3 equipment bonding conductor may be insulated or bare depending on the length of the run and the number of bends, *Rule 10–803(3)*. Note that these conductors are in parallel and must be sized on the basis of the overcurrent protective device in protecting the circuit conductors.

FIGURE 2-33 Sizing bonding conductors for the electric boiler feeder. Note that these conductors are in parallel and must be sized on the size of the feeder conductors.

When a fuse–switch combination serves as the service disconnect, the fuses must have adequate interrupting capacity to interrupt the available fault current (*Rule 14–012*), and must be capable of opening any fault currents exceeding the interrupting rating of the switch during any time that the GFP system will not cause the switch to open.

GFP is not required on

- Delta-connected three-phase systems

- Ungrounded, wye-connected three-phase systems

- 120/240-volt single-phase systems rated at less than 2000 amperes

- 120/208-volt three-phase four-wire systems rated at less than 2000 amperes

- Systems over 750 volts, for example, 2400/4160 volts

- Service disconnecting means rated at less than 1000 amperes

The time of operation as well as the ampere setting of the GFP device must be considered carefully to ensure that the continuity of the electrical service is maintained. The time of operation of the device includes: (1) the sensing of a ground fault by the GFP monitor, (2) the monitor signalling the disconnect switch to open, and (3) the actual opening of the contacts of the disconnect device (either a switch or a circuit breaker). The total time of operation may result in a time lapse of several cycles or more. (See Units 17 and 18.)

GFP circuit devices were developed to overcome a major problem in circuit protection: the low-value phase-to-ground arcing fault, Figure 2-34. The amount of current that flows in an arcing phase-to-ground fault can be low when compared to the rating or setting of the overcurrent device. For example, an arcing fault can generate a current flow of 600 amperes. Yet, a main breaker rated at 1600 amperes will allow this current to flow without tripping, since the 600-ampere current appears to be just another *load* current. The operation of the GFP device assumes that under normal conditions the total instantaneous current in all of the conductors of a circuit will exactly balance, Figure 2-35. Thus, if a current coil is installed so that all of the circuit conductors run through it, the normal current measured by the coil will be zero. If a ground fault occurs, some current will return through the grounding system and an unbalance in the conductors will result. This unbalance is then detected by the GFP device, Figure 2-36.

The purpose of GFP devices is to sense and protect equipment against *low-level ground faults.* GFP monitors do not sense phase-to-phase faults, three-phase faults, or phase-to-neutral faults. These monitors are designed to sense phase-to-ground faults only.

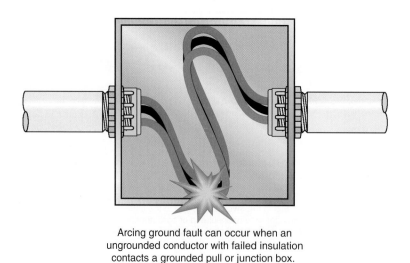

Arcing ground fault can occur when an
ungrounded conductor with failed insulation
contacts a grounded pull or junction box.

FIGURE 2-34   Ground faults.

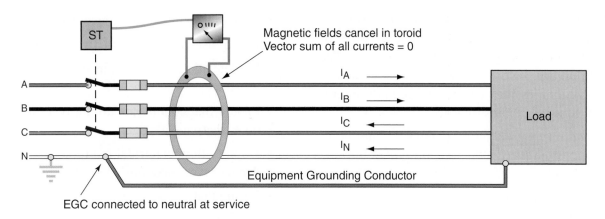

FIGURE 2-35 Normal condition.

Large (high-magnitude) ground-fault currents can cause destructive damage even though a GFP device is installed. The amount of arcing damage depends upon (1) how much current flows and (2) the length of time that the current exists. For example, if a GFP device is set for a ground fault of 500 amperes and the time setting is six cycles, then the device will need six cycles to signal the switch or circuit breaker to open the circuit, whether the ground fault is slightly more than 500 amperes or as large as 20 000 amperes. The six cycles needed to signal the circuit breaker plus the operation time of the switch or breaker may be long enough to permit damage to the switchgear.

The damaging effects of high-magnitude ground faults, phase-to-phase faults, three-phase faults, and phase-to-neutral faults can be reduced substantially by the use of current-limiting overcurrent devices.

These devices reduce both the peak let-through current and the time of opening once the current is sensed. For example, a ground fault of 20 000 amperes will open a current-limiting fuse in less than one-half cycle. In addition, the peak let-through current is reduced to a value much less than 20 000 amperes (see Units 17 and 18).

GFP for equipment, *Rule 14–102,* is not to be confused with personal ground fault protection (GFCI or GFI). (GFCI protection is covered in detail in *Electrical Wiring: Residential.*)

The GFP is connected to the normal fused switch or circuit breaker that serves as the circuit protective device. The GFP is adjusted so that it will signal the protective device to open under abnormal ground-fault conditions. The maximum setting of the GFP is 1200 amperes. In the commercial building plans, the service voltage is a 120/208 volt three-phase system.

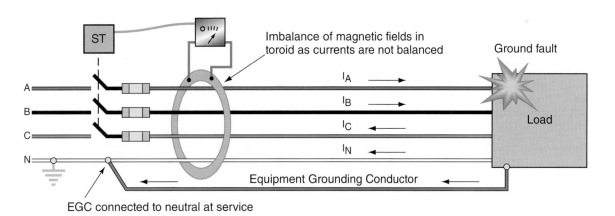

FIGURE 2-36 Abnormal condition.

The voltage to ground on this system is 120 volts. If the ampacity of the service is 2000 amperes or more, then GFP is required, *Rule 14–102(1)(b)*. The electrician can follow a number of procedures to minimize the possibility of an arcing fault. Examples of these procedures follow:

- Ensure that conductor insulation is not damaged when the conductors are pulled into the raceway.
- Ensure that the electrical installation is properly grounded and bonded.
- Ensure that locknuts and bushings are tight.
- Ensure that all electrical connections are tight.
- Tightly connect bonding jumpers around concentric and/or eccentric knockouts.
- Ensure that conduit couplings and other fittings are installed properly.
- Check insulators for minute cracks.
- Install insulating bushings on all raceways.
- Insulate all bare bus bars in switchboards when possible.
- Ensure that conductors do not rest on bolts or other sharp metal edges.
- Do not allow electrical equipment to become damp or wet either during or after construction.
- Ensure that all overcurrent devices have an adequate interrupting capacity.
- Do not work on live panels.
- Be careful when using *fish tape,* since the loose end can become tangled with the electrical panelboard.
- Be careful when working with live parts; do not drop tools or other metal objects on top of such parts.

REVIEW

Note: Refer to the *CEC* or the blueprint package as necessary.

1. A 300-kVA, dry-type transformer bank has a three-phase, 600-volt delta primary and a three-phase, 208-volt wye secondary.

   a. What is the proper size of fuse, in amperes, that must be installed in the secondary?

   _____

   _____

   _____

   _____

   b. What is kVA output if one transformer goes bad and it is necessary to use an open delta bank?

   _____

   _____

   _____

   _____

2. Draw the secondary connections for a three-phase, four-wire, wye-connected transformer bank. Label the locations where the voltage across a transformer is 120 volts.

3. A panelboard is added to an existing panelboard installation. Two knockouts—one near the top and the other near the bottom—are cut in the adjoining sides of the boxes. Indicate on the drawing the proper way of extending the phase and neutral conductors to the new panelboard. Line side lugs are suitable for two conductors.

Feeder conduit
200-ampere feeder

Panelboard interiors for
3-phase, 4-wire supply

© Cengage Learning®

4. List any five of the procedures that can be followed to minimize the possibility and/or severity of an arcing fault.

_____

_____

_____

_____

_____

_____

_____

_____

_____

_____

5. A 5-horsepower, three-phase, 230-volt motor is installed. The motor requires the installation of an equipment bonding conductor in the same conduit as the motor branch circuit. The full-load rating of the motor is 15.2 amperes. The circuit consists of No. 12 Type TW75 wire with an ampacity of 25 amperes. However, the circuit is provided with 30-ampere short-circuit protection. What size copper conductor is used for the equipment bonding conductor?

_____

6. The service of the commercial building consists of three _____ -mm (_____ -inch) conduits, each containing three _____ -kcmil phase conductors plus one No. _____ neutral conductor.

7. What is the purpose of grounding systems and enclosures?

_____

_____

_____

_____

8. What is the kVA capacity of each of the following? (Circle the correct answer.)

   a. Three 75-kVA transformers connected closed delta.

   (1) 225 kVA   (2) 150 kVA   (3) 130 kVA

   b. Two 75-kVA transformers connected open delta.

   (1) 225 kVA   (2) 150 kVA   (3) 130 kVA

   c. Three 150-kVA transformers wye-connected.

   (1) 300 kVA   (2) 450 kVA   (3) 225 kVA

9. Of the following, which is the proper size copper grounding electrode conductor for a 200-ampere service that consists of No. 3/0 RW90 (unjacketed) phase conductors? (Circle the correct answer.)

   a. No. 4 AWG   b. No. 3 AWG   c. No. 2 AWG   d. No. 6 AWG

10. A service is supplied by three 350-kcmil RW90 (unjacketed) conductors per phase. What is the minimum size copper grounding electrode conductor to be used? (Circle the correct answer.)

    a. No. 1/0 AWG   b. No. 2/0 AWG   c. No. 3/0 AWG   d. No. 6 AWG

11. Ground-fault sensing equipment is required on certain types of services. For the following systems, indicate where this equipment is used by inserting in the blanks either R (Required) or N (Not Required).

    a. 1200 A 120/208-volt three-phase system _____

    b. 1200 A 347/600-volt three-phase system _____

    c. Delta systems _____

    d. 2000 A single-phase systems _____

    e. 1200-ampere service, three-phase/four-wire wye, 277/480 volts _____

    f. 800-ampere service disconnect _____

12. The *CEC* requires that if a metal water piping system is available on the premises, the electrical system must be grounded to this water piping system. What type of wire must be used? Cite the pertinent rule.

    _____

13. Equipment is grounded and bonded so that in the event of a fault, the ground path will be able to

    _____

14. *Rule 10–806(1)* states that a grounding electrode conductor shall not be spliced. However, there are some circumstances where it may be impossible to install a grounding electrode conductor in one piece, without splicing. In such situations, *Rule* _____ does allow the grounding electrode conductor to be spliced by means of _____.

15. The engineering calculations for an 800-ampere service-entrance call for two 500-kcmil copper conductors per phase, connected in parallel. The neutral calculations show that the neutral conductors need only be No. 3 copper. Yet, the specifications and riser diagram show that the neutral conductor is No. 1/0 copper, run in parallel. The riser diagram and specifications call for two conduits, each containing three 500-kcmil phase conductors, plus one No. 1/0 neutral conductor. Explain why the neutral conductor is sized No. 1/0.

    _____

    _____

    _____

Branch Circuit and Feeders

OBJECTIVES

After completing the study of this unit, the student will be able to

- calculate the minimum allowable loading for a given area
- calculate the current in each line of a three-phase, four-wire system
- determine the correct size for the branch-circuit protection
- determine correct conductor size
- derate conductors as required
- calculate the voltage drop in single-phase circuits

The electrician often must assume the responsibility for making decisions on the sizing, protection, and installation of branch circuits. Even on well-engineered projects, the electrician will be asked to make numerous choices based on economic considerations. In the commercial building plans given in the blueprints accompanying this text, the selection of the wire type to be used and, therefore, the sizing of the conduit, the derating of the conductor, and the allowance for voltage drop are decisions to be made by the electrician.

▶ HOW EQUIPMENT TERMINAL RATINGS AFFECT CONDUCTOR SELECTION

When selecting the components for a circuit, remember that no circuit is better than its terminations, and *Rule 4–006* ensures that terminations are considered in the selection of conductors. Terminations, like other components in a circuit, have temperature ratings. Breakers, switches, and panelboards all have variable temperature ratings, and for the following example we will use common values of 75°C and 90°C. The ampacities of 90°C conductors are based on the 90°C column from *Tables 1–4* provided that all components have termination ratings of 90°C.

When a conductor rated at 90°C terminates on equipment that has been certified, tested, and marked with a maximum temperature termination rating of 75°C, the values from the 75°C column must be used. By limiting the current to the 75°C column values, we reduce the heat of the conductor and prevent nuisance tripping and other premature related failures.

For example, consider two 200 A fusible disconnects: one is connected to a 200 A nonfusible disconnect (Figure 3-1), and the other is connected to a combination panel with a main breaker, Figure 3-2. Because the breaker has a termination rating of only 75°C, column 3 of *Table 2* is used to determine the size of the conductor and therefore is one size larger than the conductor used in Figure 3-1.

200 Amp Fusible Disconnect   200 Amp Non-Fusible Disconnect
125 Amp Fuses

90°C Rated termination   90°C Rated termination

FIGURE 3-1 Conductor used #2 AWG wire type RW90 XLPE. *Table 2* 90°C column 130 amps.

200 Amp Fusible Disconnect   Insurance office
125 Amp Fuses   125 Amp combination panel

90°C Rated termination   75°C Rated termination

FIGURE 3-2 Conductor used #1 AWG wire type RW90 XLPE. *Table 2* 75°C column 130 amps. *Rule 4–006(1)(2)* and *Appendix B*.

▶ BRANCH-CIRCUIT CALCULATION

According to *Section 0* of the *Canadian Electrical Code (CEC)*, a branch circuit is that portion of the wiring system between the final overcurrent device protecting the circuit and the outlet(s). Thus, all outlets are supplied by branch circuits. An outlet is defined in *Section 0* as a point on the wiring system at which current is taken to supply equipment for a specific use. It can be seen that once the outlet requirements are tabulated, the number and sizes of the branch circuits can be determined.

Rule 8–210 governs the required calculations for the expected feeder loads and the number of branch circuits necessary. It is important to note that *Rule 8–106(2)* requires that the minimum feeder capacity must be based on the load calculated using *Rule 8–210* and *Table 14* or the actual connected load requirements, whichever is larger.

To illustrate the procedure for determining the branch circuits, the branch-circuit loads in the drugstore are divided into three categories: lighting loads, motor loads, and other loads. The *CEC* requirements for load calculations are summarized as follows:

- Use *Table 14* (Table 3–1) to find the watts per square metre for general lighting and receptacle loads or use the connected load if it is larger.

- For show window lighting, use the rating of the equipment installed, *Rule 8–210(b)*.

- The sum of the noncontinuous loading and 125% of the continuous loading shall not exceed the rating of the circuit, *Rule 8–104(1–3)*.

- Allow actual rating for specific loads, *Rule 8–210(b)*.

- For motors, see *Section 28*.

- For transformers, see *Section 26*.

- For electric heating, see *Section 62*.

- For appliances, see *Section 26*.

- For air-conditioning and refrigeration, see *Section 28*.

TABLE 3-1

CEC Table 14.

CEC TABLE 14
WATTS PER SQUARE METRE AND DEMAND FACTORS FOR SERVICES AND FEEDERS
FOR VARIOUS TYPES OF OCCUPANCY
(SEE *RULES 8–003* AND *8–210*)

Type of Occupancy	Watts Per Square Metre	Demand Factor, %	
		Service Conductors	Feeders
Stores, Restaurants	30	100	100
Offices			
First 930 m²	50	90	100
All in excess of 930 m²	50	70	90
Industrial and Commercial	25	100	100
Churches	10	100	100
Garages	10	100	100
Storage Warehouses	5	70	90
Theatres	30	75	95
Armouries and Auditoriums	10	80	100
Banks	50	100	100
Barber Shops and Beauty Parlours	30	90	100
Clubs	20	80	100
Court Houses	20	100	100
Lodges	15	80	100

GENERAL LIGHTING LOADS

Lighting for Display Area of Store

The minimum lighting load to be included in the calculations for a given type of occupancy is determined from *Table 14* or the actual connected load, whichever is greater. For a "store" occupancy, the table indicates that the unit load per square metre is 30 watts. Therefore, the minimum lighting load allowance for the first floor of the drugstore is:

$$(\text{Dimensions of drugstore}) \times (Table\ 14) = \text{Wattage}$$

$$(17.7\ \text{m} \times 6.7\ \text{m}) \times (30\ \text{W/m}^2) = \text{Wattage}$$

$$(118.59\ \text{m}^2) \times (30\ \text{W/m}^2) = 3558\ \text{watts}$$

Since this value is a minimum, it is also necessary to determine the connected load. The greater of the two values will become the computed lighting load, according to *Rule 80–106(2)*. The connected load for the first floor of the drugstore is:

27 Style I luminaires × 87 volt-amperes	= 2349	volt-amperes
4 Style E luminaires × 144 volt-amperes	= 576	volt-amperes
15 Style D luminaires × 74 volt-amperes	= 1110	volt-amperes
2 Style N luminaires × 60 volt-amperes	= 120	volt-amperes
Total connected load	= 4155	volt-amperes

As presented in detail in a later unit, the illumination (lighting) in the main area of the drugstore is provided by fluorescent luminaires equipped with two lamps and a ballast required to operate the lamps. According to the Luminaires Schedule (Drawing E7), this lamp–ballast combination will use 75 watts of power and has a load rating of 87 volt-amperes. (Watts are what we pay for; volt-amperes are what we design for.) It will be the convention in this text to use the word "load" when referring to the watts or the volt-ampere rating of the load. The total connected load, 4155 volt-amperes, is used to determine the computed loading according to *Rule 8–106(2)*. In past practice, it has been common to round luminaire volt-ampere values upward when tabulating the load.

Now, with increasing utilities costs, clients often want an accurate estimate of anticipated energy costs. This requires an accurate presentation of the loads.

Storage Area Lighting

The minimum load allowance for the storage space is:

$$(\text{Dimensions of drugstore storage}) \times (Table\ 14) = \text{Wattage}$$

$$(12.905\ \text{m}) \times (6.7\ \text{m}) \times (Table\ 14) = \text{Wattage}$$

$$(86.463\ \text{m}^2) \times (30\ \text{W/m}^2) = 2594\ \text{watts}$$

The connected load for the storage area (basement) is:

$$9\ \text{Style L luminaires} \times 87\ \text{volt-amperes} = 783\ \text{volt-amperes}$$

Show Window Lighting

Rule 8–210(b) requires the use of the actual connected load for the purposes of determining branch-circuit and feeder sizing. The connected load for the show window is:

3 receptacle outlets × 500 volt-amperes	= 1500	volt-amperes
1 lighting track × 1650 volt-amperes	= 1650	volt-amperes
Total connected load	= 3150	volt-amperes

Determining the Lighting Loading

All of the lighting in the drugstore area has now been tabulated, and the next step is to determine the wattage values that comply with *Rule 8–210*. Adding the drugstore area of 118.59 m² and the drugstore storage area of 86.463 m² gives a total area of 205 m². The minimum demand watts according to *Rule 8–210* and *Table 14* are:

General store area	3558	watts
Storage area	2594	watts
Total minimum allowance	6152	watts

The total connected lighting load for the drug-store is:

General lighting		4155	watts
Storage area lighting		783	watts
Receptacles	21 @ 120VA =	2520	watts
Total connected load		7458	watts

Because the connected load is greater than the minimum allowance, the connected load, which is 7458 watts, will be used in the calculations for the branch circuits and feeders.

MOTOR LOADS (APPLIANCES AND AIR-CONDITIONING AND REFRIGERATION EQUIPMENT)

Rule 28–700 should be referred to for air-conditioning and refrigeration equipment that incorporates hermetic refrigerant motor-compressors.

Units 9 and 11 of this text contain much more additional information relative to the *CEC* requirements for appliance and motor applications.

The drugstore is provided with air-conditioning equipment, Figure 3-3. The data on the nameplate are as follows:

- Voltage: 208 volts, three-phase, three-wire, 60 hertz
- Hermetic refrigerant compressor-motor: rated load current 20.2 amperes, 208 volts, three-phase
- Evaporator motor: full load current 3.2 amperes, 208 volts, single-phase
- Condenser motor: full load current 3.2 amperes, 208 volts, single-phase
- Minimum circuit ampacity: 31.65 amperes
- Overcurrent protection: 50 amperes, time-delay fuse
- Locked rotor current: 140 amperes

It is important to note that when the nameplate specifies a maximum size fuse, the equipment must be protected by fuses; using circuit breakers is not

FIGURE 3-3   Internal load connections of a typical air-conditioner.

approved and would be a violation of *Rule 2–100(4)*. When the nameplate specifies the maximum size fuse or circuit breaker, then either may be used as the branch-circuit protection.

BRANCH CIRCUITS

The actual number of branch circuits to be installed in the drugstore is determined by reference to the specifications, or to the Panelboard Summary shown on Drawing E7 of the blueprints. The number of branch circuits is based on information recorded in Table 3-2, the drugstore feeder loading schedule. The following discussion will illustrate the process used to arrive at the final number of circuits.

The specifications (see Appendix) for this commercial building direct the electrician to install copper conductors of a size not smaller than No. 12 AWG (American Wire Gauge). The maximum load on a No. 12 AWG copper conductor is limited by the maximum allowable overcurrent protective device, which is 20 amperes, unless stated otherwise in the *CEC, Rule 14–104 (6)*.

Continuous loads limit the branch-circuit loads to 80% of the rating of the overcurrent protection device. For the 20 ampere, 120-volt circuit, this would limit the load to $(120 \times 20 \times 0.8) = 1920$ volt-amperes. All of the lighting load in Table 3-2 is considered continuous.

(1110 + 576 + 2349 + 120 + 783 = 4938 W)		
4938 W/1920 W per branch circuit =	3	branch circuits
Add:		
Receptacle circuits (5–15 R receptacles); 1440 W per branch circuit (2520W/1440W) =	2	branch circuits
Roof circuit =	1	branch circuit
Track lighting =	2	branch circuits
Show window rec. =	1	branch circuit
Sign circuit =	1	branch circuit
Cooling circuit =	1	branch circuit
Minimum required for drugstore =	11	branch circuits

Other factors, such as the switching arrangement and convenience of installation, are also important considerations in determining the actual number of circuits used. For example, in the drugstore, 17 circuit breakers are scheduled.

CONSIDERATIONS WHEN SIZING WIRE

When sizing wire, it is important to consider the following five factors:

1. The type of installation and location
2. The type of equipment rating, 100% or 80%, *Rule 8–104, Appendix B*
3. The number of current-carrying conductors and the ambient temperature
4. The voltage drop on the feeder or branch circuit
5. Temperature limitations

Each of these considerations provides specific derating factors that must be compared in order to find the correct size wire to be installed. The factor that requires the **largest size** wire is the determining factor in deciding the wire size to be installed, *Rule 8–104(7)*.

1. *The type of installation and location.* The four types of installation are:
   - Free air
   - Direct burial, Rule 8–104(8)
   - Conduit
   - Cable

   The type used depends on the location—whether it is wet, damp, dry, corrosive, flammable, or hazardous, for example.

2. *The type of equipment rating, 100% or 80%, Rule 8–104(5,6) and Appendix B.* This is divided into two main subdivisions (refer to Table 3-3):

   A. For 100%-rated equipment, Rule 8–104(5), the maximum continuous load is:
   - 85% of the circuit rating for free air installations (the circuit rating is determined by the lesser of the overcurrent device or the wire ampacity, Rule 8–104(1))
   - 100% of the circuit rating for conduit or cable installations

   B. For 80%-rated equipment, *Rule 8–104(6)*, the maximum continuous load is:
   - 70% of the circuit rating for free air installations
   - 80% of the circuit rating for conduit or cable installations

   Most equipment is 80%-rated equipment. This means that it is not marked for 100% operation and is capable of supplying a continuous load at 80% of its marked and/or rated ampacity.

EXAMPLE

A 100-ampere switch that is 80% rated is capable of supplying a continuous load of 80 amperes.

3. *The number of current-carrying conductors and the ambient temperature.* See *Rule 4–004.* Two tables specify derating factors that apply when current-carrying conductors are in contact with each other: *Table 5B* is for free air installations with two to four conductors in contact, and *Table 5C* is for conduit or cable installations where more than three current-carrying conductors are in contact with each other.

TABLE 3-2

Drugstore Feeder Loading Schedule.

ITEM	COUNT	VA/UNIT	DEMAND LOAD	INSTALLED LOAD	DEMAND FACTOR		USE	
					FEEDER	SERVICE	FEEDER	SERVICE
MINIMUM BASIC LOAD								
(CEC Rule 8–210)	205 m²	30 W/m²	6 150		1	1	6 150	6 150
INSTALLED BASIC LOADS								
Style D luminaires	15	74		1 110			1 110	1 110
Style E luminaires	4	144		576			576	576
Style I luminaires	27	87		2 349			2 349	2 349
Style N luminaires	2	60		120			120	120
Style L luminaires	9	87		783			783	783
General receptacles	21	120		2 520			2 520	2 520
Basic installed loads							7458	7458
BASIC LOAD	(Use the greater of the demand or installed loads, *Rule 8–106(2)*.)						7458	7458
SPECIAL LOADS								
Show window rec.	3	500	1 500				1 500	1 500
Track lighting	15	110	1 650				1 650	1 650
Sign	1	1 200	1 200				1 200	1 200
Roof receptacle	1	1 440	1 440				1 440	1 440
Motors	**Volts**	**FLA**	**Phase**					
Cooling unit								
• compressor	208	20.2	3	7 269	1.25		9 086	7 269
• evaporator motor	208	3.2	1	666			666	666
• condenser motor	208	3.2	1	666			666	666
Total load				8 601			23 666	21 849

$$I = \frac{P}{E \times 1.73} = \frac{23\ 666}{208 \times 1.73} = 65.76 \text{ A}$$

Assuming continuous loads, using 75°C rated equipment (*Rule 4–006*) and 90°C RW90-XLPE copper conductors (*Table 2*, 75°C column), the ampere rating of the circuit would be:

$$\frac{65.76}{0.8} = 82.2 \text{ A}$$

The feeder conductors would be four No. 4 AWG RW90-XLPE copper conductors in a 35-mm raceway, supplied by a 100-ampere disconnect c/w 90-ampere fuses. 90 amp fuse allowed as per *Rule 14–104(a)* and *Table 13* of *CEC*. The 35-mm raceway feeds the 90-amp circuit breaker in the drugstore panel. In the above calculations of conductor size, you will notice that both 75°C and 90°C type conductors of No. 4 AWG meet the rule requirements of *CEC Rule 4–006*.

It is important to note that the ampacity of conductors size 1/0 AWG and larger in underground installations is not solely determined from CEC *Tables 1–4*. Conductor ampacity is determined by first referring to *Diagrams B4–1* to *B4–4* in *Appendix B* for an appropriate installation configuration. Once the installation configuration has been determined, refer to *Tables D8* to *D11B* in *Appendix D* for the appropriate ampacity. Be sure to read the notes at the end of the tables to ensure that the proper table is being used.

The ampacity of the conductor shall be the lower of this value or the value calculated

TABLE 3-3

Derating Factors Based on Equipment Rating, *Rule 8–104(5,6).*

TYPE OF INSTALLATION	CODE AMPACITY TABLE	DERATING FACTORS FOR LISTED AMPACITY OF CONDUCTORS	
	Rule 8–104, App. B	Percent of Continuous Operation of Equipment	
		100%	80%
Free Air	Cu *Table 1*	*8–104(5)(b)*	*8–104(6)(b)*
	Al *Table 3*	0.85	0.70
Conduit or Cable	Cu *Table 2*	*8–104(5)(a)*	*8–104(6)(a)*
	Al *Table 4*	1.00	0.80
Direct Burial (Free Air Rating)	Cu *Table 1*	*8–104(8)*	*8–104(8)*
	Al *Table 3*	0.85	0.70
Direct Burial (Cables in Contact or Inadequate Spacing)	Cu *Table 2*	*8–104(8)*	*8–104(8)*
	Al *Table 4*	0.85	0.70

from CEC *Tables 1–4* with application of *Rule 8–104(5,6).*

Another consideration is the ambient temperature, especially the maximum probable air temperature surrounding the conduit or cable. This is important due to the positive temperature coefficient of resistance and the temperature rating of the insulation of the wires. *Table 5A* provides the derating factors for various ambient temperatures.

4. *The voltage drop on the feeder or branch circuit.* The *CEC* permits a total voltage drop on a feeder and branch circuit to be 5%. However, it permits up to 3% on either the feeder or branch circuit. If the feeder voltage drop is 3%, the branch circuit voltage drop cannot be greater than 2%. If the branch circuit voltage drop is 3%, the feeder voltage drop cannot be greater than 2%. *Rule 8–102* and *Table D3* provide the information necessary to ensure safe operating levels of voltage. The voltage drop is usually a maximum of 2% on the feeder and 3% on the branch circuit.

Table D3 may be used to size the wire where long runs will have a significant voltage drop, to ensure that the maximum percent is not exceeded.

To make a safe and proper installation, the electrician will select a wire that will satisfy each of the various derating factors.

5. *Temperature limitations.* The ampacity of a conductor must be based on the equipment's maximum conductor termination temperature. The maximum allowable ampacity shall be based on the corresponding temperature column from *Tables 1–4, CEC 4–006(1).*

EXAMPLE

A No. 12-2 AC90 copper cable is used to feed an electric space heating unit located 40 metres from the panel. The area is rated a dry location. The heater draws 16 amperes at 240 volts. It is fed from a 20-ampere, two-pole 80% rated (not marked 100%) circuit breaker. The maximum ambient temperature is 50°C (122°F).

The branch circuit must satisfy all five of the considerations discussed above.

1. *The type of installation and location.* The installation is in cable, therefore ampacity will be taken from *Table 2* of the *CEC* (Table 3-4). *Table 19* approves Type AC90 for dry locations.

2. *The type of equipment rating, 100% or 80%, Rule 8–104, Appendix B.* The breaker is not marked as 100% rated, therefore the maximum

continuous current that it is rated to carry is 80% of its rating or 20 amperes × 0.8 = 16 amperes.

Rule 8–104(6)(a) allows the maximum continuous load to be 80% of the circuit rating.

The No. 12 wire from *Table 2*, column 4, has a 30-ampere allowable ampacity and the breaker is rated at 20 amperes, therefore the rating of the circuit is 20 amperes, *Rules 8–104(1)* and *14–104(2)*.

$$\text{Maximum continuous load} = 20 \text{ amperes} \times 0.8$$
$$= 16 \text{ amperes}$$

3. *The number of current-carrying conductors and the ambient temperature.* There are two current-carrying conductors. Referring to *Table 5C* (Table 3–5), the correction factor is 1.00. For an ambient temperature over 30°C, *Table 5A* (Table 3–6) gives the correction factor. Column 4 of *Table 5A* at 50°C gives a correction factor of 0.82.

Therefore, the maximum current permitted on the No. 12 wire is: 30 amperes × 1.00 × 0.82 = 24.6 amperes, *Rules 4–004(1)(b), 4–004(8)*, and *14–104(2)*.

4. *The voltage drop on the feeder or branch circuit.* The maximum voltage drop permitted on a branch circuit is 3%. *Table D3* gives a method of calculating the maximum length of wire in a two-wire circuit with the voltage drop on the circuit permitted by the *CEC*. This may be up to 3%, *Rule 8–102(1)*.

The calculation involves:

- AWG of the wire

- current on the wire (load)

- distance of the conductor run (e.g., 40 metres from panel to outlet)

- rated conductor temperature

- percent of allowable ampacity

- type of conductor, Cu or Al (see *Table D3, Note 5*)

- percent voltage drop on feeder or branch circuit (2% or 3%)

- voltage applied to circuit

The following formula may be used to find the maximum length of a run of wire for a given voltage drop.

$$L = \text{TD3 dist.} \times \% \text{ V.D.} \times \text{D.C.F.} \times \text{volts}/120$$

where:

L = Maximum length of two-wire copper conductor run in metres.

TD3 dist. = The distance shown in *Table D3* for the size of wire (AWG heading on top of table) and the actual load in amperes (Current, Amperes column).

% V.D. = The maximum percent voltage drop allowed on the circuit. *Note:* This value should be shown as a whole number, not as a percentage.

D.C.F. = Distance correction factor. This is taken from the table in *Table D3, Note 3*. The correction factor depends on the rated conductor temperature and the percentage of allowable ampacity (% allowable ampacity = Actual load amperes/ Allowable conductor ampacity × 100).

Volts = The voltage at which the circuit operates (e.g., 120, 208, 240).

Using our example of No. 12 AC90 with a 3% voltage drop, baseboard heater 40 metres away, drawing 16 amps at 240 volts, the maximum length of run is:

$$L = \text{TD3 dist.} \times \% \text{ V.D.} \times \text{D.C.F.} \times \text{volts}/120$$
$$= 6.1 \times 3 \times 1.04 \times 240/120 = 38 \text{ m}$$

Since the outlet is 40 metres from the panel, the wire size must be increased to a No. 10 copper conductor, *Rule 8–104(7)*. The No. 10-2 AC90 copper cable is the smallest size that will satisfy *all* of the derating requirements.

5. *Temperature limitations.* All correction factors being applied as required by 4-004 must correspond to the maximum termination temperature marked on the equipment.

TABLE 3–4

CEC Table 2.

TABLE 2
ALLOWABLE AMPACITIES FOR NOT MORE THAN THREE COPPER CONDUCTORS, RATED NOT MORE THAN 5000 V AND UNSHIELDED, IN RACEWAY OR CABLE (BASED ON AN AMBIENT TEMPERATURE OF 30°C*)
(SEE *RULES 4–004, 8–104, 12–2210, 12–2260, 12–3034, 26–142, 42–008, AND 42–016 AND TABLES 5A, 5C, 19, 39, AND D3.*)

	Allowable Ampacity[†][††]					
				110°C[‡]	125°C[‡]	200°C[‡]
Size, AWG or kcmil	60°C[‡]	75°C[‡]	90°C[‡**]	See Note	See Note	See Note
14[§]	15	20	25	25	30	35
12[§]	20	25	30	30	35	40
10[§]	30	35	40	45	45	60
8	40	50	55	65	65	80
6	55	65	75	80	90	110
4	70	85	95	105	115	140
3	85	100	115	125	135	165
2	95	115	130	145	155	190
1	110	130	145	165	175	215
0	125	150	170	190	200	245
00	145	175	195	220	235	290
000	165	200	225	255	270	330
0000	195	230	260	290	310	380
250	215	255	290	320	345	—
300	240	285	320	360	385	—
350	260	310	350	390	420	—
400	280	335	380	425	450	—
500	320	380	430	480	510	—
600	350	420	475	530	565	—
700	385	460	520	580	620	—
750	400	475	535	600	640	—
800	410	490	555	620	660	—
900	435	520	585	655	700	—
1000	455	545	615	690	735	—
1250	495	590	665	745	—	—
1500	525	625	705	790	—	—
1750	545	650	735	820	—	—
2000	555	665	750	840	—	—
Col. 1	Col. 2	Col. 3	Col. 4	Col. 5	Col. 6	Col. 7

Courtesy of CSA Group

* See *Table 5A* for the correction factors to be applied to the values in Columns 2 to 7 for ambient temperatures over 30°C.

† The ampacity of aluminum-sheathed cable is based on the type of insulation used on the copper conductors.

‡ These are maximum allowable conductor temperatures for one, two, or three conductors run in a raceway, or two or three conductors run in a cable, and may be used in determining the ampacity of other conductor types listed in *Table 19*, which are so run, as follows: From *Table 19* determine the maximum allowable conductor temperature for that particular type, then from this table determine the ampacity under the column of corresponding temperature rating.

§ See *Rule 14–104(2)*.

** For mineral-insulated cables, these ratings are based on the use of 90°C insulation on the emerging conductors and for sealing. Where a deviation has been allowed in accordance with *Rule 2–030*, mineral-insulated cable may be used at a higher temperature without decrease in allowable ampacity, provided that insulation and sealing material approved for the higher temperature is used.

†† See *Table 5C* for the correction factors to be applied to the values in Columns 2 to 7 where there are more than three conductors in a run of raceway or cable.

Note: These ampacities apply to bare wire or under special circumstances where the use of insulated conductors having this temperature rating is acceptable.

TABLE 3-5

CEC Table 5C.

TABLE 3-5

CEC Table 5C.

TABLE 5C
AMPACITY CORRECTION FACTORS FOR
TABLES 2 AND 4
(SEE *RULES 4–004* AND *12–2210* AND *TABLES 2* AND *4.*)

Number of Conductors	Ampacity Correction Factor
1–3	1.00
4–6	0.80
7–24	0.70
25–42	0.60
43 and up	0.50

Courtesy of CSA Group

DETERMINING CONDUCTOR SIZE AND TYPE

The insulation on a conductor may become damaged as a result of extremely high temperatures. The insulation can soften, melt, and finally break down, causing grounds, shorts, and possible equipment damage and personal injury.

The source of this damaging heat comes from the surrounding room temperature (ambient temperature) and the heat generated by the current in the wire. The heat generated in the wire can be calculated by

$$\text{Watts} = I^2R$$

TABLE 3–6

CEC Table 5A.

TABLE 5A
CORRECTION FACTORS APPLYING TO TABLES 1, 2, 3, AND 4 (AMPACITY CORRECTION FACTORS FOR AMBIENT TEMPERATURES ABOVE 30°C)
(SEE *RULES 4–004(8)*, *12–2210* AND *12–2260* AND *TABLES 1 TO 4, 57,* AND *58.*)

Ambient Temperature, °C	Correction Factor								
	Insulation Temperature Rating °C								
	60	75	90	105*	110*	125*	150*	200*	250*
35	0.91	0.94	0.96	0.97	0.97	0.97	0.98	0.99	0.99
40	0.82	0.88	0.91	0.93	0.94	0.95	0.96	0.97	0.98
45	0.71	0.82	0.87	0.89	0.90	0.92	0.94	0.95	0.97
50	0.58	0.75	0.82	0.86	0.87	0.89	0.91	0.94	0.95
55	0.41	0.67	0.76	0.82	0.83	0.86	0.89	0.92	0.94
60	—	0.58	0.71	0.77	0.79	0.83	0.87	0.91	0.93
65	—	0.47	0.65	0.73	0.75	0.79	0.84	0.89	0.92
70	—	0.33	0.58	0.68	0.71	0.76	0.82	0.87	0.90
75	—	—	0.50	0.63	0.66	0.73	0.79	0.86	0.89
80	—	—	0.41	0.58	0.61	0.69	0.76	0.84	0.88
90	—	—	—	0.45	0.50	0.61	0.71	0.80	0.85
100	—	—	—	0.26	0.35	0.51	0.65	0.77	0.83
110	—	—	—	—	—	0.40	0.58	0.73	0.80
120	—	—	—	—	—	0.23	0.50	0.69	0.77
130	—	—	—	—	—	—	0.41	0.64	0.74
140	—	—	—	—	—	—	0.29	0.59	0.71
	Col. 1	Col. 2	Col. 3	Col. 4	Col. 5	Col. 6	Col. 7	Col. 8	Col. 9	Col. 10

Courtesy of CSA Group

* These ampacities are applicable only under special circumstances where the use of insulated conductors having this temperature rating is acceptable.

Notes: (1) These correction factors apply to *Tables 1, 2, 3,* and *4.* The correction factors in column 2 also apply to *Table 57.*

(2) The ampacity of a given conductor type at higher ambient temperatures is obtained by multiplying the appropriate value from *Table 1, 2, 3,* or *4* by the correction factor for that higher temperature.

Thus, the more current-carrying conductors that are installed in a single raceway or cable, the more damaging is the cumulative effect of the heat generated by the conductors. This heat must be kept to a safe, acceptable level by either restricting the number of conductors in a raceway or cable, or limiting the amount of current that a given size conductor may carry; *Rules 4–004 and 12–2210*. This means that various derating factors and correction factors must be applied to the ampacity of the conductor being installed. *Ampacity* is defined as "the current in amperes that a conductor can carry continuously under the condition of use without exceeding its temperature rating."

When called upon to connect electrical loads such as lighting, motors, heating, refrigeration, or air-conditioning equipment, the electrician must have a working knowledge of how to select the proper sizes and types of conductors to be installed. Installing the right size wire will ensure that the voltage at the terminals of the equipment is not below the minimums set out in the *CEC*. It also ensures that the insulation will be able to withstand the temperature conditions that exist. The conduit must also be the correct size to facilitate pulling in the conductors without damage; *Rule 12–1014*.

A number of things must be considered when selecting wires. The *CEC* must be followed carefully. It contains terms such as *total demand load, calculated load, intermittent load, noncontinuous load, connected load, load current rating, maximum allowable load current, rated load current, derating factors, correction factors, ambient temperature,* and *ampacity*. These terms have a specific meaning and are very important in the selection of conductors. The electrician should become familiar with each of them.

Table 19 lists the types of wire and cable that are available for use. A person selecting a conductor should always refer to this table to determine if the conductor insulation is appropriate for the conditions of installation; *Rule 4–008*.

Tables 1, 2, 3, and *4* are referred to regularly by electricians, engineers, and electrical inspectors when information about wire sizing is needed. These tables show the allowable ampacities of insulated conductors. The temperature limitations, the types of insulation, the material the conductor is made of (copper, copper-clad aluminum, aluminum), the conductor size in AWG or in kcmil (thousand

circular mils), and the ampacity of the conductor are indicated in the appropriate table. Correction factors for high ambient temperatures are shown in *Table 5A*. See Table 3-6 in the text.

Table 2 (Table 3-4) is used most since it is referred to when the copper conductors are installed in a raceway or cable. All of these tables can be found in the middle section of the *CEC*.

◗ CORRECTION FACTORS FOR AMBIENT TEMPERATURE

The ampacities as shown in *Table 1, 2, 3,* or *4* must be multiplied by the appropriate correction factor from *Table 5A*.

For example, consider a No. 3 TW75 conductor, with an initial ampacity of 100 amperes, being installed in a boiler room where the ambient temperature is expected to reach 42°C. The maximum permitted load current for this conductor would be:

$$100 \text{ amperes} \times 0.82 = 82 \text{ amperes}$$

◗ DERATING FACTORS FOR MORE THAN THREE CURRENT-CARRYING CONDUCTORS IN ONE RACEWAY

Conductors must be derated according to *Rule 4–004, Rule 12–2210,* and *Table 5C*:

- If more than three current-carrying conductors are installed in a raceway or cable Figure 3-4(A).

- When single conductors or cable assemblies are stacked or bundled, as in a cable-tray for lengths over 600 mm (2 ft) without spacing, Figure 3-4(B).

- If the number of cable assemblies (e.g., non-metallic sheathed cable) are run together for distances over 600 mm (2 ft), Figure 3-4(C); *Rule 4–004(13)*.

It is important to note that the *CEC* refers to current-carrying conductors for the purpose of derating

FIGURE 3-4   Bundled and stacked conductors.

when more than three conductors are installed in a raceway or cable. Following are the basic rules:

- **Do** count all current-carrying wires.

- **Do** count neutrals of a three-wire circuit or feeder when the system is four-wire, three-phase, wye connected, Figure 3–5.

- **Do** count neutrals of a four-wire, three-phase, wye-connected circuit or feeder when the major portion of the load is electric discharge lighting (fluorescent, mercury vapour, high-pressure sodium, etc.), data processing, and other loads

where third harmonic (180 Hz) currents flow in the neutral, Figure 3-6.

- **Do not** count neutrals of a three-wire, single-phase circuit or feeder where the neutral carries only the unbalanced current of the "hot" phase conductors, Figure 3-7; *Rule 4–004(3)*.

- **Do not** count equipment bonding conductors that are run in the same conduit with the circuit conductors, Figure 3-8; *Rule 4–006(6)*. However, the bonding conductors must be included when calculating conduit fill; *Rule 12–1014, Appendix B*.

FIGURE 3-5   The neutral of a three-wire circuit supplied by a four-wire, three-phase, wye-connected system will carry the vector sum of the "hot" conductors of the circuit. All three of these conductors are considered to be current-carrying conductors. See *Rule 4–004(4)*.

FIGURE 3-6   The neutral of a four-wire, three-phase circuit that carries only the unbalanced current from other conductors as in the case of normally balanced circuits of 3 or more conductors, shall not be considered a current carrying conductor. See *Rule 4–004(3)*

FIGURE 3-7   The neutral in the diagram is not to be counted as a current-carrying conductor. See *Rule 4–004(3)*.

- **Do not** derate conductors in short sections of raceway 600 mm or less in length, Figure 3-9.

- **Do not** count the noncurrent-carrying "travellers" on a three-way or four-way switching arrangement, Figure 3-10. The actual conductor count is used for purposes of conduit fill calculations. Only one of the travellers is carrying current at any one time, so only one conductor need be counted when determining the ampacity adjustment.

▶ FEEDERS

The discussion so far has concentrated on the branch circuit. This is a necessary first step in determining the size of the feeder that will carry the combined load of all the branch circuits in the designated area. The specific requirements for feeders are given in *Rule 8–210* for buildings other than schools, hospitals, and residential buildings (including hotels, motels, and dormitories). Much of the information has been presented in the discussion of branch circuits.

FIGURE 3-8 In this diagram, there are three current-carrying conductors. The equipment grounding conductor must be included when determining conduit fill. The equipment grounding conductor does not have to be counted when determining conductor ampacity.

FIGURE 3-9 It is not necessary to derate the conductors in short sections of raceway as illustrated in the diagram, but all of the conductors must be included when calculating conduit fill; *Rule 4–004(13)*.

FIGURE 3-10 Counting conductors in a three-way switch installation.

FIGURE 3-11   Service, feeder, and branch circuit conductors.

The Canadian Standards Association (CSA) defines a feeder as "any portion of an electrical circuit between the service box or other source of supply and the branch circuit overcurrent devices." Figure 3-11 illustrates the differences between service, feeder, and branch circuit conductors.

A feeder:

- Must have an ampacity no less than the sum of the loads of the branch circuits supplied by the feeder, except as reduced for demand factors.

- May be reduced in size for loads that are not operated simultaneously, such as air-conditioning and heating loads.

- Neutral may be sized only for the unbalanced portion of the load. See *Rule 4–024(1)*. See Unit 8 for a discussion of neutral currents.

- Does not need to be larger than the conductors for the service supplying the feeder.

The demand factors listed in *Table 14* are allowed because in these specific cases it is highly unlikely that all the loads would be energized at the same time, *Rule 8–210(a)*.

The same logic applies to noncoincident loads as to the demand factors, *Rule 8–106(3)*.

The neutral of a system carries only the unbalanced current in many load conditions. In such cases, it need not be the same size as the ungrounded conductors, *Rule 4–024,* but in no case may a neutral service conductor be smaller than the size set forth in *Rule 10–812* for the grounding electrode conductor. See *Rule 4–024(3)(b)*.

▶ OVERCURRENT PROTECTION AND CIRCUIT RATING

The details for the selection of overcurrent protective devices are covered in Units 17, 18, and 19, and throughout the text.

The rating of a circuit depends upon the rating of the overcurrent protection device or the ampacity of the conductor, **whichever is less.** See *Rule 8–104(1)* and Figure 3-12.

Electrical equipment is rated to carry continuous loads at either 100% of its rating or 80% of its rating. A standard 15-ampere circuit breaker is permitted to carry 80% of its rating (12 amps) when the circuit conductors are rated to carry 15 amps, are run in a raceway or cable, and have an insulation temperature rating of 60°C to 90°C.

A load is considered to be continuous if, in normal operation, it will be energized for more than one hour in any two-hour period. See *Rule 8–104(3)*. Larger

For motors of design B, C, or D, conductors with 75°C insulation are permitted (75°C terminals required at OC protective device).

FIGURE 3-12 Where a larger conductor is installed for any reason, such as to minimize voltage drop, the ampere rating of the overcurrent device determines the branch-circuit classification. In this example, 30-ampere conductors have been installed and connected to the 20-ampere overcurrent device. This circuit is classified as a 20-ampere circuit, *Rule 8–104(1)*.

TABLE 3-7

Maximum Overcurrent Protection (Refer to *Rule 14–104(2)*).

WIRE SIZE	COPPER	ALUMINUM
No. 14 AWG	15 amperes	(not permitted)
No. 12 AWG	20 amperes	15 amperes
No. 10 AWG	30 amperes	25 amperes

loads are considered continuous if they are energized for more than three hours in any six-hour period.

The minimum rating of a branch circuit when standard-rated equipment is used is 125% of the full-load current if the loads are continuous. If the loads are not continuous, then the rating of the branch circuit is 100% of the non-continuous loads. For combined loads, the minimum rating of the circuit is the non-continuous loads plus 125% of the continuous loads.

Conductor overcurrent protection is covered in *Rule 14–104*. The basic requirement is that conductors shall be protected at their ampacity, but several conditions exist when the overcurrent protection can exceed the ampacity. Commonly used maximum overcurrent protection is shown in Table 3-7.

When the ampacity of a conductor does not match that of a standard-size fuse or circuit breaker, the conductor is considered adequately protected by installing the next higher standard-size overcurrent protection device up to 600 amperes. If the rating is above 600 amperes, then the next lower standard size must be selected *(Rule 14–104(a))*.

VOLTAGE DROP

Rule 8–102 requires that the voltage drop in a branch circuit, or in a feeder, not exceed 3% of the system voltage. Also, the total voltage drop, including both the branch circuit and the feeder, must not exceed a total of 5% of the system voltage, Figure 3-13.

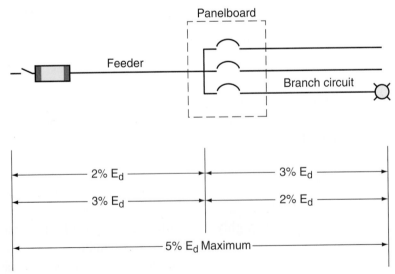

FIGURE 3-13 For light, power, and lighting loads, it is required that the combined voltage drop shall not exceed 5% on the feeder and branch-circuit wiring. As in the above diagram, if the feeder has a 2% voltage drop, then the branch circuit shall have not over a 3% voltage drop. See *Rule 8–102(1)*.

To determine the voltage drop, two factors must be known: (1) the resistance of the conductor, and (2) the current in the conductor. For example, the drugstore window display receptacle outlets are to be loaded to 1500 volt-amperes or 12.5 amperes. The distance from the panelboard to the centre receptacle is 25.9 m (85 ft). The resistance of a No. 12 uncoated copper conductor is 6.33 ohms/km (1.93 ohms per 1000 ft). Therefore the total resistance is:

(6.33 ohms/1000 m) $\times$ 2 $\times$ 25.9 m = 0.328 ohm

(1.93 ohms/1000 ft) $\times$ 2 $\times$ 85 ft = 0.328 ohm

The factor of 2 in the preceding equation is included since both the phase conductor and the neutral conductor carry current; thus, the resistance of each must be included. The voltage drop is now defined by Ohm's law, $E = IR$.

E = 12.5 amperes $\times$ 0.3281 ohm = 4.1 volts

The allowable voltage drop, according to *Rule 8–102*, is:

120 volts $\times$ 0.03 = 3.6 volts (Note that 0.03 = 3%)

Accordingly, it is necessary to use No. 10, which would have a voltage drop of 2.57 volts, for this circuit.

This method of determining the voltage drop neglects any inductance in the circuit. However, inductance usually is insignificant at the short distances that are encountered in commercial wiring. When it is desired to calculate the voltage drop, taking into consideration both the resistance and the reactance, available tables and charts can be used to simplify the calculations. Tables 3-8 and 3-9 show how to determine conductor sizes and voltage loss. Voltage drop is an important consideration because the higher temperature conductors now installed in electrical systems permit the use of a smaller size of conductor for a given load. These smaller conductors often result in excessive voltage loss in the circuit conductors.

▶ DETERMINING VOLTAGE DROP (VOLT LOSS) USING TABLES

With higher current ratings on new insulations, it is extremely important to keep *volt loss* in mind.

How to Figure Volt Loss

The calculation for volt loss is as follows:

$$\text{Volt loss} = \frac{L \times I \times T.V.}{304\ 800}$$

where:

L = length of conduit in metres

I = current in amperes

T.V. = value shown in table for kind of conduit and size of wire to be used.

EXAMPLE

Calculate the volt loss for No. 6 copper wire in 55 metres (180.4 ft) of steel conduit—three-phase, 40-ampere load at 80% power factor.

$$\text{Volt loss} = \frac{55 \times 40 \times 745}{304\ 800}$$

$$= 5.38 \text{ volts}$$

(For a 240-volt circuit, the percentage voltage drop is 5.38/240 $\times$ 100%, or 2.24%.)

These tables take into consideration *reactance on AC circuits* as well as resistance of the wire. Remember on short runs to check that the size and type of wire indicated has sufficient ampere capacity.

How to Select Size of Wire

The calculation to arrive at the correct size of wire is

$$\frac{V.L.}{L \times I} \times 304\ 800$$

where

V.L. = Volt loss

L = Length in metres of one wire

I = Current in amperes

TABLE 3-8

Copper Conductors—Ratings and Volt Loss.

| CONDUIT | WIRE SIZE | AMPACITY | | | DIRECT CURRENT | VOLT LOSS | | | | | | | | | |
| | | 60°C Wire | 75°C Wire | 90°C Wire | | Three Phase (60 Cycle, Lagging Power Factor) | | | | | Single Phase (60 Cycle, Lagging Power Factor) | | | | |
						100%	90%	80%	70%	60%	100%	90%	80%	70%	60%
Steel Conduit	14	20*	20*	25*	6140	5369	4887	4371	3848	3322	6200	5643	5047	4444	3836
	12	25*	25*	30*	3860	3464	3169	2841	2508	2172	4000	3659	3281	2897	2508
	10	30	35*	40*	2420	2078	1918	1728	1532	1334	2400	2214	1995	1769	1540
	8	40	50	55	1528	1350	1264	1148	1026	900	1560	1460	1326	1184	1040
	6	55	65	75	982	848	812	745	673	597	980	937	860	777	690
	4	70	85	95	616	536	528	491	450	405	620	610	568	519	468
	3	85	100	110	490	433	434	407	376	341	500	501	470	434	394
	2	95	115	130	388	346	354	336	312	286	400	409	388	361	331
	1	110	130	150	308	277	292	280	264	245	320	337	324	305	283
	0	125	150	170	244	207	228	223	213	200	240	263	258	246	232
	00	145	175	195	193	173	196	194	188	178	200	227	224	217	206
	000	165	200	225	153	136	162	163	160	154	158	187	188	184	178
	0000	195	230	260	122	109	136	140	139	136	126	157	162	161	157
	250	215	255	290	103	93	123	128	129	128	108	142	148	149	148
	300	240	285	320	86	77	108	115	117	117	90	125	133	135	135
	350	260	310	350	73	67	98	106	109	109	78	113	122	126	126
	400	280	335	380	64	60	91	99	103	104	70	105	114	118	120
	500	320	380	430	52	50	81	90	94	96	58	94	104	109	111
	600	355	420	475	43	43	75	84	89	92	50	86	97	103	106
	750	400	475	535	34	36	68	78	84	88	42	79	91	97	102
	1000	455	545	615	26	31	62	72	78	82	36	72	84	90	95
Non-magnetic Conduit (Lead-Covered Cables or Installation in Fibre or Other Non-magnetic Conduit, etc.)	14	20*	20*	25*	6140	5369	4876	4355	3830	3301	6200	5630	5029	4422	3812
	12	25*	25*	30*	3860	3464	3158	2827	2491	2153	4000	3647	3264	2877	2486
	10	30	35*	40*	2420	2078	1908	1714	1516	1316	2400	2203	1980	1751	1520
	8	40	50	55	1528	1350	1255	1134	1010	882	1560	1449	1310	1166	1019
	6	55	65	75	982	848	802	731	657	579	980	926	845	758	669
	4	70	85	95	616	536	519	479	435	388	620	599	553	502	448
	3	85	100	110	470	433	425	395	361	324	500	490	456	417	375
	2	95	115	130	388	329	330	310	286	259	380	381	358	330	300
	1	110	130	150	308	259	268	255	238	219	300	310	295	275	253
	0	125	150	170	244	207	220	212	199	185	240	254	244	230	214
	00	145	175	195	193	173	188	183	174	163	200	217	211	201	188
	000	165	200	225	153	133	151	150	145	138	154	175	173	167	159
	0000	195	230	260	122	107	127	128	125	121	124	147	148	145	140
	250	215	255	290	103	90	112	114	113	110	104	129	132	131	128
	300	240	285	320	86	76	99	103	104	102	88	114	119	120	118
	350	260	310	350	73	65	89	94	95	94	76	103	108	110	109
	400	280	335	380	64	57	81	87	89	89	66	94	100	103	103
	500	320	380	430	52	46	71	77	80	82	54	82	90	93	94
	600	355	420	475	43	39	65	72	76	77	46	75	83	87	90
	750	400	475	535	34	32	58	65	70	72	38	67	76	80	83
	1000	455	545	615	26	25	51	59	63	66	30	59	68	73	77

Courtesy of Cooper Bussmann

* The overcurrent protection for conductor types marked with an asterisk (*) shall not exceed 15 amperes for No. 14 AWG, 20 amperes for No. 12 AWG, and 30 amperes for No. 10 AWG copper; or 15 amperes for No. 12 AWG and 25 amperes for No. 10 AWG aluminum and copper-clad aluminum after any correction factors for ambient temperature and number of conductors have been applied.

TABLE 3-9

Aluminum Conductors—Ratings and Volt Loss.

CONDUIT	WIRE SIZE	AMPACITY			DIRECT CURRENT	VOLT LOSS									
		60°C Wire	75°C Wire	90°C Wire		Three Phase (60 Cycle, Lagging Power Factor)					Single Phase (60 Cycle, Lagging Power Factor)				
						100%	90%	80%	70%	60%	100%	90%	80%	70%	60%
Steel Conduit	12	20*	20*	25*	6360	5542	5039	4504	3963	3419	6400	5819	5201	4577	3948
	10	25	30*	35*	4000	3464	3165	2836	2502	2165	4000	3654	3275	2889	2500
	8	30	40	45	2520	2251	2075	1868	1656	1441	2600	2396	2158	1912	1663
	6	40	50	60	1616	1402	1310	1188	1061	930	1620	1513	1372	1225	1074
	4	55	65	75	1016	883	840	769	692	613	1020	970	888	799	708
	3	65	75	85	796	692	668	615	557	497	800	771	710	644	574
	2	75	90	110	638	554	541	502	458	411	640	625	580	529	475
	1	85	100	115	506	433	432	405	373	338	500	499	468	431	391
	0	100	120	135	402	346	353	334	310	284	400	407	386	358	328
	00	115	135	150	318	277	290	277	260	241	320	335	320	301	278
	000	130	155	175	252	225	241	234	221	207	260	279	270	256	239
	0000	155	180	205	200	173	194	191	184	174	200	224	221	212	201
	250	170	205	230	169	148	173	173	168	161	172	200	200	194	186
	300	190	230	255	141	124	150	152	150	145	144	174	176	173	168
	350	210	250	280	121	109	135	139	138	134	126	156	160	159	155
	400	225	270	305	106	95	122	127	127	125	110	141	146	146	144
	500	260	310	350	85	77	106	112	113	113	90	122	129	131	130
	600	285	340	385	71	65	95	102	105	106	76	110	118	121	122
	750	320	385	435	56	53	84	92	96	98	62	97	107	111	114
	1000	375	445	500	42	43	73	82	87	89	50	85	95	100	103
Non-magnetic Conduit (Lead-Covered Cables or Installation in Fibre or Other Non-magnetic Conduit, etc.)	12	20*	20*	25*	6360	5542	5029	4490	3946	3400	6400	5807	5184	4557	3926
	10	25	30*	35*	4000	3464	3155	2823	2486	2147	4000	3643	3260	2871	2480
	8	30	40	45	2520	2251	2065	1855	1640	1423	2600	2385	2142	1894	1643
	6	40	50	60	1616	1402	1301	1175	1045	912	1620	1502	1357	1206	1053
	4	55	65	75	1016	883	831	756	677	596	1020	959	873	782	688
	3	65	75	85	796	692	659	603	543	480	800	760	696	627	555
	2	75	90	100	638	554	532	490	443	394	640	615	566	512	456
	1	85	100	115	506	433	424	394	360	323	500	490	455	415	373
	0	100	120	135	402	346	344	322	296	268	400	398	372	342	310
	00	115	135	150	318	277	281	266	247	225	320	325	307	285	260
	000	130	155	175	252	225	234	223	209	193	260	270	258	241	223
	0000	155	180	205	200	173	186	181	171	160	200	215	209	198	185
	250	170	205	230	169	147	163	160	153	145	170	188	185	177	167
	300	190	230	255	141	122	141	140	136	130	142	163	162	157	150
	350	210	250	280	121	105	125	125	123	118	122	144	145	142	137
	400	225	270	305	106	93	114	116	114	111	108	132	134	132	128
	500	260	310	350	85	74	96	100	100	98	86	111	115	115	114
	600	285	340	385	71	62	85	90	91	91	72	98	104	106	105
	750	320	385	435	56	50	73	79	82	82	58	85	92	94	95
	1000	375	445	500	42	39	63	70	73	75	46	73	81	85	86

Courtesy of Cooper Bussmann

* The overcurrent protection for conductor types marked with an asterisk (*) shall not exceed 15 amperes for No. 14 AWG, 20 amperes for No. 12 AWG, and 30 amperes for No. 10 AWG copper; or 15 amperes for No. 12 AWG and 25 amperes for No. 10 AWG aluminum and copper-clad aluminum after any correction factors for ambient temperature and number of conductors have been applied.

Look under the column applying to the type of current and power factor for the figure nearest, but not above, your result and you have the size of wire needed.

EXAMPLE

Calculate the correct size of copper wire in 55 metres of steel conduit for a three-phase, 40-ampere load at 80% power factor. Maximum volt loss from local code is not to exceed 5.5 volts.

$$\frac{5.5}{55 \times 40} \times 304\ 800 = 762$$

Select the number from the table, three phase at 80% factor power, that is nearest but not greater than 762. This number is 745, which indicates the size of wire needed: No. 6 AWG.

ENERGY SAVING CONSIDERATIONS

The rapidly escalating cost of energy encourages the consideration of installing a one-size-larger conductor than is required by the *CEC*. This is particularly true for the smaller size conductors (i.e., Nos. 14, 12, 10, and 8).

The watt loss (heat) in the copper conductors can be calculated as follows:

$$\text{Watts} = I^2R$$

Consider the energy cost comparisons for a typical circuit with a 16-ampere load at a distance of 30.5 m (100 ft) from the panel. The resistance of No. 12 copper is 6.33 milliohms per metre (0.193 ohm per 100 ft), and the resistance of No. 10 copper is 3.97 milliohms per metre (0.121 ohm per 100 ft).

EXAMPLE

Using the No. 12 copper wire, the total watt loss in the conductor for 10 hours per day for 250 days is

$$\begin{aligned}
\text{Watts} &= I^2R \\
&= 16 \times 16 \times 0.006\ 33 \times 2 \times 30.5 \\
&= 98.8
\end{aligned}$$

kWh per year

$$= \frac{98.8 \times 10\text{hours per day} \times 250\text{ days per day}}{1000}$$

$$= 247$$

Cost of watt loss @ $0.10/kWh equals $24.70 per year.

EXAMPLE

Using the No. 10 copper wire, the total watt loss in the conductor is:

$$\begin{aligned}
\text{Watts} &= I^2R \\
&= 16 \times 16 \times 0.003\ 97 \times 2 \times 30.5 \\
&= 62.0
\end{aligned}$$

kWh per year

$$= \frac{62.0 \times 10\text{hours per day} \times 250\text{ days per day}}{1000}$$

$$= 155$$

Cost of watt loss @ $0.10/kWh equals $15.50 per year.

Comparison

Suppose that the list price of No. 12 copper Type T90 conductor is $45.00 per 300 metres. Therefore the cost of 61 m (200 ft) is $9.15.

Suppose that the list price of No. 10 copper Type T90 conductor is $75.00 per 300 metres. Therefore the cost of 61 m (200 ft) is $15.25.

The difference in wire cost is $6.10. The estimated saving is $24.70 − $15.50 = $9.20 per year. The investment in No. 10 copper Type T90 conductors would be returned in less than eight months.

Connections for the wires on the wiring devices terminal could be more time consuming. The electrician might prefer to attach No. 12 pigtails to the terminals of wiring devices and then splice these to the No. 10 circuit conductors.

If the electrical contractor installs 3000 metres of No. 10 copper instead of 3000 metres of No. 12

copper, a considerable energy saving will be experienced. However, the contractor must consider whether or not the *CEC* requires the use of larger-size conduit. (See *Table 6*.) The larger-size conductors also keep voltage drop to a minimum, which permits the connected electrical equipment to operate more efficiently. Reduction of heat produced by the wiring conductors will not only decrease air-conditioning costs for the building, but also increase heating costs. The savings in energy consumption through the use of larger conductors are worthy of consideration.

▶ ALUMINUM CONDUCTORS

The conductivity of aluminum is not as great as that of copper for a given wire size. For example, checking *Table 1,* the ampacity for copper No. 12 Type R90 is 40 amperes, whereas the ampacity for aluminum taken from *Table 3* for No. 12 Type R90 is 35 amperes. As another example, a No. 8 Type T90 Nylon copper wire (*Table 2*) has an ampacity of 55 amperes, whereas a No. 8 Type T90 Nylon aluminum (*Table 4*) or copper-clad aluminum wire has an ampacity of only 45 amperes. It is important to check the ampacities listed in *Tables 1, 2, 3,* and *4* and the footnotes to these tables.

Resistance is an important consideration when installing aluminum conductors. An aluminum conductor has a higher resistance compared to a copper conductor for a given wire size, which therefore causes a greater voltage drop.

Common Connection Problems

Some common problems associated with aluminum conductors when not properly connected may be summarized as follows:

- A corrosive action is set up when dissimilar wires come in contact with one another when moisture is present.
- The surface of aluminum oxidizes as soon as it is exposed to air. If this oxidized surface is not broken through, a poor connection results.

When installing aluminum conductors, particularly in large sizes, an inhibitor is brushed onto the aluminum conductor, then the conductor is scraped with a stiff brush where the connection is to be made. The process of scraping the conductor breaks through the oxidation, and the inhibitor keeps the air from coming into contact with the conductor. Thus, further oxidation is prevented. Aluminum connectors of the compression type usually have an inhibitor paste factory-installed inside the connector.

- Aluminum wire expands and contracts to a greater degree than does copper wire for an equal load. This is another possible cause of a poor connection. Crimp connectors for aluminum conductors are usually longer than those for comparable copper conductors to provide greater contact area for the connection.

▶ PROPER INSTALLATION PROCEDURES

Proper, trouble-free connections for aluminum conductors require terminals, lugs, and/or connectors that are suitable for the type of conductor being installed. See Table 3-10.

Records indicate that the majority of failures in the use of aluminum wire have occurred because of poor installation. The electrician should take special care to make sure that all connections are thoroughly tight. Screwdrivers whose blade is rotated by pushing on the handle should not be used for installation of terminals, since they often give a false impression as to the tightness of the connection.

Terminals on receptacles and switches must be suitable for the conductors being attached. Some equipment, such as the connections to circuit breakers, is required by CSA and the manufacturer to be tightened with a torque wrench to a specified torque. The manufacturer's literature should be consulted and strictly followed when torquing requirements are set forth. *Table D–6* provides recommended tightening torques for wire binding screws.

TABLE 3-10

Conductors Permitted to Connect to Terminals.

TYPE OF DEVICE	MARKING ON TERMINAL OR CONNECTOR	CONDUCTOR PERMITTED
15- or 20-ampere receptacles and switches	CO/ALR	aluminum, copper, copper-clad aluminum
15- and 20-ampere receptacles and switches	NONE	copper, copper-clad aluminum
30-ampere and greater receptacles and switches	AL/CU	aluminum, copper, copper-clad aluminum
30-ampere and greater receptacles and switches	NONE	copper only
Screwless pressure terminal connectors of the push-in type	NONE	copper or copper-clad aluminum
Wire connectors	AL/CU	aluminum, copper, copper-clad aluminum
Wire connectors	NONE	copper only
Wire connectors	AL	aluminum only
Any of above devices	COPPER OR CU ONLY	copper only

Wire Connections

When splicing wires or connecting a wire to a switch, luminaire, circuit breaker, panelboard, meter socket, or other electrical equipment, some type of wire connector is required.

Wire connectors are known in the trade by such names as screw terminal, pressure terminal connector, wire connector, Wing-Nut™, Wire-Nut™, Marrette™, split-bolt connector, pressure cable connector, solderless lug, soldering lug, solder lug, and others. Soldering-type lugs are not often used today. Solderless connectors, designed to establish connections by means of mechanical pressure, are quite common. Examples of some types of wire connectors, and their uses, are shown in Figure 3-14.

As with the terminals on wiring devices (switches and receptacles), wire connectors must be marked *AL* when they are to be used with aluminum conductors only. This marking is found on the connector itself or on or inside the shipping carton.

Connectors marked *AL/CU* are suitable for use with aluminum, copper, or copper-clad aluminum conductors. This marking is found on the connector itself or on or inside the shipping carton.

Connectors not marked *AL* or *AL/CU* are for use with copper conductors only.

Unless specifically stated on or inside the shipping carton, or on the connector itself, conductors made of copper, aluminum, or copper-clad aluminum may not be used in combination in the same connector. Combinations, when permitted, are usually connectors rated for *AL/CU* and a joint compound is used to prevent the oxide film. This is usually limited to dry locations only.

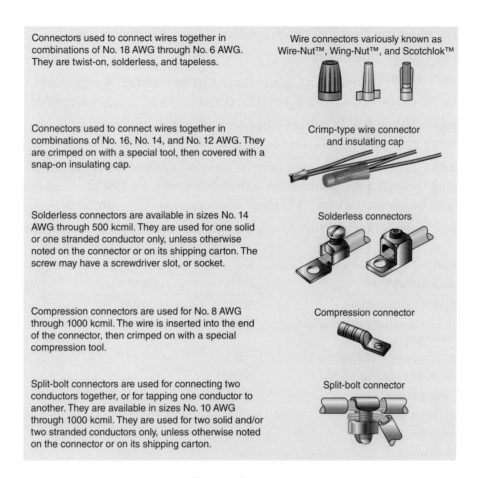

Connectors used to connect wires together in combinations of No. 18 AWG through No. 6 AWG. They are twist-on, solderless, and tapeless.

Wire connectors variously known as Wire-Nut™, Wing-Nut™, and Scotchlok™

Connectors used to connect wires together in combinations of No. 16, No. 14, and No. 12 AWG. They are crimped on with a special tool, then covered with a snap-on insulating cap.

Crimp-type wire connector and insulating cap

Solderless connectors are available in sizes No. 14 AWG through 500 kcmil. They are used for one solid or one stranded conductor only, unless otherwise noted on the connector or on its shipping carton. The screw may have a screwdriver slot, or socket.

Solderless connectors

Compression connectors are used for No. 8 AWG through 1000 kcmil. The wire is inserted into the end of the connector, then crimped on with a special compression tool.

Compression connector

Split-bolt connectors are used for connecting two conductors together, or for tapping one conductor to another. They are available in sizes No. 10 AWG through 1000 kcmil. They are used for two solid and/or two stranded conductors only, unless otherwise noted on the connector or on its shipping carton.

Split-bolt connector

FIGURE 3-14   Types of wire connectors.

REVIEW

Note: Refer to the *CEC* or the blueprint package as necessary.

1. For each of the following, indicate the unit load that is to be included in the branch-circuit calculations.

   a. A restaurant _____

   b. A school room _____

   c. A corridor in a school _____

   d. A corridor in a dwelling _____

2. A 208-volt, single-phase load rated at 7280 volt-amperes is located 30 m (100 ft) from the panelboard. The branch circuit is installed in a conduit to serve this load.

   a. The current in each conductor is _____ amperes.

   b. The minimum Type T90 Nylon conductor size for each conductor is _____ AWG.

   c. The voltage drop will be _____%.

3. When Type R90 conductors are installed in an ambient temperature of 60°C, the ampacity, as shown in *Table 2,* must be corrected by a _____ multiplier.

4. When four Type RW75 current-carrying conductors are installed in one raceway, the ampacity is derated by a factor of _____.

5. Four No. 1 Type RW75 copper current-carrying conductors are installed in a conduit running through a room where the ambient temperature is 45°C. The maximum load that can be connected to each of these conductors is _____ amperes.

6. Complete the following chart to select conductors in compliance with the *CEC.*

CONDUCTOR TYPE	RW75	RW90 XLPE	T90 NYLON
Ambient temperature	27°C	33°C	27°C
Number of current-carrying conductors	2	4	3
Noncontinuous loading	0	19 A	0
Continuous loading	13 A	0	100 A
Derating factor			
Conductor size			
Conductor ampacity			
Terminations	60°C	75°C	90°C
Circuit ampacity			
Minimum overcurrent protection*			
Maximum overcurrent protection*			

*Note that *Rule 14–104(2)* may apply.

7. The ampacity for aluminum conductors, for a given size and type, is less than that for copper conductors. Refer to *Table 3* or *4* and complete the following chart.

CONDUCTOR SIZE & TYPE	COPPER AMPACITY	ALUMINUM AMPACITY
No. 12 AWG RW75		
No. 10 AWG RW75		
No. 3 AWG RW75		
500 kcmil RW75		

8. Is it permissible for an electrician to connect aluminum, copper, or copper-clad aluminum conductors together in the same connector? Yes _____ No _____

9. Terminals of switches and receptacles marked *CO/ALR* are suitable for use with _____, _____, and _____ conductors.

10. Wire connectors marked *AL/CU* are suitable for use with _____, _____, and _____ conductors.

11. (Answer yes or no.) A wire connector bearing no marking or reference to *AL*, *CU*, or *ALR* is suitable for use with

   a. copper conductors only. _____

   b. aluminum conductors only. _____

   c. copper and aluminum conductors. _____

12. Draw a line to match the proper *CEC* section to the specific subject covered.

   a. *Section 28*   1. Grounding

   b. *Section 26*   2. Fuses

   c. *Section 28*   3. Appliances

   d. *Section 10*   4. Motors

   e. *Section 14*   5. Air-conditioning equipment

13. When terminating aluminum wire to a device with a wire binding terminal screw, describe the correct termination procedure and provide the rule number. _____

   _____

   _____

   _____

   _____

14. The branch-circuit rating for an appliance shall not be less than the _____ rating of the appliance.

15. The branch-circuit overcurrent protective device for a typical electric water heater must not exceed 125% of the appliance's ampere rating. The appliance is marked "7.5 kW, 240 volts, single phase." What is the maximum overcurrent protective device?

   _____

16. A disconnect switch is not capable of being locked in the open position 12 metres from the motor and the motor controller. The motor is visible from the disconnect switch.

   a. Does the *CEC* consider this to be acceptable? _____

   b. Give the reference that supports your answer to part (a). _____

   c. How should the installation be changed to meet code? _____

   _____

17. Appliances that have motors plus other loads (e.g., resistance heaters) are required to be marked with their (minimum, maximum) overcurrent protection and (minimum, maximum) supply circuit conductor ampacity. Circle the correct answers.

18. An electric bake oven is rated at 7.5 kW, 208 volts, three phase. The terminals in the appliance, the disconnect switch, and the panel are suitable for 75°C conductors. Determine the following. Show your work.

   a. Current draw

   b. Minimum supply conductor ampacity

   c. Size of Type R90 (unjacketed) conductors

   d. Ampere rating of fuses for branch-circuit protection

   e. Size (ampere rating) of disconnect switch

   f. Conduit size

UNIT 4

House Circuits (Owner's Circuits)

OBJECTIVES

After completing the study of this unit, the student will be able to

- diagram the proper connections for photocells and timer-controlled lights

- list and describe the main parts of an electric boiler control

- understand how to connect a sump pump circuit

Each occupant of the commercial building is responsible for electric power used within that occupant's area. It is the owner's responsibility to provide the power to light the public areas and areas not accessible to the tenants, and to provide heat for the entire building. The circuits supplying the power to those areas and the devices for which the building owner is responsible are referred to as house circuits or owner's circuits.

▶ LOADING SCHEDULE

The loading schedule can be divided into three parts (see Table 4-1):

1. The boiler feed, which has a separate 800-ampere main switch in the service equipment;

2. The emergency power system, including the equipment and devices that must continue to operate if the utility power fails;

TABLE 4-1

Owner's Area Calculation.

ITEM	COUNT	VA/UNIT	DEMAND LOAD	INSTALLED LOAD	DEMAND FACTOR		USE	
					Feeder	Service	Feeder	Service
MINIMUM BASIC LOAD								
(CEC Rule 8–210)	66 m²	25 W/m²	1 650 W		1	1	1 650	1 650
INSTALLED BASIC LOADS								
Style E luminaires	4	144		576			576	576
Style H luminaires	4	87		348			348	348
Style J luminaires	3	150		450			450	450
Style K luminaires	3	192		576			576	576
Style L luminaires	6	87		522			522	522
Style N luminaires	4	60		240			240	240
GFCI receptacles ext.	2	120		240			240	240
General receptacles	8	120		960			960	960
Installed basic loads							3 672	3 672
BASIC LOAD	(Use the greater of the demand or installed loads, *Rule 8–106(2)*.)						3 672	3 672
SPECIAL LOADS								
Telephone outlets	2	1 250					2 500	2 500
Outdoor sign	1	1 060					1 060	1 060
Motors	**Volts**	**FLA**	**Phase**					
Sump pump	120	9.4	1	1 128	1.25		1 410	1 128
Circulating pump	120	4.4	1	528			528	528
Circulating pump	120	4.4	1	528			528	528
Circulating pump	120	4.4	1	528			528	528
Circulating pump	120	4.4	1	528			528	528
Circulating pump	120	4.4	1	528			528	528
Special loads subtotal							4 050	3 768
Total load							11 522	11 240

$$I = \frac{P}{E \times 1.73} = \frac{11\ 522}{208 \times 1.73} = 32.02A$$

Assuming all loads are continuous, using 75°C rated equipment (*Rule 4–006*), and RW90-XLPE copper conductors from *Table 2*, the ampere rating of the circuit would be

$$\frac{32.02}{0.8} = 40.11A$$

The feeder would be four No. 8 RW90-XLPE copper conductors in a 27-mm conduit, supplied by a 60-A disconnect c/w 45-A fuses.

3. The remaining circuits that do not require connection to the emergency system.

FIGURE 4-1   An occupancy (motion) sensor.

CONTROL CIRCUITS

The sump pump lighting circuits are considered to be the owner's responsibility and can be divided into five groups, according to the method used to control the circuit:

- Continuous operation
- Manual control
- Automatic control
- Timed control
- Photocell control

Continuous Operation

The luminaires installed at the top of each of the stairways to the second floor of the commercial building are in continuous operation. Since the stairways have no windows, it is necessary to provide artificial light even in the daytime. The power to these luminaires is supplied directly from the panelboard.

Manual Control

A conventional lighting control system supplies the remaining second-floor corridor lights as well as several other miscellaneous lights. All of the lights on manual control can be turned on as needed.

Automatic Control

The sump pump, in the Motors section of Table 4-1, is an example of equipment that is operated by an automatic control system. No human intervention is necessary—the control system starts and stops the pump so as to keep the liquid in the sump between the preset maximum and minimum levels. Another example is a lighting circuit controlled by a motion detector, Figure 4-1.

Timed Control

The lights at the entrance to the commercial building are controlled by a time clock located near the house panel in the boiler room. The circuit is connected, as shown in Figure 4-2. When the time clock is first installed, it must be adjusted to the correct time. Thereafter, it automatically controls the lights. The clock motor is connected to the emergency panel so that the correct time is maintained in the event that the utility power fails. In the case of daylight saving time, a manual adjustment is required: forward in the spring and backward in the autumn.

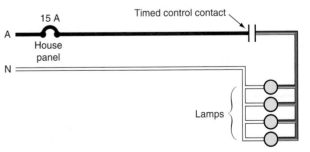

FIGURE 4-2   Timed control of lighting.

FIGURE 4-3 Photocell control of lighting.

Photocell Control

The three lights located on the exterior of the building are controlled by individual photocells, Figure 4-3. Control consists of a light-sensitive photocell and an amplifier that increases the photocell signal until it is sufficient to operate a relay that controls the light. The identified conductor (white) of the circuit must be connected to the photocell to provide power for the amplifier and relay.

◗ SUMP PUMP CONTROL

The sump pump is used to remove water entering the building because of sewage line backups, water main breakage, minor flooding due to natural causes, or plumbing system damage within the building. Since the sump pump is a critical item, it is connected to the emergency panel. The pump motor is protected by a manual motor starter, Figure 4-4, and is controlled by a float switch, Figure 4-5. When the water in the sump rises, a float is lifted that mechanically completes the circuit to the motor and starts the pump, Figure 4-6. When the water level falls and the float lowers, the pump shuts off.

One type of sump pump available commercially is controlled by a water-pressure switch. The water pushes against a flexible neoprene gasket to operate a small integral switch within the submersible pump. This type of sump pump does not use a float.

FIGURE 4-4 Manual motor controller for sump pump.

FIGURE 4-5 Float switch.

◗ BOILER CONTROL

An electrically heated boiler supplies heat to all areas of the commercial building. The boiler has a full-load rating of 200 kW. As purchased, the boiler

FIGURE 4-6 Sump pump control.

is completely wired except for the external control wiring, Figure 4-7. A heat sensor, plus a remote capillary tube and bulb, Figure 4-8, is mounted so that the bulb is on the outside of the building. The bulb must be mounted where it will not receive direct sunlight and must be spaced at least 13 mm (1/2 in.) from the brick wall. If these mounting instructions are not followed, the bulb will give inaccurate readings. The heat sensor is adjusted so that it closes when the outside temperature falls below a set point, usually 18°C (65°F). The sensor remains closed until the temperature increases

FIGURE 4-7 Boiler control diagram.

FIGURE 4-8 Modulating temperature controller.

Courtesy of Honeywell

FIGURE 4-9 Heating control circuit.

to a value above the differential setting, approximately 21°C (70°F). Once the outside temperature causes the heat sensor to close, the boiler automatically maintains a constant water temperature and is always ready to deliver heat to any zone in the building that requires it.

A typical heating control circuit is shown in Figure 4-9, while Figure 4-10 is a schematic drawing of a typical connection scheme for the boiler heating elements.

The following points summarize the wiring requirements for a boiler:

- *Section 62* covers the rules for fixed electric space-heating equipment.

- *Rule 62–114(7)(8)* states the conductor and overcurrent requirements.

- The requirements for heating equipment that use fossil fuels are covered in *Subsection 26–800.*

- Heating equipment of this size will be supplied from its own branch circuit or feeder.

- A disconnect means must be installed to disconnect the ungrounded supply conductors.

- The disconnecting means for fossil-fueled heating equipment must be located so that a person operating the disconnect will not have to walk near or past the heating equipment in order to operate the disconnect, *Rule 26–808(1)(2).*

- The branch-circuit conductor sizes are to be based on 125% of the rated load.

- Overload protective devices are based on the service factor of the motor and are to be sized at 115% or 125% of the rated full load current, *Rule 28–306.*

It should be noted that the boiler for the commercial building is not sized for a particular heat loss; its selection is based on certain electrical requirements. All of the building occupants have a heating thermostat located within their particular areas. The thermostat operates a relay (Figure 4-11) which controls a circulating pump (Figure 4-12) in the hot water piping system serving the area. The circuit to the circulating pump is supplied from the emergency panel so that the water will continue to

Two 300 kcmil per phase

Main lugs rated for 75 °C: connect to two 300 kcmil per phase conductors, run to 800-ampere switch *Rule 62–114(7)*

Heating elements

Relays controlling groups of heating elements in steps or stages through the sequencing control

Groups of elements fused at 70 amperes. Load approximately 56 amperes, which is 80% of the 70 ampere fuse rating. If an element shorts or grounds out, the fuse blows, and only that group loses power. The remaining elements continue to heat.

Boiler: 200 kW, 208 volt, 3-phase, 3-wire

© Cengage Learning®

FIGURE 4-10   Boiler heating elements.

circulate if the utility power fails (Figure 4-13). In subzero weather, this continuous circulation prevents the freezing of the boiler water, which can damage the boiler and the piping system. A thermostat controls a solenoid that opens and closes a zone valve to allow water to flow through the heater unit.

FIGURE 4-11   Relay in CSA Type 1 enclosure.

Courtesy Schneider Electric

FIGURE 4-12   Water circulating pump.

FIGURE 4-13   Installation of Panel "EF" relays, thermostats, conduits, wireway, and motor controller to connect circulating pumps for each occupant.

◼ EMERGENCY AND EXIT LIGHTING

The requirement to install safety systems such as emergency and exit lights is based on Parts 3 and 9 of the national/provincial building code and municipal bylaws. The installation of fire alarm systems is covered by *Section 32* of the *CEC*. The installation of emergency and exit lights is covered by *Section 46* of the *CEC*.

Appendix G of the *CEC* lists references to the electrical installation of fire safety systems required by the *National Building Code of Canada.*

Emergency Lighting

Emergency light fixtures will normally have incandescent or fluorescent lamps, and LED lamps. They must have at least two lamps so that if one lamp fails, the area will not be left in total darkness. They may be supplied by a central emergency power system or from individual units.

Emergency lighting is required to provide a lighting level of 10 lux for a period of one-half to two hours (depending on the occupancy and the height of the building) from the time the normal power is lost. Many installations exceed these requirements.

Installation Requirements

Emergency lighting circuits must be kept independent of all other wiring and must not enter boxes, fittings, or raceways containing the conductors of any other system, except that they may enter transfer switches and light fixtures supplied from both the normal and emergency power supplies, *Rule 46–108*. (See Figure 5-10, page 101.)

Emergency lighting circuits in buildings of noncombustible construction must be installed in a metallic raceway or cable. Rigid PVC and ENT may be used if embedded in 50 mm of concrete or masonry.

CENTRAL SUPPLY

See Unit 5 for a discussion of emergency power systems.

UNIT EQUIPMENT

An emergency lighting unit consisting of lamps, battery, and charger (Figure 4-14) is commonly used in such applications as churches, schools, and offices where a central emergency supply is not available. *Subsection 46-300 CEC* covers the installation requirements of unit equipment.

Some installation requirements of unit equipment are

- Units must be mounted at a height of 2.0 m A.F.F. if possible.

- The receptacle supplying power to the unit must be mounted at a height of 2.5 m A.F.F. and not be farther than 1.5 m from the unit.

- If remote lamps are used, the voltage drop on the circuit supplying the lamps must not exceed 5% of the unit output voltage. See *Table D4, CEC,* for an example of determining voltage drop on unit equipment.

Courtesy of Eaton's Cooper Lighting Business

FIGURE 4-14   Emergency lighting unit.

REVIEW

Note: Refer to the *CEC* or the blueprint package as necessary.

1. What size conduit(s) and wire are required to supply electric power to the boiler?

   _____

   _____

2. An overload protective device with a rating of _____ ampere(s) should be installed in the manual motor controller for the sump pump. (Show calculations.)

3. There are _____ circulating pumps.

4. The total load connected to the house panel is _____ volt-ampere(s).

5. The outside temperature sensor should not be installed _____.

6. Find the maximum trip rating for an inverse time circuit breaker used to provide short-circuit protection to a circulating pump motor branch circuit. (Show calculations.)

7. List four methods used to control lighting circuits originating in the house panel. Indicate a positive or negative feature for each type of control.

   a. _____

   b. _____

   c. _____

   d. _____

8. Where must the disconnect means for the boiler be located?

   _____

9. Conductors supplying an electrically heated boiler must have an ampacity of at least _____% of the rated load as required by *CEC, Rule* _____.

10. The sump pump operation is controlled by a(n) _____.

11. What is the actual current in each of the conductors serving the boiler when all elements are on? (Show calculations.)

Emergency Power Systems

OBJECTIVES

After completing the study of this unit, the student will be able to

- select and size an emergency power system
- install an emergency power system

Many provincial and local codes require that equipment be installed in public buildings to ensure that electric power is provided automatically if the normal power source fails. The electrician should be aware of the special installation requirements for these emergency systems. *Section 46* of the *Canadian Electrical Code (CEC)* covers emergency system requirements.

SOURCES OF POWER (*RULES 46–200 TO 46–212*)

Battery-Powered Lighting (*Rule 46–202(1)(a)*)

When the need for emergency power is confined to a definite area, such as a stairway, and the power demand in this area is low, self-contained battery-powered units are a convenient and efficient means of providing power. In general, these units are wall mounted and are connected to the normal source by permanent wiring methods. Under normal conditions, this regular source powers a regulated battery charger to keep the battery at full power. When the normal power fails, the battery is automatically connected to one or more lamps that provide enough light to the area to permit its use, such as lighting a stairway sufficiently to allow people to leave the building. Battery-powered units are commonly used in stairwells, hallways, shopping centres, supermarkets, and other public structures.

Central Battery Power

If the power demand is high (excluding the operation of large motors), central battery-powered systems are available. These systems usually operate at 32 volts and can service a large number of lamps.

SPECIAL SERVICE ARRANGEMENTS

If separate utility power sources are available (*Rule 6–102(1)*, Figure 5-1), emergency power can be provided at a minimum cost. If protection is required to cover power failures within the building only, then a connection may be made ahead of the main switch, Figure 5-2. However, all of the equipment must have an adequate interrupting rating for the available short-circuit current (see Unit 18).

Emergency source

Regular source

© Cengage Learning®

FIGURE 5-1 Separate services, *Rule 6–102(1)*.

To protect the smaller conductors, cable limiters may be installed at the point where the smaller conductors for the systems are tapped ahead of the main

Emergency source

Switches and fuses must have an interrupting capacity not less than the available short-circuit current at their supply terminals

Main service disconnect means

© Cengage Learning®

FIGURE 5-2   Possible connection ahead of service disconnect means, *Rule 32–208*.

EMERGENCY GENERATOR SOURCE (RULE 46-202(3))

The generator for the commercial building is located in the boiler room and marked "MG" on drawing E2 and E5. Generator sources may be used to supply emergency power. The selection of such a source for a specific installation involves consideration of the following factors:

- Engine type
- Generator capacity
- Load transfer controls

A typical generator for emergency use is shown in Figure 5-3.

Engine Types and Fuels

The type of fuel to be used in the driving engine of a generator is an important consideration in the installation of the system. Fuels that may be used are low pressure gas, natural gas, gasoline, and diesel fuel. Factors affecting the selection of the fuel to be used include its availability and local regulations governing its storage. Natural gas and gasoline engines differ only in the method of supplying the fuel; therefore, in some installations, one of these fuels may be used as a standby alternative for the other, Figure 5-4.

An emergency power source that uses gasoline and/or natural gas usually has lower installation and

Courtesy of Kohler Power Systems

© Cengage Learning®

FIGURE 5-3   Typical generator for emergency use.

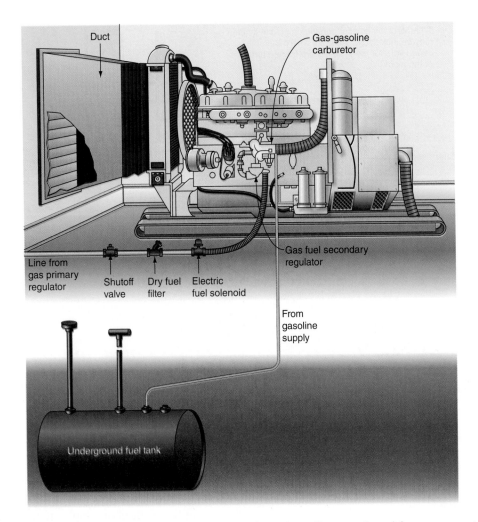

FIGURE 5-4   Typical installation of a natural gas/gasoline engine-driven generator.

operating costs than a diesel-powered source. However, the problems of fuel storage can be a deciding factor in the selection of the type of emergency generator. Gasoline is not only dangerous, but also becomes stale after a relatively short period of time and thus cannot be stored for long periods. If natural gas is used, the Btu content must be greater than 1100 Btu per cubic foot. Diesel-powered generators require less maintenance and have a longer life. A diesel system is usually selected for installations having large power requirements, since the initial costs closely approach the costs of systems using other fuel types. A typical installation using gasoline is shown in Figure 5-5.

Cooling

Smaller generator units are available with either air or liquid cooling systems. Units having a capacity greater than 15 kilowatts generally use liquid cooling. For air-cooled units, it is recommended that the heated air be exhausted to the outside. In addition, a provision should be made to bring in fresh air so that the room where the generator is installed can be kept from becoming excessively hot. Typical installation of an air-cooled emergency generator system is shown in Figure 5-6.

Generator Voltage Characteristics

Generators having any required voltage characteristic are available. A critical factor in the selection of a generator for a particular application is that it must have the same voltage output as the normal building supply system. For the commercial building covered in this text, the generator selected

To muffler

Condensation trap

Flexible line

Drain cock

If exhaust line must be pitched upward, construct a trap of pipe fittings at point of rise

Single-thimble diameter 300 mm larger than or double ventilated thimble 150 mm larger than exhaust line

Holes in end

Inner sleeve

|225 mm| |225 mm|

Exhaust line passing through wall or partition

To load transfer switch box

Support

Large radius type bends

Exhaust line to muffler

To floor drain

Flexible exhaust connection

Note:
The flow of heated air and cooling air may be regulated by automatically controlled shutters

Flexible section

Air outlet duct

Cool air

Canvas section

Important:
Cooling air inlet must be at least 1 1/2 times larger than radiator duct outlet area on radiator cooled models.

Fill pipe

Hooded vent

Fuel line

Fuel return line (from primer tank or "day" tank if used)

Main fuel tank

Restricted outlet

Plug

"Day" tank
Note:
"Day" tank must be located at the same level as the generator

Be sure that a return line is installed from tank to main fuel tank

Flexible line between tank and solid line

To ground

To ignition coil terminal

Return line should pitch downward all the way to supply tank. Avoid dips which trap liquid fuel

Line from fuel pump

Fuel solenoid valve (where used)

Line to carburator

FIGURE 5-5 Typical installation (gasoline).

provides 120/208-volt, wye-connected, three-phase, 60-hertz power.

Capacity (*Rule 46–100*)

It is an involved—but extremely important—task to determine the correct size of the engine-driven power system so that it has the minimum capacity necessary to supply the selected equipment. If the system is oversized, additional costs are involved in the installation, operation, and maintenance of the system. However, an undersized system may fail at the critical time when it is being relied upon to provide emergency power. To size

Automatic
load transfer
control, *Rule 14–612*

Meter

Incoming services

Main fuse and switch

Large radius elbow

Flexible section

Flexible fuel line

Concrete base

Fuel return line

Louvre
near ceiling

To load

Thimble

Muffler

Condensation
trap

Duct to remove
heated engine air

Canvas
section
Air in

Automatic louvres

Hooded vent

Batteries

Fill pipe

Fuel tank

Note: If revolving-field-type
plant is installed in smaller
than approximately
10.9 m³ compartment,
duct generator-heated
air to outside.

Do not join with duct for
engine-heated air.

FIGURE 5–6   Typical installation (pressure).

the emergency system, it is necessary to determine all of the equipment to be supplied with emergency power. If motors are to be included in the emergency system, then it must be determined if all of the motors can be started at the same time. This information is essential to ensuring that the system has the capacity to provide the total starting inrush kVA required.

The starting kVA for a motor is equivalent to the motor's locked rotor kVA, a value that is independent of the voltage characteristics of the motor. Thus, a 5-horsepower Code E motor requires a generator capacity of (see Table 5-1).

$$5 \text{ hp} \times 4.99 \text{ kVA/hp} = 24.95 \text{ kVA}$$

If two motors are to be started at the same time, the emergency power system must have the capacity

to provide the sum of the starting kVA values for the two motors.

For a single-phase motor rated at less than $\frac{1}{2}$ horsepower, Table 5-2 lists the approximate kVA values that may be used if exact information is not available. The power system for the commercial building must supply the following maximum kVA:

Five $\frac{1}{6}$-horsepower C-type motors 5 × 1.85 kVA	= 9.25 kVA
One $\frac{1}{2}$-horsepower Code L motor $\frac{1}{2}$ × 9.99 kVA	= 4.99 kVA
Lighting and receptacle load from emergency panel schedule	= 4.51 kVA
Total	18.75 kVA

TABLE 5-1

Locked-Rotor Indicating Code Letters.

TABLE 430–7(B)
LOCKED-ROTOR INDICATING CODE LETTERS

Code Letter	Kilovolt-Amperes per Horsepower With Locked Rotor		
A	0	—	3.14
B	3.15	—	3.55
C	3.55	—	3.99
D	4.0	—	4.49
E	4.5	—	4.99
F	5.0	—	5.59
G	5.6	—	6.29
H	6.3	—	7.09
J	7.1	—	7.99
K	8.0	—	8.99
L	9.0	—	9.99
M	10.0	—	11.19
N	11.2	—	12.49
P	12.5	—	13.99
R	14.0	—	15.99
S	16.0	—	17.99
T	18.0	—	19.99
U	20.0	—	22.39
V	22.4	—	and up

Courtesy NEMA

Source: Reprinted from *NEMA MG 1-2006 Motors and Generators*, by permission of the National Electrical Manufacturers Association. © Copyright 2006 by the National Electrical Manufacturers Association. All rights including translation into other languages, reserved under the Universal Copyright Convention, the Berne Convention for the Protection of Literary and Artistic Works, and the International and Pan American Copyright Conventions.

Thus, the generator unit selected must be able to supply this maximum kVA load as well as the continuous kVA requirement for the commercial building.

Continuous kVA requirement:	
Lighting and receptacle load	4.510 kVA
Motor load	3.763 kVA
Total	8.273 kVA

A check of manufacturers' data shows that a 12-kVA unit is available having a 20-kVA maximum rating for motor starting purposes and a 12-kVA

TABLE 5-2

Approximate Locked-Rotor kVA Values for Fraction Horsepower Motors.

APPROXIMATE KVA VALUES

HP	Type	Locked Rotor kVA
1/6	C	1.85
1/6	S	2.15
1/4	C	2.50
1/4	S	2.55
1/3	C	3.0
1/3	S	3.25

C = Capacitor-start motor
S = Split-phase motor

continuous rating. A smaller generator may be installed if provisions, such as time delays on the contactors, can be made to prevent the motors from starting at the same time.

Derangement Signals

According to *Rule 46–210,* a *derangement signal,* that is, a signal device having both audible and visual alarms, should be installed outside the generator room in a location where it can be readily and regularly observed. The purposes of a device such as the one shown in Figure 5-7 are to indicate any malfunction in the generator unit, any load on the system, or the correct operation of a battery charger.

Courtesy of Kohler Power Systems

FIGURE 5-7 Audible and visual signal alarm annunciator.

FIGURE 5-8   Automatic transfer control panel.

Automatic Transfer Equipment (*Rule 46–206(1)*)

Equipment must be provided to start the engine of the emergency generator if the main power source fails and to transfer the supply connection from the regular source to the emergency source. These operations can be accomplished by a control panel such as the one shown in Figure 5-8. A voltage-sensitive relay is connected to the main power source. This relay (transfer switch) activates the control cycle when the main source voltage fails, Figure 5-9. The generator motor is started when the control cycle is activated. As soon as the motor reaches the correct speed, a set of contactors is energized to disconnect the load from its normal source and connect it to the generator output.

Wiring

The branch-circuit wiring of emergency systems must be separated from the standard system except for the special conditions noted. Figure 5-10 shows a typical branch-circuit installation for an emergency system.

Relay	1	Voltage sensitive
Relay	2	Voltage and frequency sensitive
Coil	L	Powered from regular source
Coil	E	Powered from emergency source

Contactors L and E are mechanically interlocked so that both cannot be energized at the same time.

FIGURE 5-9   Transfer switch.

Emergency system circuits
must be separated from
standards circuits
Rule 46–108

Fixture on emergency circuit
Rule 46–106

(A) Connected to normal circuit

Barrier not permitted to be used to separate
emergency and normal circuits. Must be a
separate box. *Rule 46–108(4,5)*

Normal branch-
circuit wiring

Separate emergency system
branch-circuit wiring

© Cengage Learning®

FIGURE 5-10   Branch-circuit wiring for emergency systems and normal systems.

Emergency systems must be controlled by automatic transfer equipment, *Rule 46–206(1)*. An automatic light-activated device (photocell) may be used to separately control lights not required during daylight hours, *Rule 46–206(2)*.

Under certain conditions, emergency circuits are permitted to enter the same junction box as normal circuits, *Rule 46–108(4)*. Figure 5-11 shows an exit light that contains two lamps: one connected to the normal circuit, and the other connected to the emergency circuit.

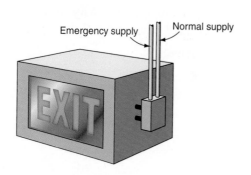

Emergency supply   Normal supply

FIGURE 5-11   Both normal and emergency supply may enter the same junction box, *Rule 46–108(4)(b)*.

REVIEW

Note: Refer to the *CEC* or the blueprint package as necessary.

1. List three different methods of providing an emergency power source.

    a. _____ , b. _____ , c. _____

2. A $7\frac{1}{2}$-horsepower, three-phase, 230-volt Code F motor has a starting rating ranging between _____ kVA and _____ kVA. (Show calculations.)

_____

_____

_____

3. If two $7\frac{1}{2}$-horsepower, three-phase, 230-volt Code F motors are to be started at the same time, the kVA capacity of the generator required is _____ kVA. (Show calculations.)

_____

_____

_____

4. If two $7\frac{1}{2}$-horsepower, three-phase, 230-volt Code F motors are to be operated by a generator, the minimum size generator (in kVA capacity) that can be used with selective control of the motors is _____ kVA. (Show calculations.)

_____

_____

_____

5. Name three conditions that can be indicated by a derangement signal device.

    a. _____, b. _____, c. _____

6. Emergency circuit wiring must be kept (together with) (separate from) normal circuit wiring. (Circle the correct answer.)

7. Three-way, four-way, and series-connected switches (are) (are not) permitted to control emergency lighting. (Circle the correct answer.)

UNIT **6**

Fire Alarm and Safety Systems

OBJECTIVES

After completing the study of this unit, the student will be able to

- define the four stages of a fire
- list the four classes of fire
- list four fire signatures
- explain the purpose of a fire alarm system
- determine types of occupancy in which a fire alarm system is required
- list components of passive systems of fire prevention
- identify and define manual and automatic initiation devices
- list types of fire suppression systems
- determine minimum zoning requirements of commercial buildings for fire alarm systems
- list components of emergency standby systems for fire alarm systems
- identify the *Section 32* requirements for the installation of commercial fire alarm systems

SPECIAL TERMINOLOGY

The following special terminology applies to fire alarm systems:

- *Ancillary device*—A device from another protection system that is actuated by the fire alarm system.

- *Annunciator*—A device used to visually display signals received from a fire alarm system.

- *Class A circuit*—A Class A circuit (return loop circuit) begins at a pair of terminals in the control panel, makes a continuous loop through all the devices connected to the circuit, and terminates at a second pair of terminals in the control panel. A Class A circuit does not require the placement of a separate end-of-line resistor. The end-of-line resistor is already in the panel, connected to the second set of terminals.

- *Class B circuit*—A Class B circuit (terminated circuit) starts at a pair of terminals in the control panel, has one continuous path connecting the devices on the circuit, and terminates at an end-of-line resistor that may be mounted in the field beyond the last device or in the panel.

- *Electrical supervision*—The ability to detect a fault condition that would interfere with the normal operation of the circuit.

- *Fire detector*—A device that can sense a fire condition. It includes heat and smoke detectors.

- *Trouble signal*—An audible and visual signal resulting from failure of equipment or faults on circuits.

- *Alarm signal*—An audible signal used to advise occupants of a building of an (emergency) fire condition.

- *Single-stage fire alarm system*—A system that will cause an alarm signal to sound immediately on all the audible alarm devices of the system when any pull station or automatic detector is operated.

- *Two-stage fire alarm system*—A system in which the operation of the first pull station or automatic detector initiates an alert signal. This is followed by an alarm signal if the alert signal is not acknowledged within five minutes.

FOUR STAGES OF A FIRE

In order to understand the purpose of a fire alarm system, it is necessary to understand what fire is and how it works.

A fire consists of four main stages: (1) incipient stage; (2) smouldering stage; (3) flame stage; and (4) heat stage. In each of these stages there are identifying characteristics that may be detected by various means. Refer to Figure 6-1. The earlier the detection of a fire, the less the damage and the quicker it may be extinguished.

1. *Incipient Stage:* There is no smoke, and the heat being generated is almost undetectable. The fire is just beginning to appear.

2. *Smouldering Stage:* Smoke is now visible and various toxic gases are present. The fire is now readily detectable. It is important to note that life cannot be sustained where the smoke level in the air exceeds 6%. Most fire fatalities are caused by smoke inhalation and subsequent damage to lung tissue. Even at this stage, some self-contained breathing apparatus (SCBA) is necessary to sustain life.

3. *Flame Stage:* Flame is now present but may be invisible, depending on the type of material being consumed.

4. *Heat Stage:* In this final stage, structural steel can buckle and the shattering of glass windows adds to the hazards. Uncontrolled heat and rapidly expanding air complete the development of the fire.

Fire produces characteristic signatures, namely the presence of aerosols, the formation of smoke, visible light from flames, and the radiation of heat. See Figure 6-2. Fires produce smoke and toxic gases that can overwhelm occupants if they are sleeping. Most fatalities are caused from the inhalation of smoke and toxic gases rather than from burns. Fire is the third leading cause of accidental death in Canada.

Heavy smoke reduces visibility and causes disorientation on the part of those trying to escape. Thus, there is a great need for clear and easily understood signs to indicate the exits from the building and the location of fire-fighting equipment.

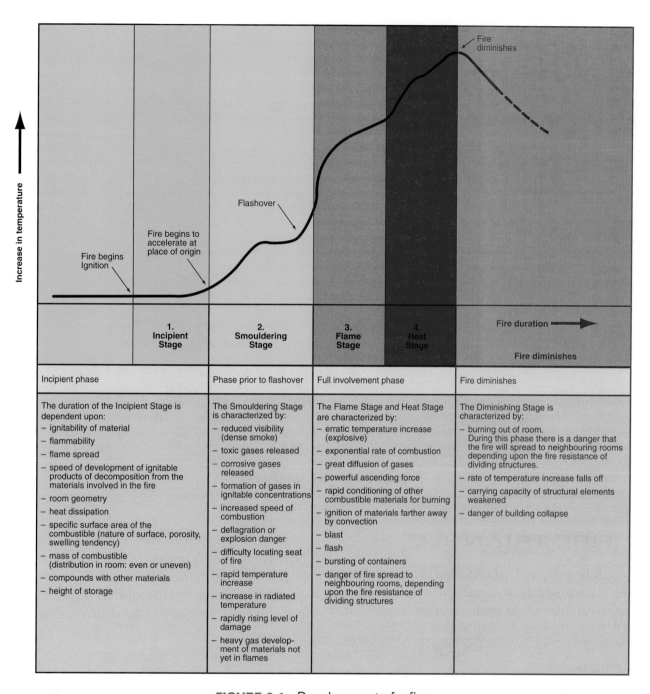

FIGURE 6-1   Development of a fire.

FOUR CLASSES OF FIRE

The four classes into which fire can be divided are:

- Class A—Wood, paper, rags
- Class B—Vapour, air, gas, paint
- Class C—Electrical
- Class D—Combustible metals

A manual fire extinguisher will list the classes of fire that it is designed to extinguish on a label that is attached to the cylinder that holds the extinguishing agent.

Different extinguishing agents are required for the various classes of fire. Water under pressure is the most common. It deprives the fire of oxygen as it expands and turns into steam. Powder, foam, carbon dioxide, and halon are used in specialized situations.

		Fire signatures (by-products of fires)			
		Aerosols: Invisible smoke (small charged particles)	Visible smoke	Flames	Heat
Stages of development of a typical class A fire	1. Incipient stage	Yes	No	No	No
	2. Smouldering stage	Varies	Yes	No	No
	3. Flame stage	Varies	Varies	Yes	Yes
	4. Heat stage	Varies	Varies	Yes	Yes
Type of detector used to sense the fire		Ionization smoke detector	Photoelectric or beam smoke detector	Flame detector	Heat detector

FIGURE 6-2 Fire signatures.

FIRE TRIANGLE

Successful combustion requires all of the components of the fire triangle: heat, oxygen, and fuel. There is a fourth part, called the chain reaction component, which assists the fire; this will be discussed later.

Fire will not be able to start or grow if any of the following occurs:

1. Fuel is removed or controlled.

2. Oxygen is removed or controlled.

3. Heat is removed or controlled.

4. Chain reaction is removed or controlled.

The fire may be extinguished by smothering, because this will remove oxygen to below the critical level. It can also be extinguished by dilution of the combustible material or by cooling the material below the fire point.

Fire is a chemical reaction involving the three basic elements and the resultant continuing chain reaction. An unimpeded chain reaction is essential for its progress. The fire triangle of fuel, oxygen, and heat thus evolves into the fire tetrahedron, Figure 6-3. Any type of fire fighting or extinguishing revolves around removing one component of the chain reaction.

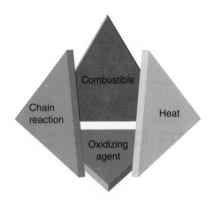

A combustible, heat, and an oxidizing agent must combine in a chain reaction for fire to occur.

FIGURE 6-3 Fire tetrahedron.

PURPOSE OF FIRE ALARM SYSTEMS

The primary purpose of fire alarm systems is to protect lives. The secondary purpose is to protect property and to initiate fire fighting in the event of a fire.

It should be noted that the best way to fight a fire is to prevent it from starting by avoiding circumstances that could initiate a fire. Modern early warning fire detection devices enable fires to be detected and acted upon before they can threaten life and property.

Once the alarm has been initiated, every occupant of the building can be automatically warned so that safe, controlled evacuation of the building can occur. In addition, the system can have the provision to automatically signal a monitoring company, which can then contact the local fire department.

Whenever a fire occurs with significant loss of life, the ensuing investigation will endeavour to find the cause and make recommendations to upgrade the fire code and building code. This will help prevent these situations from recurring.

TYPES OF OCCUPANCY REQUIRING A FIRE ALARM

A fire alarm system is required if any of the following conditions are met:

- If the maximum permitted occupancy is greater than 300

- If the occupant load for any major occupancy is exceeded

- If the occupant load is greater than 150 above or below the first storey

- If the building, including basements and levels below grade, is more than three storeys in height

- If there is a contained use area, such as is found in jails or mental health centres

- If there is an impeded egress zone

- If the building contains a school, college, or daycare facility with an occupancy greater than 40

- If the building contains a licensed beverage establishment or a restaurant with an occupant load of more than 150

- If the building contains a residential occupancy with sleeping accommodations of more than 10 occupants

- If the building contains most industrial occupancies

- If the building contains an open-air seating area for more than 300 persons

These requirements are intended for guidance only and the authority having jurisdiction (AHJ) should be contacted to ensure that local building code and fire code regulations are complied with. The AHJ is usually the fire department for the city or region.

CLASSIFICATION OF MAJOR OCCUPANCIES

Major occupancies are divided into six main categories, with the following designations:

A—Assembly, for example, theatres, arenas, restaurants, churches, schools
B—Institutional, for example, hospitals, jails
C—Residential, for example, homes, hotels
D—Business, for example, offices
E—Mercantile, for example, shopping plazas
F—Industrial, for example, flour mills, garages, dairies

Categories A, B, and F are further subdivided, as shown in Figure 6-4. For further information, refer to Part 3 of the *National Building Code of Canada (NBC)*.

Determining Occupant Load

Every building is designed for a certain occupant load by the architects. This is calculated on a square metre per person basis. For example, in an office, the occupant load is 9.3 m^2 per person.

GROUP	DIVISION	DESCRIPTION OF MAJOR OCCUPANCIES
A	1	Assembly occupancies intended for the production and viewing of the performing arts
A	2	Assembly occupancies not elsewhere classified in Group A
A	3	Assembly occupancies of the arena type
A	4	Assembly occupancies in which provision is made for the congregation or gathering of persons for the purpose of participating in or viewing open-air activities
B	1	Institutional occupancies in which persons are under restraint or incapable of self-preservation because of security measures not under their control
B	2	Institutional occupancies in which persons because of mental or physical limitations require special care or treatment
C	—	Residential occupancies
D	—	Business and personal services occupancies
E	—	Mercantile occupancies
F	1	High hazard industrial occupancies
F	2	Medium hazard industrial occupancies
F	3	Low hazard industrial occupancies

FIGURE 6-4 Classification of major occupancies.

AREAS WHERE FIRE DETECTORS ARE TO BE INSTALLED

- Storage rooms that are outside dwelling units
- Service rooms that are outside dwelling units
- Janitor's rooms
- Elevator shafts
- Rooms used to store hazardous materials

AREAS WHERE HEAT DETECTORS ARE TO BE INSTALLED

- In every room in an area of a building classified as Group A Division 1, or Group B other than sleeping rooms
- Except in a hotel, in every suite, and every room not located within a suite, in portions of buildings classified as Group C and buildings more than three storeys in height

- In a hotel, in every room in a suite, and in every room not located within a suite, other than saunas and swimming pools or refrigerated areas

AREAS WHERE SMOKE DETECTORS ARE TO BE INSTALLED

The areas where smoke detectors are required are covered in the *NBC*. These are summarized as follows:

- Group A Division 1—every corridor
- Group B—every sleeping room including jail cells
    —every corridor or hallway serving as a means of egress from sleeping rooms
- Group C—every public corridor
    —every exit stair shaft
    —every room and corridor serving a contained use area
    —in the vicinity of draft stops, as per *Section 3.2.8.7* of the *NBC*

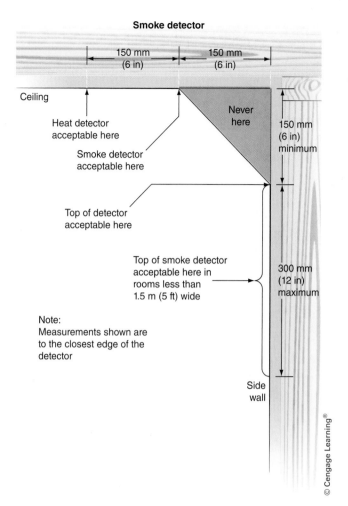

FIGURE 6-5   Ceiling- or wall-mounted detectors.

For the specific location on a ceiling or wall where the detector should be placed, refer to Figure 6-5.

▶ TYPES OF SYSTEMS

There are three building systems that protect occupants of buildings and the structure itself from fire.

1. Passive systems
2. Detection/signalling systems
3. Suppression systems

Passive Systems

Passive systems include a variety of building components and choices made during the design and construction of a building. Many of these choices are mandated by the building and fire safety codes. These would include fire walls, separation between hazardous locations, fire resistance ratings, and flame spread ratings together with smoke control and fire stops.

There must be approved exit systems to give occupants a means of emergency egress from a building.

High-rise buildings pose special problems because of the stack effect. Created by wind and temperature differences, the stack effect can cause stairwells and corridors to fill with smoke during a fire.

Detection/Signalling Systems

A fire alarm system that is able to detect a fire and then warn the occupants and signal the fire department is another type of fire protection.

Fire alarm systems fall into two categories: (1) local systems and (2) proprietary systems. Most are local fire alarm systems. A local fire alarm system has three main components: (1) manual and automatic initiating and detection devices, (2) audible alarm signal devices, and (3) a central control unit to process the information. Refer to Figure 6-6.

An annunciator panel at a remote location could be connected to the fire alarm system to indicate trouble or alarm in a zone. See Figure 6-7.

Alarm detection and initiation can be manual or automatic. Manual pull stations provide for manual alarm initiation, Figures 6-8A and 6-8B.

Heat and smoke detectors form a major category of automatic detection systems. Heat detectors can be either fixed temperature or a combination of fixed temperature and rate-of-rise. A dual action, fixed temperature and rate-of-rise heat detector is shown in Figure 6-9. Smoke detectors can be of the ionization, photoelectric, or beam type. A typical ceiling-mount smoke detector is shown in Figure 6-10.

Alarm signalling is usually in the form of bells, horns, and sirens. In areas where there is a high noise level or there are hearing-impaired occupants, visual signals such as strobe lights are used in addition to bells. Many modern systems use voice paging and tone signals on one-way speaker systems.

Detection or initiating circuits

Signalling or indicating circuits

Remote annunciators

Primary electrical power supply

Control equipment

Off-site alarm connection

Ancillary devices

FIGURE 6-6 Simple fire alarm system.

Courtesy of Honeywell

FIGURE 6-7 Typical back-lit annunciator.

Suppression Systems

The various fire extinguishing methods can be placed into four groups:

- Diluting
- Cooling
- Smothering
- Inhibiting chain reaction

Among these methods there are a variety of extinguishing agents. Water is the most common because it is inexpensive and easily available. Wet pipe sprinklers constitute 97% of sprinkler system

Courtesy of Honeywell

FIGURE 6-8A Manual fire alarm pull station.

installations, provide a very effective method of fire suppression, and are mandatory in new construction of almost all commercial buildings in Canada. The sprinkler system interconnects with

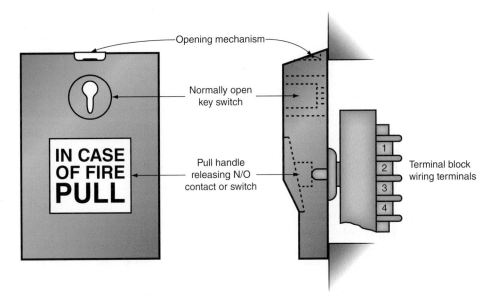

FIGURE 6-8B   Typical pull station assembly.

FIGURE 6-9   Rate-of-rise heat detector.

the fire alarm system. Figure 6-11 shows a typical sprinkler valve with initiating devices that are supervised by the fire alarm system. Various components of the sprinkler system could be monitored for the following indications of trouble:

- Position of the supply valve
- Loss of power to the fire pump
- Low temperature in the supply tank
- High or low air pressure in pressure tanks
- Failure of automatic fire pump controls

FIGURE 6-10   Smoke detector.

Courtesy of Honeywell

FIGURE 6-11   Sprinkler system initiating device.

Other extinguishing systems include carbon dioxide, dry chemical, foam, and halon gas.

SMOKE DETECTORS

Two types of smoke detectors are commonly used: photoelectric and ionization.

Photoelectric Type

The photoelectric type of smoke detector has a light sensor that measures the amount of light in a chamber. When smoke is present, light is reflected off of the smoke in the chamber, triggering the alarm. Refer to Figure 6-12A. This type of sensor detects smoke from burning materials that produce great quantities of white or grey smoke, such as furniture, mattresses, and rags. This type of detector is less effective for gasoline and alcohol fires, which do not produce heavy smoke. Additionally, it is not as effective with black smoke, which absorbs light rather than reflecting it.

Ionization Type

In an ionization type of detector, a low-level radioactive source supplies particles that ionize the air in the smoke chamber of the detector. Plates in this chamber are oppositely charged. Because the air is ionized, an extremely small amount of current (millionths of an ampere) flows between the plates. Smoke entering the chamber impedes the movement of the ions, reducing the current flow, which causes an alarm to go off. This type of detector is able to sense products of combustion (POC) in the incipient stage of a fire and is sometimes referred to as a POC detector. It can detect airborne particles (aerosols) as small as five microns in diameter.

The ionization type of detector is effective for detecting small amounts of smoke, as in gasoline and alcohol fires, which are fast-flaming. Also, it is more effective in detecting dark or black smoke than the photoelectric type of detector. A comparison of the effectiveness of the photoelectric smoke detector and the ionization smoke detector is shown in Figure 6-12B.

Features of Smoke Alarms

Smoke alarms may contain an indicator light to show that the unit is functioning properly. They may also have a test button. When the button is pushed, it is the circuitry and alarm that are tested, not the smoke-detecting ability.

Heat Detectors

Heat detectors are available that sense a specific *fixed* temperature, such as 57°C (135°F) or 90°C (194°F). Available also are *rate-of-rise* heat detectors that sense rapid changes in temperature (6°C [14°F] per minute), such as those caused by flash fires. Fixed and rate-of-rise temperature detectors are available as a combination unit.

The spacing of heat detectors shall be as recommended by the manufacturer because each detector is capable of sensing heat within a limited area in a given time limit.

Wiring Requirements

Class A and Class B Circuits.   There are two classes of fire alarm system circuits—Class A and Class B. Class B is a two-wire system that has an end-of-line resistor and represents 80% to 90% of installed systems. A single break or ground fault in the wiring will initiate a trouble signal, but an alarm may not be sent in to the control panel.

FIGURE 6-12A   Light scattering principle of smoke detector.

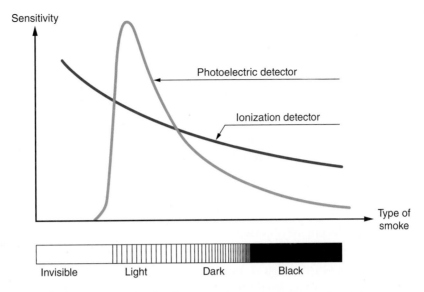

FIGURE 6-12B   Ionization vs. photoelectric smoke detector.

A Class A fire alarm system is a four-wire system that originates and terminates in a terminal strip in the panel. It does not require an end-of-line resistor, since an end-of-line resistor is already mounted on a module located inside the panel. A single break or ground fault in the wiring will not prevent an alarm from being sounded, and a trouble signal will be initiated as soon as the ground fault or break in the wiring occurs.

Class B Circuits.   Figure 6-13 shows a typical Class B initiating circuit. Figure 6-14 shows typical Class A and Class B signal circuits.

Because of the vital importance of smoke and fire detectors in preventing loss of life, it is suggested that the student obtain a copy of Underwriters Laboratories of Canada (ULC) standards.

Also, read very carefully the application and installation manuals that are prepared by the manufacturers of the smoke and heat detectors.

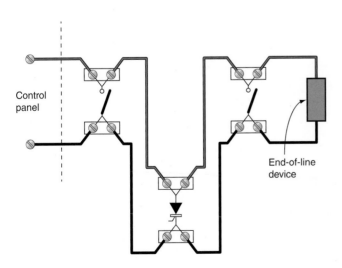

FIGURE 6-13   Class B detection circuit.

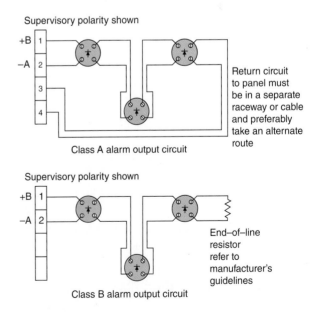

FIGURE 6-14   Class A & B signal circuits.

ZONING REQUIREMENTS OF FIRE ALARM SYSTEMS

If a building is not more than three storeys in height and has a total floor area of not more than 2000 m^2, then the fire alarm system does not require zoning. Even though this is the case, it is recommended that the fire alarm system installed in any building be zoned to provide quick location of a trouble or alarm signal in the building.

Figure 6-15 shows a typical placement of a five-storey building for the alarm initiating circuits. These would include smoke, heat, and rate-of-rise detectors in the required locations. In addition to this would be the signal circuit placement to ensure adequate sound coverage to alert the occupants of an alarm.

FIRE CODES AND STANDARDS

Underwriters Laboratories of Canada

The ULC provides standards for the installation, verification, and inspection of fire alarm systems. These standards are:

- CAN/ULC S524-06, "Standard for the Installation of Fire Alarm Systems"

- CAN/ULC S537-04, "Standard for the Verification of Fire Alarm Systems"
- CAN/ULC S536-04, "Standard for the Inspection and Testing of Fire Alarm Systems"

Minimum requirements are stated in the standards.

Fire warning devices commonly used in a commercial building are heat detectors, rate-of-rise detectors, and smoke detectors. These must be connected to a central fire alarm control panel that will process the information and send an output signal to alarm bells or buzzers.

In some provincial jurisdictions, it is mandatory that qualified technicians install, verify, and maintain fire alarm systems. The certification to become a fire alarm technician is available through a variety of government-approved courses including, for example, college courses, local electrical union (IBEW) courses, and Canadian Fire Alarm Association (CFAA) courses.

FIGURE 6-15 Typical zoning of a five-storey building.

In recent years the standards of ULC, together with the *NBC* and associated provincial and municipal codes, have made mandatory the installation of smoke detectors and heat and/or smoke detectors in commercial buildings that were built before these standards were in effect. A sufficient number of detectors must be installed so as to provide full coverage. For more information, see ULC website, www.ulc.ca.

Canadian Fire Alarm Association

Courtesy of the Canadian Fire Alarm Association

Organized in 1973, the Canadian Fire Alarm Association (CFAA) has the mission to maximize the effectiveness and use of fire alarm systems in the protection of life and property in Canada. Its vision is to be recognized nationally as the definitive resource for fire alarm related information, education, qualifications, expertise, and industry support and to do this, it maintains chapters throughout Canada. The CFAA offers fire alarm technician certification courses, and various other technical seminars, and has regularly published informative journals several times a year since the 1980s. For more information, see the Canadian Fire Alarm Association website, www.cfaa.ca.

▶ CEC REQUIREMENTS (SECTION 32)

Section 32 covers the installation requirements of fire alarm systems and fire pumps. This section is an amendatory section, which means that its requirements override other sections of the *Canadian Electrical Code (CEC)* with regard to this type of installation.

- *Rule 32–102* covers the wiring methods required for the installation of conductors in a fire alarm system.

- *Rule 32–106* requires that wiring to dual terminals must have each wire independently terminated to each terminal.

- *Rule 32–108* requires that the feed to the fire alarm system be as close to the load side of the service box as practicable, and that the disconnect be red and be lockable in the "On" position.

It is recommended that electricians familiarize themselves with the requirements of this section of the *CEC*.

▶ FIRE PUMPS

Fire pumps and associated equipment may be connected to a separate service, according to *Rule 32–204*. This service may be placed at a different location from the main service, but must be labelled in a "conspicuous, legible and permanent manner" with the words "FIRE PUMP."

Consideration must be given to running the power feeds to fire pumps underground, *Rule 32–202, Appendix B*. Underground feeds are better protected from hazards such as flood, vandalism, or explosion, and so have fewer failures. All conductors must be copper, *Rule 32–100*. The wiring must be installed in a metal raceway or be an armoured cable type listed in *Table 19*. It may also be installed in PVC or EMT if embedded in at least 50 mm of concrete.

Overcurrent Protection for Fire Pumps

Overcurrent protection is based on the locked-rotor current (LRC) of the motor(s) in addition to the rated current of any control or power equipment on the circuit, *Rule 32–206(4)* and *Appendix B*. No separate overload or overheating protection is required for fire pump motors; this protection is achieved by the overcurrent device, *Rule 32–210*. In a fire situation, the fire pump is required to operate until the motor fails, therefore reduced protection is permitted by *Rule 32–210*.

To round out this unit's treatment of fire alarm and safety systems, see Figure 6-16 for a fire alarm control panel, Figure 6-17 for equipment typically

FIGURE 6-16 Fire alarm control panel.

EXAMPLE

For a 10-horsepower, 575-volt, three-phase, Class B insulation, totally enclosed, nonventilated motor, the following is a calculation of disconnect, fuse, and wire size:

FLA = 11 amperes, *Table 44*

11 amperes × 6 = 66 amperes LRC, *Rule 32–206(4)* and *Appendix B* of the *CEC*

Minimum overcurrent device is 70 amperes and minimum disconnect size is 100 amperes.

Minimum wire ampacity is 11 × 1.25 = 13.75 amperes, *Rule 32–200*.

Minimum wire size is No. 14 Type TWN75 copper, *Table 2*. Use $\frac{1}{2}$-in (16-mm) rigid or EMT conduit, *Table 6*.

From junction box to motor, use No. 14 Type R90 silicone wire, *Rule 28–104(1)*, *Table 37*, and *Table 19*, *Note 10*.

associated with a fire alarm system, and two depictions of fire protection symbols that you might see in relation to mechanical (Figure 6-18) and electrical (Figure 6-19) systems.

NFW-50
Fire Alarm
Control Panel

Spool CL

N-ANN-80

N-ANN-S/PG
Printer Gateway

N-ANN-IO
Graphic LED
Module

N-ANN-LED

N-ANN-RLY
Relay Module

NP-100
Photo Detector

NI-100
Ionization Detector

NH-100 Series
Heat Detector

NDM-100
Dual Monitor Module

P2R
Strobe

NP-100T, NP-A100
Photo/Thermal

Power Supply
FCPS-24S6
FCPS-24S8

P2R
Strobe

NMM-100
Monitor Module

NZM-100-6,
NMM-100-10
Multiple-Circuit
Interface Module

P2R
Strobe

NZM-100
2-Wire Detector
Monitor Module

DNR
Duct Detector

NFW-50

Addressable fire alarm
control panel

NOT-BG12LX
Addressable
Manual Pull Station

FIGURE 6-17 Typical equipment for fire alarm system.

MECHANICAL FIRE PROTECTION SYMBOLS	
PREFERRED SYMBOL	DESCRIPTION
	Pressure gauge
	Pressure reducing valve
	Check valve
	Gate valve
	Globe valve
	Fire hose cabinet
	Fire hose rack
M	Multi purpose extinguisher
W	Water type extinguisher
C	CO_2 type extinguisher
D	Dry chemical extinguisher
	Fire department input connection
	Supply air duct
	Return air duct
	Fire damper
— G —	Gas line
— F —	Fire line
— SP —	Main sprinkler supply line
	Sprinkler line & upright head
	Sprinkler line & pendent head
	Sprinkler line & sidewall head
	Post hydrant
	Box hydrant
FD	Floor drain
	Funnel floor drain
	Roof drain
	Catch basin

FIGURE 6-18   Mechanical fire protection symbols.

ELECTRICAL FIRE PROTECTION SYMBOLS	
PREFERRED SYMBOL	DESCRIPTION
	Fire alarm manual station
K	Fire alarm key-operated station
	57° C fixed temperature heat detector
	57° C R. of R. & fixed temp. Heat detector
	88° C fixed temperature heat detector
	57° C rate of compensation heat detector
	Ionization type smoke detector
	Photoelectric type smoke detector
	Water flow switch
	Valve supervisory switch
	Fire alarm visual signal appliance
	Fire alarm audible signal appliance
S	Cone speaker
S	Horn speaker
T	Emergency telephone
	Remote alarm indicator
ANN	Fire alarm system annunciator
FAC	Fire alarm system control unit
CACF	Central alarm and control facility
VC	Voice communication control unit
DGP	Data gathering panel
TB	Terminal box
	Magnetic door holder
	Combination door holder/closer unit
	Electric release lock

FIGURE 6-19   Electrical fire protection symbols.

REVIEW

Note: Refer to the *CEC* or the blueprint package where necessary.

1. List the four stages of a fire.

    a. _____

    b. _____

    c. _____

    d. _____

2. Draw lines to connect the four classes of fire with the appropriate material.

    Class A   Electrical

    Class B   Paper

    Class C   Combustible metals

    Class D   Paint

3. Name two types of smoke detectors.

    _____

    _____

4. Which type of smoke detector is most sensitive to black smoke?

    _____

5. What is the minimum clearance from the corner of a ceiling for a smoke detector?

    _____

6. List six types of circuits that may be connected to a fire alarm panel.

    a. _____

    b. _____

    c. _____

    d. _____

    e. _____

    f. _____

7. How does a rate-of-rise detector work?

    _____

    _____

    _____

8. With reference to Figure 6-15, indicate

    a. The minimum number of detection or initiating circuits required

        _____

b. Where the remote annunciator panel will be located

_____

9. For a 15-horsepower, 575-volt, three-phase, Class H, totally enclosed, nonventilated fire pump, calculate:

a. Wire size

_____

b. Minimum size of overcurrent device

_____

c. Minimum size of disconnect

_____

d. Minimum size of raceway

_____

10. Draw a diagram of a Class B initiating circuit.

Hazardous Materials in the Workplace

OBJECTIVES

After completing the study of this unit, the student will be able to

- describe the Workplace Hazardous Materials Information System (WHMIS)
- explain the responsibilities of suppliers and employers
- describe a typical Material Safety Data Sheet (MSDS)
- explain the responsibilities of employees
- provide a list of relevant health hazards
- describe the major classes of hazardous materials
- provide a list of symbols for hazardous materials

▶ WORKPLACE HAZARDOUS MATERIALS INFORMATION SYSTEM (WHMIS)

The Workplace Hazardous Materials Information System (WHMIS) is a Canada-wide legislated system of providing information on workplace hazardous materials and how to safely use, store, and handle them.

Many of the materials in today's workplaces can cause injury, illness, or death due to fire, shock, explosion, or other serious causes if they are not handled properly. WHMIS describes the dangers of these workplace materials and how workers can protect themselves from the hazards associated with them. WHMIS legislation also makes employers responsible for providing their workers with workplace-specific training and education.

The legislation was passed at the federal level (as Bill C-70) and states that workers "have the right to know" about hazardous materials to which they are exposed on the job. The legislation at the provincial level incorporates the federal legislation and introduces the concept of the Material Safety Data Sheet (MSDS) and the public's right to know concerning hazardous material in the public domain.

The major components of WHMIS legislation are as follows:

- Warning labels must be placed on containers of hazardous materials.

- Hazards of each material must be listed on an MSDS available in the workplace.

- Employee education programs concerning WHMIS are mandatory in the workplace.

Compliance with WHMIS legislation falls into three categories and requires action by suppliers, employers, and employees.

1. *Responsibility of the supplier.* Suppliers have to determine which of their products intended for use in the workplace in Canada are hazardous materials under WHMIS. In addition, as a condition of sale or importation, suppliers must provide information about the products that are hazardous materials. This information is provided in the form of a regulation label (Figure 7-1) and an MSDS.

(A) Product identifier (common name/chemical name)

(B) Hazard symbol

(C) Risk phrase(s)

(D) First aid measures

(E) Supplier identifier

(F) Reference to Material Safety Data Sheet (MSDS)

(G) Precautionary measures

(H) Border

FIGURE 7-1   Supplier label.

2. *Responsibility of the employer.* Employers must ensure that any container of potentially hazardous material received at the workplace has the required label. In addition, any hazardous material transferred from a supply container into another container must have a label fixed to the new container indicating its contents.

An MSDS for each of these materials must be readily accessible at the worksite and be given to the joint health and safety committee. An MSDS must be provided for any hazardous material produced at the workplace as well.

The employer also has the responsibility of training the employees in all matters related to WHMIS and MSDSs. The training program is to be reviewed on an annual basis in consultation with the health and safety committee.

3. *Responsibility of the employee.* The employee is required to participate in the training program and to use the information learned to maintain a safe and healthy workplace.

HEALTH

Hazardous materials are in use in all aspects of our lives. This includes many commonly used cleaning supplies, paints, chemicals, solvents, and glues.

TYPICAL HAZARDS

- *Personal exposure (health).* Certain chemicals can cause harmful effects to the body. These can include burns, eye irritation, and allergic reactions. In addition, the breathing of chemical vapours into the body can harm various organs.

- *Reactivity.* Certain chemicals—when mixed with other chemicals or exposed to heat, water, or air—may burn, explode, or form toxic vapours. For example, mixing chlorine bleach with acid-based cleaners will produce poisonous chlorine gas. Breathing this will have a serious effect on the body.

- *Flammability.* Chemicals that may catch fire with improper use must be identified as being flammable.

WORKPLACE CONTROLS

The harmful effects of hazardous materials in the workplace can be minimized by the provision of proper storage areas with adequate ventilation, the implementation of safe work procedures, and the proper disposal of the hazardous materials.

PERSONAL PROTECTIVE EQUIPMENT

Appropriate personal protective equipment must be worn by employees when workplace controls are not possible. This may include adequate eye and face protection, gloves, proper footwear, safety mats, protective clothing, and breathing protection. It is mandatory that employees wear the appropriate personal protective equipment provided for them.

This is extremely important when the employee may be exposed to a hazardous material. For example, when taking an oil sample from a transformer that may contain PCBs or topping up liquid levels in batteries in battery rooms, the appropriate safety equipment must be worn by the employee. If an employee does not comply with this requirement, disciplinary action must be mandatory.

All staff must follow basic personal hygiene procedures, which include washing hands after chemical use and not eating or drinking where chemicals are used or stored.

MATERIAL SAFETY DATA SHEET (MSDS)

Employees have the right to know about the products they are using on the job. The information is found on the MSDS for each product. See Figure 7-2 for an example of an MSDS. The MSDS must include the following information:

a. *Product identifier:* the common name and the chemical name, for example, household bleach/ sodium hypochlorite

b. *Hazard symbols:* symbols that indicate materials that are flammable, corrosive, explosive, or poisonous (e.g., flammable symbol in Figure 7-1)

c. *Physical data:* included in the MSDS

d. *Fire and explosion data:* includes the flash point of gases or vapours of the product

e. *Reactivity:* description of how the product reacts with other substances (e.g., household bleach when mixed with acid forms chlorine gas—one of the types of corrosive poison gas that caused such devastation in World War I)

f. *Toxicological properties:* describes the health effects of exposure by ingestion, inhalation, or direct skin or eye contact

MATERIAL SAFETY DATA SHEET
<u>IPEX 100</u> PVC CONDUIT SOLVENT CEMENT (Low VOC)

Page : 1

SLUYTER COMPANY LTD.

375 Steelcase Road East
Markham, Ontario L3R 1G3 Canada
Tel (905) 475-6011 Fax (905) 475-3119

SECTION 01: CHEMICAL PRODUCT AND COMPANY IDENTIFICATION

```
MANUFACTURER----------------------SLUYTER COMPANY LTD.
                                  375 Steelcase Road East
                                  Markham, Ontario L3R 1G3
                                  Canada
                                  Tel (905) 475-6011
PRODUCT NAME----------------------IPEX 100 PVC CONDUIT SOLVENT CEMENT — LOW VOC
MANUFACTURED FOR-------------------Ipex Electrical Inc. 2441 Royal Windsor Drive, Mississauga,
                                  ON L5J 4C7
PRODUCT USES----------------------For welding plastic pipes and fittings.
CHEMICAL FAMILY-------------------Poly Vinyl Chloride based cement.
```

SECTION 02: COMPOSITION / INFORMATION ON INGREDIENTS

INGREDIENTS / %	CAS / TLV	LD/50, ROUTE, SPECIES	LC/50, ROUTE, SPECIES
TETRAHYDROFURAN 35 — 50	109-99-9 200 ppm	1650 mg/kg Oral (Rat)	18000 ppm 4 hours Inhalation (Rat)
METHYL ETHYL KETONE 35 — 50	78-93-3 200 ppm	2737 mg/kg Oral (Rat)	23500 mg/m3 8 Hours Inhalation (Rat)
ACETONE 20 - 40	67-64-1 750 ppm	9750 mg/kg Oral (Rat)	16000 ppm 4 hours Inhalation (Rat)
CYCLOHEXANONE 3 - 8	108-94-1 25 ppm	1620 mg/kg Oral (Rat)	8000 ppm 1 Hour Inhalation (Rat)

SECTION 03: HAZARDS IDENTIFICATION

```
ROUTE OF ENTRY:---------------------
        SKIN CONTACT-----------------Can cause moderate skin irritation,
                                    defatting and dermatitis.
        SKIN ABSORPTION--------------No data available.
        INHALATION-------------------See "EFFECTS OF CHRONIC EXPOSURE", below.
        INHALATION CHRONIC-----------Breathing of high vapour concentrations
                                    could cause dizziness, headache or even
                                    unconsciousness. May be anesthetic which
                                    could result in other central nervous
                                    system effects.
        INGESTION--------------------Can cause gastro-intestinal irritation,
                                    nausea, vomiting and diarrhea. Small
                                    amounts of liquid aspirated into
                                    respiratory system could cause severe
                                    health effects (e.g. Bronchopneumonia or
                                    Pulmonary Edema).
        EYE CONTACT------------------Contains materials that are severely
                                    irritating to the eyes.
EFFECTS OF ACUTE EXPOSURE----------Refer to "ROUTE ENTRY" section.
EFFECTS OF CHRONIC EXPOSURE--------May cause damage to the central nervous
                                  system. Prolonged or repeated skin contact
                                  may cause drying or cracking of the skin.
```

Ref: 00000319

Preparation Date : September 6, 2012

EMERGENCY - CALL CANUTEC (613) 996-6666 (Collect)

FIGURE 7-2 Material Safety Data Sheet.

MATERIAL SAFETY DATA SHEET
IPEX 100 PVC CONDUIT SOLVENT CEMENT (Low VOC)

Page : 2

SECTION 04: FIRST AID MEASURES

EYE CONTACT------------------------Check for and remove any contact lenses.
Immediately flush with water for a minimum
of 20 minutes and get medical attention.

SKIN CONTACT-----------------------Remove contaminated clothing. Wash
affected area with water and soap. Seek
medical attention if irritation occurs or persists.

INHALATION------------------------Remove patient to fresh air. If not
breathing, trained personnel should
administer artificial respiration. Get
medical attention.

INGESTION-------------------------Do NOT induce vomiting. Get immediate medical attention.

ADDITIONAL INFORMATION------------Contact your local poison control centre.

SECTION 05: FIRE FIGHTING MEASURES

FLAMMABILITY-----------------------Flammable.

UNDER WHAT CONDITIONS--------------Extremely flammable liquid. Dangerous fire
hazard when exposed to heat, flame or
temperatures above the flash point. As
vapours are heavier than air, they may
travel to a source of ignition and flash back

SPECIAL PROCEDURES-----------------A self-contained breathing apparatus is
required for fire fighting personnel. Use
water spray to cool fire exposed surfaces
and to protect personnel.

FLASH POINT (METHOD)--------------- -5ºC TAG Closed Cup.

AUTO IGNITION TEMPERATURE----------321ºC.

UPPER FLAMMABLE LIMIT (% VOL)------11.80.

LOWER FLAMMABLE LIMIT (% VOL)------2.20.

EXTINGUISHING MEDIA----------------Alcohol foam, CO2 or dry chemical.

HAZARDOUS COMBUSTION PRODUCTS------Carbon Dioxide. Carbon Monoxide. Hydrogen Chloride. Smoke.

SENSITIVITY TO MECHANICAL----------Unknown.
IMPACT

SENSITIVITY TO STATIC--------------May be sensitive.
DISCHARGE

SECTION 06: ACCIDENTAL RELEASE MEASURES

LEAK/SPILL-------------------------Ventilate. Remove all sources of ignition,
open flames, sparks and heaters. Wear
protective gear (See SECTION 8).Large
spills must be collected for disposal.
Small spills can be wiped. Use a
non-combustible absorbent inorganic
material. Prevent run-off into drains,
sewers and other waterways.

SECTION 07: HANDLING AND STORAGE

HANDLING PROCEDURES----------------Avoid skin and eye contact. Avoid breathing
vapours. Use adequate ventilation. Keep away
from heat, sparks and open flame.

STORAGE NEEDS----------------------Store away from all sources of heat and
ignition. Store in well ventilated
area. Keep container closed and out of
reach from children and pets when not in use.

Ref: 00000319

Preparation Date : September 6, 2012

EMERGENCY - CALL CANUTEC (613) 996-6666 (Collect)

FIGURE 7-2 Material Safety Data Sheet   (continued).

Courtesy Sluyter Company Ltd.

MATERIAL SAFETY DATA SHEET
<u>IPEX 100</u> PVC CONDUIT SOLVENT CEMENT (Low VOC)

Page : 3

SECTION 08: EXPOSURE CONTROLS / PERSONAL PROTECTION

```
PROTECTIVE EQUIPMENT:---------------
        EYE/TYPE----------------------Safety glasses.
        RESPIRATORY/TYPE--------------None required for normal use if adequate
                                      ventilation is maintained. Use NIOSH/MSHA
                                      approved respirator if the TLV is exceeded.
        GLOVES/ TYPE------------------Wear impervious gloves (Neoprene or Rubber).
        CLOTHING/TYPE-----------------Not applicable.
        FOOTWEAR/TYPE-----------------Not applicable.
        OTHER/TYPE--------------------Eye bath and safety shower.
VENTILATION REQUIREMENTS-----------Natural or mechanical (Explosion Proof)
                                      ventilation to keep vapour levels well
                                      below the TLV.
```

SECTION 09: PHYSICAL AND CHEMICAL PROPERTIES

```
PHYSICAL STATE---------------------Liquid.
ODOUR------------------------------Ketone odour.
SPECIFIC GRAVITY-------------------0.90 - 1.00.
ODOUR THRESHOLD (ppm)--------------Not available.
VAPOUR PRESSURE--------------------145 mmHg @ 20ºC.
VAPOUR DENSITY (AIR=1)------------- >1.00.
EVAPORATION RATE-------------------6.00 (NBUAC = 1).
BOILING POINT (deg C)--------------65ºC.
pH---------------------------------Not applicable.
SOLUBILITY IN WATER (% W/W)--------Negligible.
COEFFICIENT OF WATER\OIL-----------Not available.
DISTRIBUTION
FREEZING POINT--------------------- <0ºC.
MELTING POINT (deg C)--------------Not applicable.
MOLECULAR WEIGHT-------------------Not applicable.
```

SECTION 10: STABILITY AND REACTIVITY

```
INCOMPATIBILITY--------------------Strong acids and strong bases. Chlorinated
                                    solvents. Oxidizing agents.
REACTIVITY CONDITIONS--------------Excessive heat, sparks and open flame.
HAZARDOUS PRODUCTS OF--------------Carbon Dioxide. Carbon Monoxide. Hydrogen
DECOMPOSITION                       Chloride. Peroxides.
```

SECTION 11: TOXICOLOGICAL INFORMATION

```
EXPOSURE LIMIT OF MATERIAL---------See "HAZARDOUS INGREDIENTS" in SECTION 2.
IRRITANCY OF MATERIAL--------------Moderate.
SENSITIZING CAPABILITY OF----------Not available.
MATERIAL
CARCINOGENICITY OF MATERIAL--------Reversible liver damage at high dosages is
                                    possible.
TERATOGENICITY---------------------No information is available and no adverse
                                    teratogenicity effects are anticipated.
MUTAGENICITY-----------------------No information is available and no adverse
                                    mutagenicity effects are anticipated.
REPRODUCTIVE EFFECTS---------------Not available.
SYNERGISTIC MATERIALS--------------Not available.
```

Ref: 00000319

Preparation Date : September 6, 2012

EMERGENCY - CALL CANUTEC (613) 996-6666 (Collect)

FIGURE 7-2 Material Safety Data Sheet (continued).

Courtesy Sluyter Company Ltd.

MATERIAL SAFETY DATA SHEET

<u>IPEX 100</u> PVC CONDUIT SOLVENT CEMENT (Low VOC)

SECTION 12: ECOLOGICAL INFORMATION

ENVIRONMENTAL----------------------Not available. Can be dangerous if allowed to enter drinking water intakes. Product has an unaesthetic appearance and can be a nuisance. Do not contaminate domestic or irrigation water supplies, lakes, streams, ponds and rivers.

BIODEGRADABILITY--------------------Not available. The solvent portion of this product is biodegradable and vaporizes rapidly.

VOC INFORMATION--------------------This product emits VOC's (volatile organic compounds in use. Always ensure that the use of this product complies with local VOC Emission Regulations, where they exist. The VOC level is <u>475 grams/litre</u> (SCAQMD Test Method 316A)

SECTION 13: DISPOSAL CONSIDERATIONS

WASTE DISPOSAL---------------------Spilled material and water rinses are classified as chemical waste. To be disposed of in accordance with current Local, Provincial and Federal regulations.

SECTION 14: TRANSPORT INFORMATION

T.D.G. CLASSIFICATION--------------
(A)------------------------------(A) In containers up to 5 litres - shipped as CONSUMER COMMODITY. If the shipment exceeds 500 kg in weight, shipped as CONSUMER COMMODITY - ADHESIVES CLASS 3.

(B)------------------------------(B) In containers over 5 litres - ADHESIVES CLASS 3 UN1133 P.G. II.

SECTION 15: REGULATORY INFORMATION

CPR COMPLIANCE---------------------This product has been classified in accordance with the hazard criteria of the CPR and the MSDS contains all the information required by the CPR.

WHMIS CLASSIFICATION---------------Class B Div.2 Flammable Liquid Class D Div.2B Toxic Material.

SECTION 16: OTHER INFORMATION

IMPORTANT:------------------------The information on this Material Safety Data Sheet is furnished without warranty, expressed or implied. All the information appearing herein is based upon data obtained from manufacturers and/or recognized technical sources. While the information is believed to be accurate, we make no representations for the accuracy or sufficiency.

Ref: 00000319

Preparation Date : September 6, 2012

EMERGENCY - CALL CANUTEC (613) 996-6666 (Collect)

FIGURE 7-2 Material Safety Data Sheet (continued).

Courtesy Sluyter Company Ltd.

g. *Preventive measures:* describes protective clothing required when using the product

h. *First-aid measures:* describes emergency and first-aid procedures (e.g., certain products require vomiting to be induced, while other products, such as acids, indicate that vomiting should not be induced)

i. *Date and source of MSDS:* used to determine whether the information is the most current (MSDSs older than three years must be updated)

> *PVC GLUE*
> — *Wear safety gloves.*
> — *Use in well-ventilated area.*
> — *Refer to MSDS for further information.*

FIGURE 7-3 Workplace label.

◗ SUPPLIER LABEL

Each package of a product must be provided with a label in both official languages. This label has a distinctive slashed border and depicts the applicable hazard symbols for the product. The supplier of the product should be identified along with an address and telephone number, and the information should be highlighted so that it stands out from the background colour of the container. For an example of a supplier label, see Figure 7-1.

Employers must ensure that all hazardous products bear a supplier label before they enter the workplace. If a product arrives without a label, or is placed or poured into a new container, employers are responsible for ensuring that a workplace label is attached.

◗ WORKPLACE LABEL

A label may be made up by the employer when using the product in small quantities. This label may be smaller than the supplier's label, but must contain at least the following:

a. product identification

b. preventive measures or precautions

c. a statement indicating that an MSDS is available

An example of a workplace label is shown in Figure 7-3.

◗ HAZARDOUS MATERIALS

There are six major classes of hazardous materials, which are shown with their symbols in Figure 7-4. They are as follows:

- Class A: Compressed gas
- Class B: Flammable and combustible materials
- Class C: Oxidizing materials
- Class D: Poisonous and infectious materials
- Class E: Corrosive materials
- Class F: Dangerously reactive materials

PREFERRED SYMBOL	DESCRIPTION
	Class A: Compressed gas
	Class B: Flammable and combustible materials Division 1: Flammable gases Division 4: Flammable solids Division 2: Flammable liquids Division 5: Flammable aerosols Division 3: Combustible liquids Division 6: Reactive flammable materials
	Class C: Oxidizing materials

	Class D: Poisonous and infectious materials
	
	Division 1: Materials causing immediate and serious toxic effects
	    Subdivision A: Very toxic material
	    Subdivison B: Toxic material
	
	Division 2: Materials causing other toxic effects
	    Subdivision A: Very toxic material
	    Subdivision B: Toxic material
	
	Division 3: Biohazardous infectious material

	Class E: Corrosive materials
	Class F: Dangerously reactive materials

FIGURE 7-4   Hazardous material symbols.

REVIEW

1. What does WHMIS stand for? _____

   _____

2. What does MSDS stand for? _____

   _____

3. Responsibilities regarding compliance with WHMIS legislation fall into three broad categories, and require action by:

   i. _____

   _____

   ii. _____

   _____

   iii. _____

   _____

4. Describe Bill C-70. _____

   _____

5. List all pieces of information required on a supplier label.

   _____

   _____

   _____

   _____

   _____

   _____

   _____

   _____

   _____

   _____

   _____

6. Who is responsible for ensuring that a workplace label appears on an unlabelled product?

   _____

7. What is the employee's obligation with regard to participation in WHMIS programs provided by the employer? _____

   _____

   _____

8. What do the following hazard symbols stand for?

 _____

 _____

 _____

 _____

 _____

_____

_____

_____

9. Using the MSDS in Figure 7-2, answer the following questions:

a. What are the chronic effects of exposure to this product?

_____

_____

b. How should this product be handled?

_____

_____

c. What are the storage requirements of this product?

_____

_____

d. What materials should you avoid when using this product?

_____

_____

Panelboard Selection and Installation

OBJECTIVES

After completing the study of this unit, the student will be able to

- select a panelboard, given the number of required branch circuits and their loads
- connect the branch circuits to a panelboard
- select the feeder size, given the loading of the panelboard

Changing needs for electrical energy in a commercial building will require the electrician to install new branch circuits. In addition, as the existing panelboards are filled, new panelboards must be installed. As a result, the electrician must be able to select and connect panelboards and the feeders to supply them.

PANELBOARDS

Separate feeders must be run from the main service equipment to each separate area of the commercial building. Each feeder will terminate in a panelboard that is located in the area to be served. All panelboards are Canadian Standards Association (CSA) approved. All interiors will have 225-ampere bus lugs, rated at 75°C for incoming conductors. Figure 8-1 shows two types of panelboards that are available. On the left is a combination panelboard, which has a main overcurrent device located in the cabinet. On the right is a load centre panelboard,

which requires overcurrent protection from another location, as demonstrated in Figure 8-7.

A panelboard may be defined as a single panel or group of panel units designed for assembly in the form of a single panel; such a panel will include bus bars and automatic overcurrent devices, and may or may not include switches for the control of light, heat, or power circuits. A panelboard is designed to be installed in a cabinet or cutout box placed in or against a wall or partition. This cabinet (and panelboard) is to be accessible only from the front. *Rule 14–108* describes the enclosure of overcurrent devices. Panelboards are referred to as "dead front" because no live parts are accessible without removing the cover.

While a panelboard is accessible only from the front, a switchboard may be accessible from the rear as well.

Lighting and appliance branch-circuit panelboards generally are used on single-phase, three-wire systems and three-phase, four-wire electrical systems. This type of installation provides individual branch circuits for

Jupiterimages/Thinkstock

Tetra Images/Getty Images

FIGURE 8-1   Panelboards.

lighting and receptacle outlets and permits the connection of appliances such as the equipment in the bakery.

Panelboard Construction

In general, panelboards are constructed so that the main feed bus bars run the height of the panelboard. The branch-circuit protective devices are connected to alternate main buses, as shown in Figures 8-2A and 8-2B. In an arrangement of this type, the connections directly across from each other are on the same phase, and the adjacent connections on each side are on different phases. As a result, multiple protective devices can be installed to serve the 208-volt

equipment. The numbering sequence shown in Figures 8-2A, 8-2B, and 8-3 is the system used for most panels. Figures 8-4 and 8-5 show the phase arrangement requirements of *Rule 4–038(4,5)*.

FIGURE 8-3 Panelboard circuit numbering scheme.

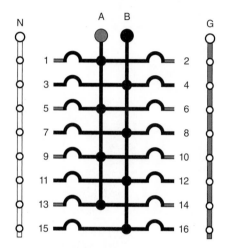

FIGURE 8-2A Lighting and appliance branch-circuit panelboard; single-phase, three-wire connections.

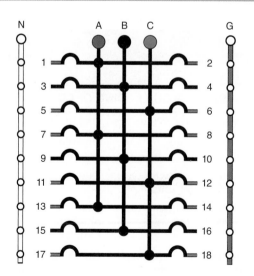

FIGURE 8-2B Lighting and appliance branch-circuit panelboard; three-phase, four-wire connections.

Top view

Front view   Side view

FIGURE 8-4 Possible phase arrangement for switchboards and panelboards. Colours of conductors based on *Rule 4–038(3)* for three-phase A, B, and C.

High leg A must be red in colour, tagged or identified by other effective means and must be isolated from ground conductor, *Rule 4–038(4)*

120/240-volt, 3-phase, 4-wire delta connected transformer with the midpoint of one transformer winding grounded. The A phase has a higher voltage to ground of 120 × 1.732 = 208.

© Cengage Learning®

FIGURE 8-5 Panelboards and switchboards supplied by four-wire, delta-connected systems shall have the A phase connected to the phase having the higher voltage to ground, *Rule 4–038(4,5)*.

Number of Circuits

The number of the overcurrent devices in a panelboard is determined by the needs of the area being served. Using the bakery as an example, there are 13 single-pole circuits and six 3-pole circuits. This is a total of 31 poles. When using a three-phase supply, the incremental number is six (a pole for each of the three phases on both sides of the panelboard). The minimum number of poles that could be specified for the bakery is 36.

Colour Code for Branch Circuits

The "hot" ungrounded conductors may be any colour *except* green or green with a yellow stripe (colours that are reserved for grounding purposes only), or white or natural grey (colours that are reserved for the grounded circuit conductors). See *Rule 4–038*.

Table 8-1 provides suggestions for colour coding of conductors.

TABLE 8-1	
Suggested Colour Coding of Conductors.	
SUGGESTED COLOUR CODING OF CONDUCTORS	
Two-wire circuits, 120 volts	One black, one white
Three-wire circuits, 120/240 volts	One black, one white, one red
Three-phase, four-wire wye circuits, 120/208 volts	One black, one white, one red, one blue
Three-phase, four-wire delta circuits, 240 volts	One black, one white, one blue, one red (for high leg A phase)
Three-phase, four-wire wye circuits, 347/600 volts	One black, one white, one red, one blue
Three-phase, three-wire delta circuits, 600 volts	One red, one black, one blue

Rule 4–038(3) requires that three-phase AC be identified with one red (Phase A), one black (Phase B), one blue (Phase C), and, where neutral is required, one white. Where different voltages exist in a building, the ungrounded conductors for each system should be identified and posted at each branch-circuit panelboard.

This situation occurs where, for instance, the building is served with 347/600 volts and step-down transformers are used to provide 120/208 volts for lighting and receptacle outlets.

▶ PANELBOARD SIZING

Panelboards are available in various sizes, including 100, 125, 200, 225, 400, and 600 amperes. These sizes are determined by the current-carrying capacity of the main bus. After the feeder capacity is determined, the panelboard size is selected to carry the demand amperes at 80% or 100% of the panel ampere rating. The number of circuits required must also be considered. Refer to Table 8-2.

Panelboard Overcurrent Protection

In many installations, a single feeder may be sized to serve several panelboards, Figure 8-6. In this situation, each panelboard must have a main overcurrent device with a trip rating not greater than the rating of the bus bars in the panelboard. In the case of the

	BREAKERS			TOTAL POLES USED	SPECIFIED POLES
	1P	2P	3P		
Drugstore	16	0	1	19	36
Bakery	13	0	6	31	36
Insurance office	19	0	1	22	36
Beauty salon	8	2	1	15	24
Doctor's office	7	2	0	11	24
Owner's	18	0	0	18	24

TABLE 8-2 Panelboard Summary.

bakery installation, a main device is not required, since the panelboard is protected by the feeder protective device, Figure 8-7. An example of a method of providing panelboard protection is shown in Figure 8-8.

▶ THE FEEDER

The size of a feeder is determined by the branch-circuit loads being served. In addition, a reasonable allowance for growth may be considered. Again using the bakery as an example, the connected load is 46 620 volt-amperes (see Table 10-1, page 159).

Since we are using standard-rated equipment and conductors listed in *Table 2*, the ampacity of

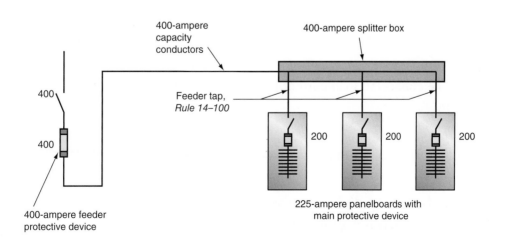

FIGURE 8-6 Panelboards with main, *Rule 14–606*.

© Cengage Learning®

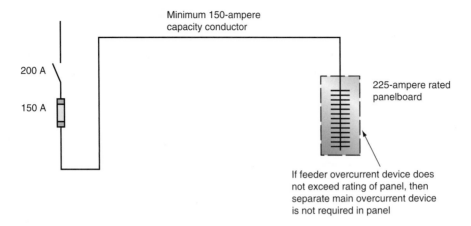

FIGURE 8-7 Panelboards without main, *Rule 14–100(a)*.

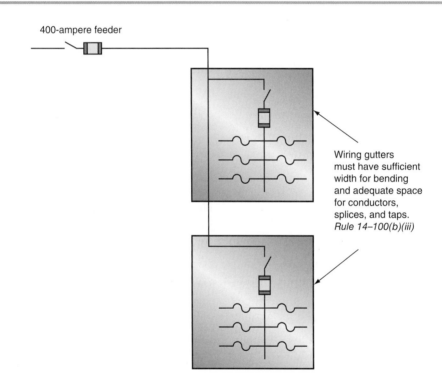

FIGURE 8-8 Feedthrough panel with adequate gutter space, *Rule 14–100(b)(iii)*.

the feeder to the bakery will be based on 125% of the connected load. The connected load current is

$$\frac{46\ 620}{208 \times 1.73} = 129.56 \text{ A}$$

The ampere rating of the feeder to the bakery would be

$$129.56 \text{ A} \times 1.25 = 161.95 \text{ A}$$

The conductor size using RW90-XLPE 600 V without jacket copper conductors would be No. 2/0 AWG. The conduit size would be 41 mm.

The conductor is sized based on 75°C termination temperature.

Neutral Current

Under certain conditions, the neutral wire of a feeder may be smaller in size than the phase wires. This reduction in size is permissible because the neutral carries only the unbalanced portion of the load. Figures 8-9 through 8-11 illustrate the use of vectors to determine the approximate current in the neutrals of a three-phase, four-wire system. However, on a

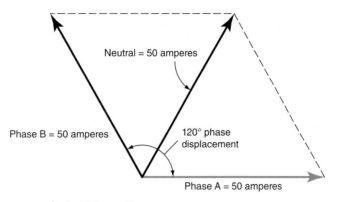

Neutral = 50 amperes

Phase B = 50 amperes

120° phase displacement

Phase A = 50 amperes

Scale: 1.25 cm = 10 amperes

FIGURE 8-9 Vector sum as shown in Table 8-3, Condition IV; loads are connected line-to-neutral.

Neutral = 43.6 amperes

120° phase displacement

Phase B = 30 amperes

Phase A = 50 amperes

Scale: 1.25 cm = 10 amperes

FIGURE 8-10 Vector sum showing current in the neutral when Phase A is carrying 50 amperes and Phase B is carrying 30 amperes; neutral is carrying approximately 43.6 amperes; loads are connected line-to-neutral.

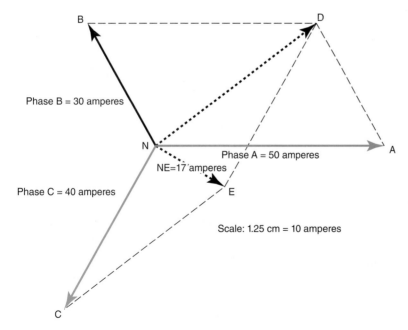

B   D

Phase B = 30 amperes

N   A

Phase A = 50 amperes

NE=17 amperes

Phase C = 40 amperes   E

Scale: 1.25 cm = 10 amperes

C

1. Draw each of the three-phase currents to scale, 120° apart.

2. Draw a parallelogram using any two phases, for example, NBDA.

3. Draw a diagonal for the parallelogram in Step 2: the diagonal is ND.

4. Draw a parallelogram using sides NC and ND; the result is parallelogram NDEC.

5. Draw a diagonal for parallelogram NDEC; the diagonal is NE.

6. Measure the length of NE. This length represents the approximate current in the neutral conductor for the three-phase, four-wire wye system. NE is approximately equal to 17 amperes. This example is for resistive loads only.

FIGURE 8-11 Drawing vectors to determine approximate current in neutral of a three-phase, four-wire wye system; loads are connected line-to-neutral.

TABLE 8-3

Load in Amperes, Neutral Unbalanced Loading*
Three-Phase, Four-Wire System.

CONDITION	PHASE A	PHASE B	PHASE C	NEUTRAL
I	50	0	0	50
II	0	50	0	50
III	0	0	50	50
IV	50	50	0	50
V	50	0	50	50
VI	0	50	50	50
VII	50	50	50	0
VIII**	50	50	50	(approx.) 50

* Table is based on resistive loads.

** For inductive loads, such as fluorescent ballasts, the neutral conductor carries approximately the same value of current as the phase conductors. In this case, the neutral is considered to be a current-carrying conductor. See *Rules 4–004(4)* and *4–024(2)(a)*.

three-phase, four-wire system, the load value is a vector sum (see Figure 8-11).

Table 8-3 shows several conditions that may exist. The neutral currents for Conditions I, II, III, and VII are obvious, but Conditions IV, V, and VI appear to be contradictory. To find the unbalanced neutral current in a three-phase, four-wire system, only the loads connected between the neutral and phase wires are considered. In addition, equal power factors are assumed. The equation used to find the unbalanced neutral current is

$$N = \sqrt{A^2 + B^2 + C^2 - AB - BC - CA}$$

where:

N = current in the neutral

A = current in Phase A

B = current in Phase B

C = current in Phase C

 EXAMPLE

Compute the neutral current for Figure 8-11.

$$N = \sqrt{30^2 + 40^2 + 50^2 - (30 \times 40) - (40 \times 50) - (50 \times 30)}$$
$$= \sqrt{900 + 1600 + 2500 - 1200 - 2000 - 1500}$$

$$= \sqrt{300}$$
$$= 17.32 \text{ amperes}$$

To calculate the neutral current when only two phases are available (from a three-phase, four-wire system), use the following equation:

$$N = \sqrt{A^2 + B^2 - AB}$$

 EXAMPLE

Compute the neutral current for Figure 8-9.

$$N = \sqrt{50^2 + 50^2 - (50 \times 50)}$$
$$= \sqrt{2500 + 2500 - 2500}$$
$$= \sqrt{2500}$$
$$= 50 \text{ amperes}$$

Another method for computing the neutral current is to tabulate the connected loads on the neutral. Do not include three-pole loads or two-pole-with-no-neutral loads.

 EXAMPLE

Assume the following volt-ampere loads.

	PHASE A	PHASE B	PHASE C
	1 305	609	1 200
	900	264	1 188
	348	1 500	540
	1 500	720	720
	900		
Total	4 953	3 093	3 648

In the worst-case scenario, only the loads on Phase A would be operating. In this case, the load on the neutral would be 4953 volt-amperes, or 41.3 amperes.

NEUTRAL SIZING (*RULE 4–024*)

Rule 4–024 sets forth a procedure for reducing the neutral size. The method is based on the loading as computed by *Section 8* (not the connected load), and makes the assumption that the unbalanced load is evenly distributed across the phases.

FEEDER LOADING SCHEDULE

The doctor's office feeder loading schedule can be found in Table 8-4.

TABLE 8-4

Doctor's Office Feeder Loading Schedule.

ITEM	COUNT	VA/UNIT	DEMAND LOAD	INSTALLED LOAD	DEMAND FACTOR		USE	
					Feeder	Service	Feeder	Service
Minimum Basic Load								
(CEC Rule 8–210)	42 m^2	50 W/m^2	2 100 W		1	0.9	2 100	1 890
Installed Basic Loads								
Style F luminaires	3	132		396			396	396
Style G luminaires	2	132		264			264	264
Style J luminaires	5	150		750			750	750
Style N luminaires	2	60		120			120	120
General receptacles	10	120		1 200			1 200	1 200
							2 730	2 730
BASIC LOAD	Use the greater of the demand or installed loads, *CEC Rule 8–106(2)*.						2 730	2 730
SPECIAL LOADS								
Equipment outlet	1	2 000		2 000			2 000	2 000
Roof receptacle	1	1 440		1 440			1 440	1 440
Water heater	1	3 800		3 800			3 800	3 800
Motors	**Volts**	**FLA**	**Phase**					
Cooling unit								
• compressor	208	16.8	1	3 494	1.25		4 368	3 494
• evap/cond. motor	208	3.7	1	770			770	770
Total Load							15 108	14 234

$$I = \frac{P}{E} = \frac{15\ 108}{208} = 72.68\ A$$

Assuming all loads are continuous, using 75°C rated equipment (*Rule 4–006*) and RW90-XLPE copper conductors from *Table 2*, 75°C column, *CEC*, the ampere rating of the circuit would be

$$\frac{72.63}{0.8} = 90.79\ A$$

The minimum size feeder would be three No. 3 AWG RW90-XLPE copper conductors in a 27-mm conduit, supplied from a 100-A disconnect c/w 100-A fuses. Conductor sized based on 75°C termination temperature.

REVIEW

Note: Refer to the *CEC* or the blueprint package as necessary.

1. Six 120-volt branch circuits supplying incandescent lighting are run from a three-phase lighting panel to a pull box. The circuits are connected to breakers at the following positions in the panel: Circuits 1, 2, 8, 17, 19, and 21. How many neutrals are required for the six branch circuits?

2. A lighting panel has 225-ampere main buses. The load on the panel is 150 amperes. What is the maximum size overcurrent protection permitted for the panelboard?

3. A neutral of a three-phase, four-wire system will carry a current of _____ amperes when the currents in the phases are as follows: Phase A, 12 amperes; Phase B, 4 amperes; and Phase C, 2 amperes. All loads are connected neutral-to-phase.

4. The preferred colours to be used for the ungrounded conductors of a three-phase, four-wire circuit are _____, _____, and _____.

5. Using the information in the Insurance Office Panel Schedule shown in the blueprints on Drawing E7, determine the maximum unbalanced current that the neutral supplying the panel will carry. Show all calculations.

The Cooling System

OBJECTIVES

After completing the study of this unit, the student will be able to

- list the parts of a cooling system

- describe the function of each part of the cooling system

- make the necessary calculations to determine the sizes of the electrical components

- read a typical wiring diagram that shows the operation of a cooling unit

The electrician working on commercial construction is expected to install the wiring of cooling systems and troubleshoot electrical problems in these systems. Therefore, it is recommended that the electrician know the basic theory of refrigeration and the terms associated with it.

REFRIGERATION

Refrigeration is a method of removing energy in the form of heat from an object. When the heat is removed, the object is colder. An energy balance is maintained, which means that the heat must go somewhere. As long as the locations where the heat is discharged and where it is removed are separate from each other, it can be said that the space where the heat was absorbed is cooled. The inside of the household refrigerator or freezer is cold to the touch, but this cold cannot be used to cool the kitchen by having the refrigerator door open. Actually, leaving the door open causes the kitchen to become hotter overall. This situation demonstrates an important principle of mechanical refrigeration: to remove heat, it is necessary to add energy to it.

Mechanical refrigeration relies primarily on the process of evaporation. This process is responsible for the cool sensation that results when rubbing alcohol is applied to the skin or when gasoline is spilled on the skin. Body heat supplies the energy required to vaporize the alcohol or gasoline, and it is the transfer of this energy from the body to the liquid that cools the skin. In refrigeration systems, such as those used in the commercial building, the evaporation process is controlled in a closed system in order to preserve the refrigerant so that it can be reused many times in what is known as the *refrigerant cycle*. As shown in Figure 9-1, the four main components of the refrigerant cycle are:

- *Evaporator.* As the refrigerant evaporates, it absorbs energy from the air around the evaporator.

- *Compressor.* This device pressurizes the refrigerant and pumps it into the condenser. Although this adds more heat, the refrigerant can be condensed readily because of the increased pressure.

© Cengage Learning®

FIGURE 9-1 The refrigerant cycle.

- *Condenser.* The compressed refrigerant condenses here as the heat is removed from it by the air blown past the condenser.
- *Expansion valve.* This control device maintains a lower pressure in the evaporator so that the refrigerant can expand and vaporize.

EVAPORATOR

The evaporator in a commercial installation normally consists of a fin tube coil, Figure 9-2, through which the building air is circulated by a motor-driven fan. A typical evaporator unit is shown in Figure 9-3. The evaporator may be located inside or outside the building. In either case, the function of the evaporator is to remove heat from the interior of the building or from an enclosed space. The chilled air is usually circulated through pipes or ductwork

to ensure a more even distribution. The window-type air conditioner, however, discharges the air directly from the evaporator coil. In general, the cooling air from the evaporator is circulated through the space to be cooled and then passed again across the cooling coil. A certain percentage of outside air is added to the circulating air to replace air lost through exhaust systems, fume hoods, or gradual leakage through walls, doors, and windows.

COMPRESSOR

The compressor serves as a pump to draw the expanded refrigerant gas from the evaporator. In addition, the compressor boosts the pressure of the gas and sends it to the condenser, Figure 9-4. This compression of the gas is necessary to condense it back into a liquid. When the temperature of the

FIGURE 9-2 Evaporator components.

Courtesy of Avantco Refrigeration

FIGURE 9-3 An evaporator with dual fans.

LoveFreedom/Shutterstock.com

FIGURE 9-4 A motor-driven reciprocating compressor.

© iStockphoto.com/Blue_Army

air or water surrounding the condenser is relatively warm, the gas pressure must be increased to ensure that the temperature around the condenser can liquefy the refrigerant.

Direct-drive compressors are usually used in large installations. For smaller installations, however, the trend is toward the use of hermetically sealed compressors. Due to several built-in electrical characteristics, these hermetic units cannot be used on all installations. (Restrictions on the use of hermetically sealed compressors are covered later in this text.)

CONDENSER

Condensers are generally available in three types: air-cooled units (Figure 9-5), water-cooled units (Figure 9-6), and evaporative cooling units in which water from a pump is sprayed on air-cooled coils to increase their cooling capacity (Figure 9-7). The function of the condenser in the refrigerant cycle is to discharge both the heat extracted by the evaporator and the heat of compression. Thus, it can be seen that keeping the refrigerator door open causes the kitchen to become hotter because the condenser transfers the combined heat load to the condensing medium. In the case of the refrigerator in a residence, the condensing medium is the room air.

Air-cooled condensers use a motor-driven fan to drive air across the condensing coil. Water-cooled condensers require a pump to circulate the water.

Once the refrigerant gas is condensed to a liquid state, it is ready to be used again as a coolant.

EXPANSION VALVE

It was stated previously that the refrigerant must evaporate or boil if it is to absorb heat. The process of boiling at ordinary temperatures can occur in the evaporator only if the pressure is reduced. The task of reducing the pressure is simplified, since the compressor draws gas away from the evaporator and tends to evacuate it. In addition, a restricted flow of liquid refrigerant is allowed to enter the high side of the evaporator. As a result, the pressure remains fairly low in the evaporator coil so that the liquid refrigerant entering through the restriction flashes into a vapour and fills the evaporator as a boiling refrigerant.

The restriction to the evaporator may be a simple orifice. However, commercial systems generally use a form of automatic expansion valve that is responsive to changes in the heat load. The expansion valve is located in the liquid refrigerant line at the inlet to the evaporator. The valve controls the rate at which the liquid flows into the evaporator. The liquid flow rate is determined by installing a temperature-sensitive bulb at the outlet of the evaporator to sense the heat gained by the refrigerant as it passes through the evaporator. The bulb is filled with a volatile liquid (which may be similar to the refrigerant). This liquid expands and

FIGURE 9-5 An air-cooled condenser.

FIGURE 9-6 A water-cooled condenser.

FIGURE 9-7 An evaporative condenser.

passes through a capillary tube connected to a spring-loaded diaphragm to cause the expansion valve to meter more or less refrigerant to the coil. The delivery of various amounts of refrigerant compensates for changes in the heat load of the evaporator coil.

◗ HERMETIC COMPRESSORS

The tremendous popularity of mechanical refrigeration for household use in the 1930s and 1940s stimulated the development of a new series of refrigerants that were nonexplosive and less toxic. These refrigerants, known as chlorofluorocarbons (CFCs), were relatively expensive at the time of their initial production, and because of their expense, it was no longer feasible to permit the normal leakage that

occurred around the shaft seals of reciprocating, belt-driven units. As a result, the hermetic compressor was developed, Figure 9-8. This unit consists of a motor–compressor completely sealed in a gas-tight, steel casing. The refrigerant gas is circulated through the compressor and over the motor windings, rotor, bearings, and shaft. The circulation of the expanded gas through the motor helps to cool the motor.

Recent research has developed a CFC-free refrigerant that, it is hoped, will reduce the harmful effects of CFCs on the ozone layer in the upper atmosphere.

The initial demand for hermetic compressors was for use in residential-type refrigerators. Therefore, most of the hermetic compressors were constructed for single-phase service. In other words, it was necessary to provide auxiliary winding and starting devices for the compressor installation. Because

Low-pressure gas in

High-pressure gas out

Motor rotor

Johannes Grau Schnittmodelle

FIGURE 9-8   The hermetic compressor.

the refrigerant gas surrounded and filled the motor cavity, it was necessary to remove the centrifugal switch commonly provided to disconnect the starting winding at approximately 85% of the full speed. This switch was removed because any arcing in the presence of the refrigerant gas caused the formation of an acid from the hydrocarbons in the gas, which attacked and etched the finished surfaces of the shafts, bearings, and valves. The acid also carbonized the organic materials used to insulate the motor winding and caused the eventual breakdown of the insulation. To replace the switch, the relatively heavy magnetic winding of a relay was connected in series with the main motor winding. The initial heavy inrush of current caused the relay to lift its magnetic core and energize the starting winding. As the motor speed increased, the main winding current decreased, causing the relay to disconnect the starting or auxiliary winding.

A later refinement of this arrangement was the use of a voltage-sensitive relay that was wound to permit pickup at values of voltage greater than the line voltage. The coil of this voltage-sensitive relay was connected across the starting winding. This connection scheme was based on the principle that the rotor of a single-phase induction motor induces in its own starting winding a voltage that is approximately in quadrature (phase) with the main winding voltage and has a value greater than the main voltage. As long as the motor was running, the voltage-sensitive relay broke the circuit to the starting winding.

Since most maintenance problems in hermetic compressors were due to starting relays and

capacitors, it was desirable to eliminate as many as possible of these devices. It was soon realized that small- and medium-size systems could make use of a different form of refrigerant metering device, thus eliminating the automatic expansion valve. Recall that this valve has a positive shutoff characteristic, which means that the refrigerant is restricted when the operation cycle is finished as indicated by the evaporator reaching the design temperature. As a result, when a new refrigeration cycle begins, the compressor must start against a head of pressure. If a small-bore, open capillary tube is substituted for the expansion valve, the refrigerant is still metered to the evaporator coil, but the gas continues to flow after the compressor stops until the system pressure is equalized. Therefore, the motor is required only to start the compressor with the same pressure on each side of the pistons.

This ability to decrease the load led to the development of a new series of motors that contained only a running capacitor (no relay was installed). This type of motor furnished sufficient torque to start the unloaded compressor and, at the same time, greatly improved the overall power factor.

COOLING SYSTEM CONTROL

Figure 9-9 shows a wiring diagram that is representative of standard cooling units.

When the selector switch in the heating–cooling control is set to COOL and the fan switch is placed

FIGURE 9-9   A cooling system control diagram.

on AUTO, any rise in temperature above the set point causes the cooling contacts (TC) to close and complete two circuits. One circuit through the fan switch energizes the evaporator fan relay (EFR), which, in turn, closes EFR contacts to complete the 208-volt circuit to the evaporator fan motor (EFM). The second circuit is to the control relay (CR) and causes the CR1 contacts to open. When the CR1 contacts open, the crankcase heater (CH) is de-energized. (This heater is installed to keep the compressor oil warm and dry when the unit is not running.) In addition, CR2 contacts close to complete a circuit to the condenser fan motor (CFM) and another circuit through low-pressure switch 2 (LP2), the high-pressure switch (HP), the thermal contacts (T), and the overload contacts (OL) to the compressor motor starter coil (CS). When the CR2 contacts close, the three CS contacts in the power circuit to the compressor motor (CM) also close.

Contacts low-pressure control (LP1) and LP2 open when the refrigerant pressure drops below a set point. When contacts LP2 open, CS is de-energized. When contacts LP1 open, CR is de-energized, the circuit to CFM opens, and the circuit to CH is completed. The high-pressure switch (HP) contacts open and de-energize the compressor motor starter (CS) when the refrigerant pressure goes above the set point of the switch

The LP1 is the normal operating control and the high-pressure control (HP) and LP2 act as safety devices. The T contacts shown in Figure 9-9 are located in the compressor motor and open when the winding temperature of the motor is too high.

The OL contacts are controlled by the OL elements installed in the power leads to the motor. The OL elements are sized to correspond to the current draw of the motor. That is, a high current causes the overload elements to overheat and open the OL contacts. As a result, the compressor starter is de-energized and the compressor stops. The evaporator fan motor (EFM) used to circulate the cooled air in the store area can be run continuously if the fan switch is turned to FAN. Actually, in many situations, it is recommended that the fan motor run continuously to keep the air in motion.

COOLING SYSTEM INSTALLATION

The owner of the commercial building leases the various office and shop areas on the condition that heat will be furnished to each area. However, tenants agree to pay the cost of operating the refrigeration system to provide cooling in their areas. Only four cooling systems are indicated in the blueprints for the commercial building since the bakery does not use a cooling system. The cooling equipment for the insurance office, the doctor's office, and the beauty salon are single-package units located on the roof. The compressor, condenser, and evaporator for each of these units are constructed within a single enclosure. The system for the drugstore is a split system with the compressor and condenser located on the roof and the evaporator located in the basement, Figure 9-10. Regardless of the type of cooling system installed, a duct system must be provided to connect the evaporator with the area to be cooled. The duct system shown diagrammatically in Figure 9-10 does not represent the actual duct system, which will be installed by another contractor.

The electrician is expected to provide a power circuit to each air-conditioning unit as shown on the plans. In addition, it is necessary to provide wiring to the thermostat in each area, and, in the case of the drugstore, wiring must be provided to the evaporator located in the basement.

ELECTRICAL REQUIREMENTS FOR AIR-CONDITIONING AND REFRIGERATION EQUIPMENT

The *Canadian Electrical Code (CEC)* rules covering air-conditioning and refrigeration equipment are found in *Section 28*. For air-conditioning and refrigeration equipment that does not incorporate hermetically sealed motor-compressor(s), *Rules 28–106, 28–200,* and *28–304* apply.

For the insurance office, an air-conditioning unit is located on the roof of the commercial building.

FIGURE 9-10 Single-package and split-system cooling units.

The data furnished with this air-conditioning unit supply the following information:

- Voltage: 208 volts, three-phase, three-wire, 60 Hz
- Hermetic refrigerant compressor-motor: Rated-load amperes 20.2 at 208 volts three-phase
- Evaporator motor: Full-load amperes 3.2 at 208 volts single-phase
- Condenser motor: Full-load amperes 3.2 at 208 volts single-phase

- Minimum circuit ampacity: 31.65 amperes[1]
- Maximum overcurrent protection: 70 amperes, time-delay fuse. 50% LRC, *Rule 28–708.*
- Locked-rotor current (LRC): 140 amperes

When the nameplate on equipment specifies fuses, then only fuses are permitted to be used as the branch-circuit overcurrent protection. Circuit breakers are not permitted.

Air-conditioning and refrigeration equipment is almost always marked with the minimum supply circuit ampacity, maximum ampere rating, and type of overcurrent protective device permitted.

---

[1]$[(20.2 \times 1.25) + 3.2 + 3.2] = 31.65$ amperes, *Rule 28–108* (see Unit 12).

▶ SPECIAL TERMINOLOGY

Understanding the following definitions is important when installing air-conditioners, heat pumps, and other equipment utilizing hermetic refrigerant motor–compressor. These terms are found in *Rule 28–010* and in various manufacturers' data.

Rated-Load Current (RLA).   The RLA is determined by the manufacturer of the hermetic refrigerant motor–compressor through testing at rated refrigerant pressure, temperature conditions, and voltage.

In most instances, the RLA is at least equal to 64.1% of the hermetic refrigerant motor–compressor's maximum continuous current (MCC).

EXAMPLE ————————————————

The nameplate on an air-conditioning unit is marked:

Compressor RLA...17.8 amperes

Locked-Rotor Current or Locked-Rotor Amperes (LRA).   This is marked on the equipment nameplate. If it is not marked, it can be calculated by multiplying the full-load current or full-load amperes (FLA) by six. If the FLA is not marked, it can be obtained from *Table 44* (Table 9-2) or *Table 45* (Table 9-1).

Branch-Circuit Selection Current (BCSC). Some hermetic refrigerant motor–compressors are designed to operate continuously at currents greater than 156% of the RLA. In such cases, the unit's nameplate is marked with "Branch-Circuit Selection Current." The BCSC will be no less than 64.1% of the maximum continuous current (MCC) rating of the hermetic refrigerant motor–compressor. (*Note:* 156% and 64.1% have an inverse relationship: 1/1.56 = 0.641, and 1/0.641 = 1.56.)

TABLE 9-1

CEC Table 45.

TABLE 45
SINGLE-PHASE AC MOTORS
(*SEE RULES 28–010 AND 28–704*)

Single-Phase AC Motors Full-Load Current in Amperes [See Notes (1) to (4)]

hp Rating	115 V	230 V
$\frac{1}{6}$	4.4	2.2
$\frac{1}{4}$	5.8	2.9
$\frac{1}{3}$	7.2	3.6
$\frac{1}{2}$	9.8	4.9
$\frac{3}{4}$	13.8	6.9
1	16	8
$1\frac{1}{2}$	20	10
2	24	12
3	34	17
5	56	28
$7\frac{1}{2}$	80	40
10	100	50

Courtesy of CSA Group

Notes:

(1) For full-load currents of 208- and 200-volt motors, increase the corresponding 230-volt motor full-load current by 10% and 15%, respectively.

(2) These values of motor full-load current are to be used as guides only. Where exact values are required (e.g., for motor protection), always use those appearing on the motor nameplate.

(3) These values of full-load current are for motors running at usual speeds and motors with normal torque characteristics. Motors built for especially low speeds or high torques may have higher full-load currents, and multi-speed motors will have full-load current varying with speed, in which case the nameplate current ratings shall be used.

(4) The voltages listed are rated motor voltages. Corresponding Nominal System Voltages are 120 and 240 volt. Refer to CSA Standard CAN3–C235.

EXAMPLE ————————————————

The maximum continuous current of a hermetic refrigerant motor–compressor is 31 amperes. The BCSC will be no less than:

31 amperes × 0.641 = 19.9 amperes

TABLE 9-2

CEC Table 44.

TABLE 44
THREE-PHASE AC MOTORS
(SEE RULES 28–010 AND 28–704)

3-Phase	AC Motor Full-Load Current in Amperes [See Notes (1), (2), (3), and (4)]											
Motor Rating hp	Induction Type, Squirrel Cage, and Wound Rotor, in Ampress							Synchronous Type, Unity Power Factor [See Note (4)], A				
	115 V	200 V	208 V	230 V	460 V	575 V	2300 V	200 V	230 V	460 V	575 V	2300 V
$\frac{1}{2}$	4.4	2.5	2.4	2.2	1	0.9	—	—	—	—	—	—
$\frac{3}{4}$	6.4	3.7	3.5	3.2	1.4	1.3	—	—	—	—	—	—
1	8.4	4.8	4.6	4.2	1.8	1.7	—	—	—	—	—	—
$1\frac{1}{2}$	12.0	6.9	6.6	6.0	2.6	2.4	—	—	—	—	—	—
2	13.6	7.8	7.5	6.8	3.4	2.7	—	—	—	—	—	—
3	19.2	11.0	10.6	9.6	4.8	3.9	—	—	—	—	—	—
5	30.4	17.5	16.7	15.2	7.6	6.1	—	—	—	—	—	—
$7\frac{1}{2}$	44	25.3	24.2	22	11	9	—	—	—	—	—	—
10	56	32.2	30.8	28	14	11	—	—	—	—	—	—
15	84	48	46.2	42	21	17	—	—	—	—	—	—
20	108	62	59.4	54	27	22	—	—	—	—	—	—
25	136	78	74.8	68	34	27	—	62	54	27	22	—
30	160	92	88	80	40	32	—	75	65	33	26	—
40	208	120	114	104	52	41	—	99	86	43	35	—
50	260	150	143	130	65	52	—	124	108	54	44	—
60	—	177	169	154	77	62	16	147	128	64	51	12
75	—	221	211	192	96	77	20	185	161	81	65	15
100	—	285	273	248	124	99	26	243	211	106	85	20
125	—	359	343	312	156	125	31	304	264	132	106	25
150	—	414	396	360	180	144	37	—	—	158	127	30
200	—	552	528	480	240	192	49	—	—	210	168	40

Courtesy of CSA Group

Notes:

(1) These values of motor full-load current are to be used as guides only. Where exact values are required (e.g., for motor protection), always use those appearing on the motor nameplate.

(2) These values of motor full-load current are for motors running at speeds typical for belted motors and motors with normal torque characteristics. Motors built for especially low speeds or high torques may require more running current, and multi-speed motors will have full-load current varying with speed, in which case the nameplate current rating shall be used.

(3) For 90 and 80% power factor, multiply the above figures by 1.1 and 1.25, respectively.

(4) The voltages listed are rated motor voltages. Corresponding Nominal System Voltages are 120, 208, 240, 480, and 600 V. Refer to CSA CAN3-C235.

Since the BCSC value, when marked, is always equal to or greater than the unit nameplate marked RLA value, the manufacturer of the air-conditioning unit must use the BCSC value, instead of the RLA value, to determine the minimum circuit ampacity (MCA) and the maximum overcurrent protection (MOP). For installation of individual hermetic refrigerant motor–compressor, where the electrician must select the conductors, the controller, the disconnecting means, and the short-circuit and ground-fault protection, the electrician must use the BCSC, if given, instead of the RLA.

Maximum Continuous Current (MCC). The MCC is determined by the manufacturer of the hermetic refrigerant motor–compressor under specific test conditions. The MCC is needed to properly design the end-use product. The installing electrician is not directly involved with the MCC.

The MCC is not on the nameplate of the packaged air-conditioning unit. The MCC is recognized as being no greater than 156% of the RLA or the BCSC, for the hermetic refrigerant motor–compressor. Except in special conditions, the overload protection must cut out if the current exceeds the 156% limit.

EXAMPLE

A hermetic refrigerant motor–compressor is marked:

Maximum Continuous Current... 31 amperes

Minimum Circuit Ampacity (MCA). The manufacturer of an air-conditioning unit is required to mark the nameplate with this value. This is what the electrician needs to know. The manufacturer determines the MCA by multiplying the RLA, or the BCSC, of the hermetic refrigerant motor–compressor by 125%. The current ratings of all other concurrent loads—such as fan motors, transformers, and relay coils—are then added to this value, *Rule 28–108.*

EXAMPLE

An air-conditioning unit's nameplate is marked:

Minimum Circuit Ampacity ... 26.4 amperes

This was derived as follows:

BCSC of 19.9 amperes × 1.25 =	24.9 amperes
plus $\frac{1}{4}$ horsepower fan motor @	1.5 amperes
MCA =	26.4 amperes

The electrician must install the conductors, the disconnect switch, and the overcurrent protection based upon the MCA value. No further calculations are needed; the manufacturer of the equipment has done it all.

Maximum Overcurrent Protective (MOP) Device. The electrician should always check the nameplate of an air-conditioning unit for this information, as the manufacturer is required to mark this value on the nameplate. This value should not exceed the calculated value of the overcurrent device as determined by *Rule 28–200,* for the highest-rated device plus all loads such as electric heaters and other motors that may operate simultaneously, *Rule 28–204(1).* This is not to exceed 50% LRC, *Rule 28–708(1).*

EXAMPLE

The nameplate of an air-conditioning unit is marked:

Maximum Time Delay Fuse ... 45 amperes

This was derived as follows:

BCSC of 19.9 amperes × 2.25 =	44.775 amperes
plus 1/4 horsepower fan motor @	1.5 amperes
Therefore, the MOP =	46.275 amperes

The 44.775 amperes calculation assumes that *Rule 28–200(e)(ii)* is used, due to nuisance tripping. Since 46.275 is the maximum, the next lower

standard size time-delay fuse, 45 amperes, must be used.

The electrician installs branch-circuit overcurrent devices with a rating **not to exceed** the MOP as marked on the nameplate of the air-conditioning unit. The electrician makes no further calculations; all calculations have been done by the manufacturer of the air-conditioning unit.

Overcurrent Protection Device Selection. How does an electrician know if fuses or circuit breakers are to be used for the branch-circuit overcurrent device? The electrician reads the nameplate and the instructions carefully. *Rule 2–024* requires that "Electrical equipment…shall be approved and shall be of a kind or type and rating approved for the specific purpose for which it is to be employed."

 EXAMPLE

If the nameplate of an air-conditioning unit reads "Maximum Size Time-Delay Fuse: 45 amperes," fuses are required for the branch-circuit protection. To install a circuit breaker would be a violation of *Rule 2–100(1)(m)*.

If the nameplate reads "Maximum Fuse or HACR Type Breaker …," then either a fuse or a HACR-type circuit breaker is permitted.

Maximum Overload Protection (MOLP). Referring to *Rule 28–710,* the maximum rating is not to exceed 140% of the RLC of the motor–compressor.

Disconnecting Means Rating. The *CEC* rules for determining the size of disconnecting means required for HVAC equipment are covered in *Rule 28–714* (see below). Because disconnect switches are horsepower rated, it is sometimes necessary to convert the horsepower rating to a locked-rotor current rating. *Rule 28–704(1)* indicates this to be six times the FLA from the nameplate of the compressor. If this information is not available, the values may be obtained from *Table 44* or *45* in the *CEC*.

Air-Conditioning and Refrigeration Equipment Disconnecting Means

- The disconnecting means for an individual hermetic motor–compressor must not be less than 115% of the rated-load current. See *Rule 28–714(1)(a).*

- The disconnecting means for an individual hermetic motor–compressor shall have an interrupting capacity or LRC rating equal to the LRC of the motor–compressor.

- For equipment that has at least one hermetic motor–compressor, and other simultaneous loads such as fans, heaters, solenoids, and coils, the disconnecting means must not be less than 115% of the largest RLA plus the sum of the currents of all of the components. See *Rule 28–714(2)(a).* In our example, the full-load current equals 31.65 amperes. Checking *Table 44,* we find that a full-load ampere rating of 31.65 amperes is slightly greater than the 30.8 amperes that a 10-horsepower, 208-volt motor draws. Thus a 15-horsepower disconnect would be required.

- The total locked-rotor current must also be checked to be sure that the horsepower rating of the disconnecting means is capable of safely disconnecting the full locked-rotor current, *Rule 28–714(2)(b).* In our example, the locked-rotor current equals 140 amperes. The equivalent LRC rating of the 15-horsepower disconnect, as determined by *Rule 28–704,* would be 277.2 amperes (46.2 amperes × 6 = 277.2 amperes). This would fall well within the requirements of *Rule 28–714(2)(b).*

If one of the selection methods (FLA or LRA) indicates a larger-size disconnect switch than the other method does, then the larger switch shall be installed.

- The disconnecting means must be within sight of, and not more than 3 metres from, the air-conditioning and refrigeration equipment. See *Rule 28–604(5).*

- The disconnecting means must be accessible. See *Rule 28–604(6).*

Air-Conditioning and Refrigeration Equipment Branch-Circuit Conductors

- For individual motor–compressor equipment, the branch-circuit conductors must have an ampacity of not less than 125% of the rated-load current or branch-circuit selection current, whichever is greater. See *Rules 28–706* and *28–106*.

- For motor–compressor(s) plus other loads, such as a typical air-conditioning unit that contains a motor–compressor, fan, heater, coils, etc., add all of the individual current values, plus 25% of that of the largest motor or motor–compressor. See *Rule 28–108*.

One can readily see the difficulty encountered when attempting to compute the actual currents flowing in Phases A, B, and C for a typical three-phase unit, Figure 9-11. Rather than attempting to add the currents vectorially, for simplicity, one generally uses the arithmetic sum of the current. However, the student must realize that an ammeter reading taken while the air-conditioning unit is operating will not be the same as the calculated current values. For the air-conditioning unit in the insurance office, the calculation is:

(20.2 + 3.2 + 3.2) amperes	=	26.60 amperes
Plus 25% of 20.2 amperes	=	5.05 amperes
Total	=	31.65 amperes

FIGURE 9-11 Load connections.

The minimum ampacity for the conductors serving the above air-conditioning unit is 31.65 amperes, with the calculation done as above. The nameplate of the air-conditioning unit also calls for a minimum circuit ampacity of 31.65 amperes. Refer to *Table 2*, which shows that No. 10 is the smallest size with adequate ampacity.

Air-Conditioning and Refrigeration Equipment Branch-Circuit Overcurrent Protection

- For individual units, the branch-circuit overcurrent protection shall not exceed 50% of the locked-rotor current (LRC). But if the 50% sizing will not allow the equipment to start, then the *CEC* permits sizing not to exceed 65% LRC. See *Rule 28–708(1)*.

- Where the circuit has more than one hermetic refrigerant motor–compressor, or one hermetic refrigerant motor–compressor and other motors or other loads, refer to *Rule 28–206*.

The branch-circuit overcurrent protection must also meet the requirements of *Rule 28–200*. Simply stated, this means that overcurrent protection must be provided that does not exceed the protection required for the motor–compressor that draws the largest amount of current, plus the sum of the rated-load current or branch-circuit selection currents of other motor–compressor, plus the ampere ratings of any other loads. In our example:

$$[(20.2 \times 1.75) + 3.2 + 3.2] \text{ amps} = 41.75 \text{ amps}$$

The nameplate for the air-conditioning unit supplying the cooling for the drugstore calls for maximum 40-ampere, dual-element, time-delay fuses. This value must not be exceeded.

If the feeder or branch circuit originates from a circuit-breaker panel, then the manufacturer's requirements can be met by installing a fusible disconnect switch near, or on, the air-conditioning equipment.

Air-Conditioning and Refrigeration Equipment Motor–Compressor Overload Protection

This topic is covered in *Rule 28–710*. The manufacturer provides the overload protection for the various internal components that make up the complete air-conditioning unit and submits the equipment to the Canadian Standards Association (CSA) for testing of the product. Overload protection of the internal components is covered in the CSA. This protection can be in the form of fuses, thermal protectors, certain types of inverse-time circuit breakers, overload relays, etc.

Rule 28–710 discusses overload protection for motor–compressor appliances. This rule requires that the motor–compressor be protected from overloads and failure to start by:

- A separate overload relay that is responsive to current flow and trips at not more than 140% of the rated-load current, or

- A time-delay fuse or inverse-time circuit breaker rated at not more than 125% of the rated-load current, or

- An integral thermal protector arranged so that its action interrupts the current flow to the motor–compressor.

It is rare for the electrician to have to provide additional overload protection for these components. The electrician's job is to make sure the equipment is fed by the proper-size conductors with the proper size and type of branch-circuit overcurrent protection and the proper horsepower rated disconnecting means.

REVIEW

Note: Refer to the *CEC* or the blueprint package as necessary.

1. Refrigeration is a method of _____ from an object.

2. Match the following items by inserting the appropriate letter from column 2 in the blank in column 1.

   a. This device raises the energy level of the refrigerant.   _____   A. Evaporator

   b. The refrigerant condenses here as heat is removed.   _____   B. Compressor

   c. This is a metering device.   _____   C. Condenser

   d. The refrigerant absorbs energy here in the form of heat.   _____   D. Expansion valve

3. Name three types of condensers.

   _____   _____   _____

4. A 230-volt, single-phase hermetic compressor with an FLA of 21 and an LRA of 250 requires a disconnect switch with a rating of _____ horsepower.

5. A motor–compressor unit with a rating of 32 amperes is protected from overloads by a separate overload relay selected to trip at not more than _____ ampere(s).

6. Should the electrician be required to furnish overload protection for a motor–compressor refrigeration unit, he or she can meet the requirements of *Rule 28–306* by selecting fuses sized at not more than _____% of the motor–compressor rated-load current.

7. Does the *CEC* permit the disconnect switch for a large rooftop air-conditioner to be mounted inside the unit? Give the appropriate reference. _____

   _____

8. Where the nameplate of an air-conditioning unit specifies the minimum ampacity required for the supply conductors, must the electrician increase this ampacity by 25% to determine the correct ampacity for the conductors that will supply the air-conditioning unit? Explain.

   _____

   _____

   _____

9. Where appliances are required to be grounded, what section of the *CEC* would you refer to for specific grounding requirements?

   _____

10. Which current rating of an air-conditioning unit is greater? (Circle the letter of the correct answer.)

    a. rated-load current

    b. branch-circuit selection current

Reading Electrical Drawings—Bakery, and Switches and Receptacles

Part I: Reading Electrical Drawings—Bakery

OBJECTIVES

After completing the study of Part I of this unit, the student will able to

- determine loading requirements
- select branch-circuit conductors

After completing the study of Part II of this unit, the student will be able to

- select switches and receptacles with the proper rating for a particular application
- install various types of receptacles correctly
- connect single-pole, three-way, four-way, and double-pole switches into control circuits

BLUEPRINTS

The electrician should check the electrical drawings long before the actual installation begins, as many decisions are required even when there is a complete set of drawings. Few plans provide full information on (1) the exact paths that the conduits or cables shall take, and (2) the actual sizes of the conductors and conduits. The circuit routings are very project-dependent and the sizings depend upon the routings.

Frequently, it is possible to install short connecting conduits between wall boxes while the slabs are being poured. Later, longer conduits may be needed for the same connections. If longer runs are necessary, the electrician may need to increase the conductor sizes and also the conduit sizes.

This section of the unit will give the student experience in determining the proper conductor size.

THE BAKERY BLUEPRINTS

Several different applications are illustrated in the bakery. See Figure 10-1. The luminaires are surface mounted as individual units. Raceway connections are made to all the luminaires. Connections are made from an electrical box located flush in the ceiling or by a raceway installed between the ends.

Special attention should be paid to the branch circuits as these plans are being reviewed. The branch-circuit panelboard summary shown in

Table 8-2 (page 135), provides important information about these circuits. However, project conditions may require the electrician to make changes.

The electrician is responsible for checking voltage drop and conductor ampacity. Once these have been determined, the electrician must check to see if any of the items on the summary need updating. A review of what can, what should not, and what cannot be changed will be useful.

- Changing the circuit load should be avoided. Changing the load could require the development of entirely new loading and panelboard schedules.

- Overcurrent protection should be carefully checked if there are changes in load or load type. Otherwise, it should not be changed.

- Reassigning phase connections can be useful if there is an advantage in grouping circuits to share neutrals.

- Continuous and unbalanced load designations are usually self-evident and should not be revised except in conjunction with other changes.

- The wire size may need to be increased if voltage drop is excessive or the conductor ampacity is derated.

- When the wire size is changed it may be possible to increase the circuit capacity.

Table 10-1 gives the bakery feeder loading schedule.

FIGURE 10-1   Electrical drawing for a bakery. *Note:* For complete blueprint, refer to Drawing E3 included with text.

TABLE 10-1

Bakery Feeder Loading Schedule.

ITEM	COUNT	VA/UNIT	DEMAND LOAD	INSTALLED LOAD	DEMAND FACTOR		USE	
					Feeder	Service	Feeder	Service
MINIMUM BASIC LOAD								
(CEC Rule 8–210)	218 m²	30 W/m²	6540		1	1	6 540	6 540
INSTALLED BASIC LOADS								
Style C luminaires	15	87		1 305			1 305	1 305
Style B luminaires	4	87		348			348	348
Style E luminaires	2	144		288			288	288
Style L luminaires	12	87		1 044			1 044	1 044
Style N luminaires	2	60		120			120	120
General receptacles	20	120		2 400			2 400	2 400
Installed basic loads							5 505	5 505
BASIC LOAD	Use the greater of the demand or installed loads, *CEC Rule 8–106(2)*.						6 540	6 540
SPECIAL LOADS								
Show win. lts. style C	3	87	261				261	261
Show win. rec.	4	720	2 880				2 880	2 880
Sign	1	1 200	1 200				1 200	1 200
Bake oven	1	16 000	16 000				16 000	16 000
Doughnut mach. Htr (See Note (1))	1	2 000	2 000				2 000	2 000
Motors	**Volts**	**FLA**	**Phase**					
Dishwasher	208	23.9	3	8 600	1.25		10 750	8 600
Disposer	208	7.44	3	2 677			2 677	2 677
Dough divider	208	2.2	3	792			792	792
Doughnut machine (See Note (1))	208	2.2	3	792			792	792
Exhaust fan	120	2.9	1	348			348	348
Multi-mixer	208	7.48	3	1 556			1 556	1 556
Multi-mixer	208	3.96	3	824			824	824
Total load							46 620	44 470

Notes: (1) Doughnut machine heater and doughnut machine motor are combined for a total load of 2792 VA.

$$I = \frac{P}{E \times 1.73} = \frac{46\,620}{208 \times 1.73} = 129.56 \text{ A}$$

Assuming all loads are continuous, using 75°C rated equipment (*Rule 4–006*), and RW90-XLPE copper conductors from *Table 2*, 75°C column, *CEC*, the ampere rating of the circuit would be

$$\frac{129.56}{0.8} = 161.95 \text{ A}$$

The minimum size feeder would be four No. 2/0 RW90-XPLE copper conductors in a 53-mm raceway, supplied by a 200-A switch c/w 175-A fuses. Conductor sized based on 75°C termination temperature.

Part II: Switches and Receptacles

During the course of the work, an electrician selects and installs numerous receptacles and switches. Therefore, it is essential that the electrician know the important characteristics of these devices, and how they are to be connected into the electrical system.

▶ RECEPTACLES

The National Electrical Manufacturers Association (NEMA) and the Canadian Standards Association (CSA) have developed standards for the physical appearance of locking and nonlocking plugs and receptacles. The differences in the plugs and receptacles are based on the ampacity and voltage rating of the device. For example, the most commonly used receptacle is the CSA 5–15R, Figure 10-2. The CSA 5–15R receptacle has a 15-ampere, 125-volt rating and has two parallel slots and a ground-pin hole. This receptacle will accept the CSA 5–15P plug only, Figure 10-3 (when "P" is used, such as in 5–15P, the "P" stands for cord end plug). The CSA 5–20R (Figure 10-4), also known as a T-slot receptacle, has its two neutral slots at right angles. This receptacle is rated at 20 amperes, 125 volts. The CSA 5–20R will accept a CSA 5–15P or 5–20P plug, *Rule 26–700, Appendix B*, Figure 10-5. A CSA 6–20R T-slot receptacle is shown in

FIGURE 10-3   CSA 5–15P. A 15-ampere, 125-volt plug.

FIGURE 10-4   CSA 5–20R. A 20-ampere, 125-volt receptacle.

FIGURE 10-5   CSA 5–20P. A 20-ampere, 125-volt plug.

Figure 10-6. This receptacle has a rating of 20 amperes at 250 volts. CSA 6–15P and 6–20P are shown in Figure 10-7. Another receptacle that is specified for the commercial building is the CSA 6–30R, Figure 10-8. This receptacle is rated at 30 amperes, 250 volts.

FIGURE 10-2   CSA 5–15R. A 15-ampere, 125-volt receptacle.

FIGURE 10-6   CSA 6–20R. A 20-ampere, 250-volt receptacle.

A   B

FIGURE 10-7   CSA 6–15P, a 15-ampere, 250-volt plug; and CSA 6–20P, a 20-ampere, 250-volt plug.

FIGURE 10-8   CSA 6–30R. A 30-ampere, 250-volt receptacle.

The CSA standards for general-purpose, non-locking and locking receptacles are shown in Figures 10-9 and 10-10. Following is the key to the catalogue numbers for these standards.

A/   Single receptacle
B/   Duplex receptacle
C/   Plug
D/   Connector
E/   Flanged inlet
F/   Flanged outlet
+   Hospital Grade
*   Available in hospital grade; add suffix HG
++   Available in isolated ground; add prefix IG

A special note should be made of the differences between the 125/250-volt (CSA 14) devices and the three-phase, 250-volt (CSA 15) devices. The connection of the 125/250-volt receptacle requires a neutral, a grounding wire, and two-phase connections. For the three-phase, 250-volt receptacle, a grounding wire and three-phase connections are required.

CSA configurations 5-20R and 6-20R are designed to accept both 15- and 20-ampere plugs. These configurations make the CSA configurations for nonlocking receptacles identical to those used in the United States. The 5–20R and 5–20RA have been reversed in *Diagram 1* as of the 2015 *CEC*. The T-slot is now the commonly used configuration and the straight blade is the alternative. This is also true for the 6–20R and the 6–20RA.

◼ HOSPITAL-GRADE RECEPTACLES

In locations where severe abuse or heavy use is expected, hospital-grade receptacles are recommended. These high-quality products meet CSA requirements as hospital-grade receptacles and are marked with a small dot, Figure 10-11.

◼ ELECTRONIC EQUIPMENT RECEPTACLES

Circuits with receptacle outlets that serve microcomputers, solid-state cash registers, or other sensitive electronic equipment should be served by receptacles that are specially designed. These

Description		15 A Receptacle	20 A Receptacle	30 A Receptacle	50 A Receptacle	60 A Receptacle
120 V	TT			TT-30R		
125 V	5	5-15R	5-20R	5-30R	5-50R	
125 V	5A		5-20RA alternate			
*250 V	6	6-15R	6-20R	6-30R	6-50R	
*250 V	6A		6-20RA alternate			
277 V AC	7	7-15R	7-20R	7-30R	7-50R	
347 V AC	24	24-15R	24-20R	24-30R	24-50R	
125/250 V	14	14-15R	14-20R	14-30R	14-50R	14-60R
3 φ 250 V	15	15-15R	15-20R	15-30R	15-50R	15-60R

(Rows 120 V through 347 V AC: 2-pole 3-wire grounding. Rows 125/250 V and 3 φ 250 V: 3-pole 4-wire grounding.)

*For configurations 6-15R, 6-20R, 6-20RA, 6-30R, and 6-50R, Y denotes the identified terminal when used on circuits derived from three-phase, 4-wire 416 V circuits.
Note: Except as noted above, in Diagrams 1 and 2,
(a) G represents the terminal for bonding to ground;
(b) W represents the identified terminal; and
(c) X, Y, and Z represent the terminals for ungrounded conductors.

FIGURE 10-9 CSA *Diagram 1.*

Courtesy of CSA Group

receptacles may be constructed for an isolated ground, Figure 10-12, may have transient voltage surge protection, may be hospital grade, or may possess any combination of these three characteristics. Where the isolated grounding is desired, an insulated grounding conductor is permitted to be run from the ground terminal at the distribution panel to the isolated terminal on the receptacle, *Rule 10–904(9).* Figure 10-13 shows the difference in wiring methods for conventional receptacles and isolated ground receptacles. These receptacles are required to have an orange triangle on their faces, except where the receptacle is orange. In this case, the triangle is black.

Surge protection will absorb high voltage surges on the line and further protect the equipment, Figure 10-14. Surge suppressors are highly recommended in regions where lightning strikes are

Description		15 A Receptacle	20 A Receptacle	30 A Receptacle	50 A Receptacle	60 A Receptacle
125 V	L5	L5-15R	L5-20R	L5-30R	L5-50R	L5-60R
125 V	ML2	ML2-15R				
125 V	SS1				SS1-50R	
250 V	L6	L6-15R	L6-20R	L6-30R	L6-50R	L6-60R
277 V AC	L7	L7-15R	L7-20R	L7-30R	L7-50R	L7-60R
347 V AC	L24		L24-20R			
480 V AC	L8		L8-20R	L8-30R	L8-50R	L8-60R
600 V AC	L9		L9-20R	L9-30R	L9-50R	L9-60R
125/250 V	L14		L14-20R	L14-30R	L14-50R	L14-60R
125/250 V	SS2				SS2-50R	
3 φ 250 V	L15		L15-20R	L15-30R	L15-50R	L15-60R
3 φ 480 V	L16		L16-20R	L16-30R	L16-50R	L16-60R
3 φ 600 V	L17			L17-30R	L17-50R	L17-60R
3 φ 208 Y / 120 V	L21		L21-20R	L21-30R	L21-50R	L21-60R
3 φ 480 Y / 277 V	L22		L22-20R	L22-30R	L22-50R	L22-60R
3 φ 600 Y / 347 V	L23		L23-20R	L23-30R	L23-50R	L23-60R

Left-side row group labels: *2-pole 3-wire grounding* (rows L5 through L9); *3-pole 4-wire grounding* (rows L14 through L17); *4-pole 5-wire grounding* (rows L21 through L23).

Note: *In Diagrams 1 and 2,*
(a) G represents the terminal for bonding to ground;
(b) W represents the identified terminal; and
(c) X, Y, and Z represent the terminals for ungrounded conductors.

FIGURE 10-10   CSA *Diagram 2.*

FIGURE 10-11   Hospital-grade receptacle.

common. Compared to the cost of replacing damaged equipment and lost data, the cost of the receptacle, probably under $50, seems small.

GROUND-FAULT CIRCUIT INTERRUPTER RECEPTACLES (*RULE 26–700(11)*)

Rule 26–700(11) requires that ground-fault circuit interrupters (GFCI) be installed in circuits where receptacles are within 1.5 metres of washbasins,

Wrap-around, locked on brass mounting strap provides additional support strength for receptacle assembly.

Green grounding screw connected directly to the grounding contacts.

Insulation barrier construction – first patented by Hubbell – isolates ground contacts from the mounting strap.

Straight blade 15A and 20A, 125V duplex receptacles are available in a choice of colours: orange, gray and ivory.

Amperage and voltage clearly indicated.

Back- and side-wiring capability provides easy installation with stranded or solid wire.

Dimensionally stable, reinforced thermoplastic polyester provides impact strength in addition to heat and flame resistance.

Impact-resistant nylon face.

Courtesy of Hubbell Canada L.P.

FIGURE 10-12   Isolated grounding receptacle.

A

Green hex-head screw on receptacle connects to grounding terminals and centre-threaded post for cover.

Mounting strap

Outlet box

Bonding jumper

Metal raceway

Normal grounding electrode system

B

Green hex-head screw on receptacle connects only to grounding terminals. Centre threaded post for cover is connected to metal yoke.

Mounting strap

Outlet box

Insulating barrier

Connect equipment grounding conductor to nearest of
• Building disconnecting means
• Separately derived system
• Service

Metal raceway

Insulated equipment grounding conductor

© Cengage Learning®

FIGURE 10-13   (A) Standard conventional receptacle and (B) isolated ground receptacle.

UL Listed to Standards 1449 and 498; CSA Certified.

Damage-alert alarm sounds when surge protection is no longer functioning...and keeps sounding until the receptacle is replaced or muting screw is utilized.

Muting screw allows damage-alert alarm to be silenced until device is replaced.

Power-on indicator light verifies instantly that power is available at the receptacle and the suppression circuit is fully functional; light off means power has been interrupted; flashing light indicates surge protection circuitry has been damaged.

Distinctive surge symbol provides quick visual identification of surge suppression receptacle. High impact nylon face resists breakage.

Automatic arounding clip attached to bridge.

Fits standard wall box.

Courtesy of Hubbell Canada L.P.

FIGURE 10-14   Surge suppressor receptacle.

bathtubs, and shower stalls. The *CEC* also requires the installation of GFCI protection on temporary wiring for construction sites and in a limited number of other applications. GFCI protection is required for the bathrooms of commercial buildings. For further information concerning GFCI application, the student is referred to *Electrical Wiring: Residential,* Seventh Canadian Edition (Nelson, 2015).

Deaths and personal injuries have been caused by electrical shock from appliances such as radios, shavers, and electric heaters. This shock hazard exists whenever a person touches both a defective appliance and a grounded conducting surface such as a water pipe or metal sink. To protect against this possibility of shock, the 15-ampere branch circuits can be protected with GFCI receptacles or GFCI circuit breakers. The GFCI receptacles are the most commonly used form of protection.

The CSA requires that Class A GFCIs trip on ground-fault currents of 4 to 6 milliamperes (0.004 to 0.006 ampere). Figures 10-15 and 10-16 illustrate the principles of how a GFCI operates.

RECEPTACLES IN ELECTRIC BASEBOARD HEATERS

Electric baseboard heaters are available with or without receptacle outlets. Figure 10-17 shows how a receptacle outlet can be installed in an electric baseboard heater. *Section 62* details the requirements for fixed electric space heating equipment.

SWITCHES (RULES 14–500 TO 14–514)

Most electricians refer to switches as toggle switches or wall switches. Refer to Figure 10-18. These switches are divided into two categories.

Category 1 contains those Alternating Current (AC)/Direct Current (DC) general-use switches that are used to control

- AC or DC circuits;
- Resistive loads not to exceed the ampere rating of the switch at rated voltage;

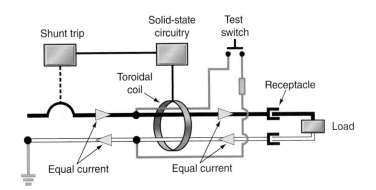

No current is induced in the toroidal coil because both circuit wires are carrying equal current. The contacts remain closed.

An imbalance of from 4 to 6 milliamperes in the coil will cause the contacts to open. The GFCI must open in approximately 25 milliseconds. Receptacle-type GFCIs have a switching contact in each circuit conductor.

© Cengage Learning®

FIGURE 10-15 Basic principle of how a ground-fault circuit interrupter operates.

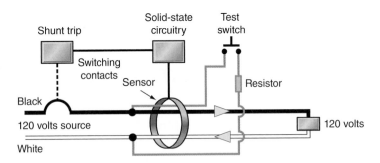

© Cengage Learning®

FIGURE 10-16 GFCI internal components and connections. Receptacle-type GFCIs switch both the black and white conductors. Note that when the test button is pushed, the test current passes through the sensor to the test button but bypasses the sensor on the way back to the opposite circuit conductor. This is how an imbalance is created and then monitored by the solid-state circuitry to signal the GFCI's contacts to open. Note that since both load currents pass through the sensor, no imbalance is present under normal receptacle use.

This receptacle outlet shall not
be connected to the heater circuit

Electric baseboard heater

FIGURE 10-17   Factory-mounted receptacle outlets on permanently installed electric baseboard heaters may be counted as the required outlets for the space occupied by the baseboard unit.

FIGURE 10-18   General-use switch.

- Inductive loads not to exceed one-half the ampere rating of the switch at rated voltage;

- Tungsten filament lamp loads not to exceed the ampere rating of the switch at rated voltage if the switch is marked with the letter "T." (A tungsten filament lamp draws a very high current at the instant the circuit is closed. As a result, the switch is subjected to a severe current surge.)

The AC/DC general-use switch normally is not marked "AC/DC." However, it is always marked with the current and voltage rating, such as "10A-125V, 5A-250V-T."

Category 2 contains those AC general-use switches that are used to control

- AC circuits only;

- Resistive, inductive, and tungsten-filament lamp loads not to exceed the ampere rating of the switch at 120 volts;

- Motor loads not to exceed 80% of the ampere rating of the switch at rated voltage.

AC general-use switches may be marked "AC only," or they may also be marked with their current and voltage ratings. A typical switch marking is "15A, 120–277V AC" or "120–277–347V AC." The 347–V rating is required on 347/600-volt systems. Refer to *Section 30* for the special requirements of 347-V switches for lighting circuits. See *Rule 14–512* for additional information pertaining to maximum voltage limitations.

Terminals of switches rated at 20 amperes or less—when marked "CU/AL" or "CO/ALR"—are suitable for use with aluminum, copper, and copper-clad aluminum conductors. Switches not marked "CU/AL" or "CO/ALR" are suitable for use with copper and copper-clad conductors only.

Many switches have screwless pressure terminals where a connection can be made by simply inserting the bared end of the conductor. These terminals may be used with copper and copper-clad aluminum conductors only. Push-in terminals are not suitable for use with ordinary aluminum conductors.

Further information on switch ratings is given in *Rules 14–508* and *14–510*.

Switch Types and Connections

Switches for branch-circuit lighting are readily available in four basic types: single-pole, three-way, four-way, and double-pole.

Single-Pole Switch. A single-pole switch is used where it is desired to control a light or group of lights, or other load, from one switching point. This type of switch is wired in series between the ungrounded ("hot") wire and the load. Figure 10-19 shows typical applications with the feed to the switch connected either at the switch or at the light. Note that the grounded wire goes directly to the load and the ungrounded wire is broken at the single-pole switch. Three diagrams are shown for each switching connection. The first diagram is a schematic drawing and is valuable when visualizing the current path. The second diagram represents the situation where a raceway will be available for installing the conductors. The third diagram illustrates the connections necessary when using armoured cable (AC) or nonmetallic-sheathed cable (NMSC), *Rule 4–036(2)*.

FIGURE 10-19 Single-pole switch connection.

Three-Way Switch. A three-way switch has a *common terminal* to which the switch blade is always connected. The other two terminals are called the *traveller terminals,* Figure 10-20. In one position, the switch blade is connected between the common terminal and one of the traveller terminals. In the other position, the switch blade is connected between the common terminal and the second traveller terminal. The three-way switch can be identified readily because it has no *On* or *Off* position. Note that *On* and *Off* positions are not marked on the switch handle in Figure 10-20. The three-way switch is also identified by its three terminals. The common terminal is darker in colour than the two traveller terminals, which have a natural brass colour. Figure 10-21 shows the application of three-way switches to provide control of the light from two points. Three-way or four-way switches shall not be considered as disconnecting means, *Rule 2–304(2).*

Four-Way Switch. A four-way switch should really be called a four-terminal switch. A four-way switch is used when a light, a group of lights, or other load must be controlled from more than two switching points. A four-way switch is similar to a three-way switch in that it does not have *On* and *Off* positions. A four-way switch is always connected between two 3-way switches. Three-way switches are connected to the source and to the load. At all other control points, four-way switches are used.

A four-way switch has four terminals. Two of these terminals are connected to the traveller conductors from one 3-way switch, and the other two terminals are connected to traveller conductors from the other three-way switch, Figure 10-23.

In Figure 10-22, terminals A1 and A2 are connected to one 3-way switch and terminals B1 and B2 are connected to the other three-way switch. In position 1, the switch connects A1 to B2 and A2 to B1. In position 2, the switch connects A1 to B1 and A2 to B2. Figure 10-23 illustrates a typical circuit in which a lamp is controlled from any one of three switching points.

Any colour other than white, grey, or green could have been used for the switch return. For the *travellers,* select a colour that is different from the switch return. In these wiring diagrams, red and black would be used for the travellers. Other colour combinations could have been used.

For cable wiring, the white conductor is used as the supply to a switch and shall be permanently re-identified as being an ungrounded conductor. The white conductor shall not be used as the switch return. In Figure 10-23, the arrangement is such that white conductors were spliced together. Care must be used to ensure that the traveller wires are connected to the proper terminals of the four-way switch. That is, the two traveller wires from one 3-way switch must be connected to the two terminals on one end of the four-way switch. Similarly, the two traveller wires from the other three-way switch must be connected to the two terminals on the other end of the four-way switch.

Double-Pole Switch. A double-pole switch is rarely used on lighting circuits. As shown in Figure 10-24, a double-pole switch can be used for those installations where two separate circuits are to be controlled with one switch. Both conductors of a circuit may be disconnected at the same time using a two-pole switch, *Rule 14–016(a).*

FIGURE 10-20 (A) Three-way switch and (B) positions.

A Circuit with 3-way switch control—feed at the switch

B Circuit with 3-way switch control—feed at the light

FIGURE 10-21   Three-way switch connections.

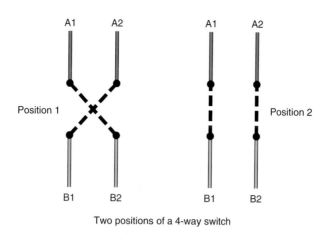

Two positions of a 4-way switch

FIGURE 10-22 Four-way switch operation.

NOTE: Four-way switches are always connected
between two 3-way switches.

FIGURE 10-23 Four-way switch connections.

Double-pole switching is required
for gasoline pump circuits, *NEC 514.11(A)*

FIGURE 10-24 Double-pole switch connections.

◗ SWITCH AND RECEPTACLE COVERS

A cover that is placed on a recessed box containing a receptacle or a switch is called a faceplate, and a cover placed on a surface mounted box is called a raised cover.

Faceplates come in a variety of colours, shapes, and materials, but for our purposes they can be placed in two categories: insulating and metal. Metal faceplates can become a hazard because they can conduct electricity; they are considered to be effectively grounded through the 6–32 screws that fasten the faceplate to the grounded yoke of a receptacle or switch. If the box is nonmetallic, the yoke must be grounded. Both switches and receptacles are available with a grounding screw for grounding the metal yoke.

It is a requirement that two screws or another approved method be used to fasten a receptacle to a raised cover. Receptacles that are installed outdoors in damp or wet locations must be covered with weatherproof covers that maintain their integrity when the receptacle is in use. These covers would be deep enough to shelter the attachment plug, *Rule 26–702.*

REVIEW

◗ PART I: READING ELECTRICAL DRAWINGS—BAKERY

Note: Use the following questions to review the loading schedule for the bakery. Use the bakery loading schedule, the bakery plans, and the *CEC* as necessary.

1. What is the basic demand load for the bakery?

   _____

   _____

2. The total connected load of the "special loads" (i.e., not basic load) is _____ volt-amperes.

3. The total connected lighting load on the main floor in the bakery is _____ volt-amperes.

4. To supply the lighting tabulated in the previous question, a total of _____ branch circuits are used.

5. The connected load on circuit numbers 1, 3, and 5 is _____ amperes.

6. The maximum allowable load on branch circuit 20 will be _____ volt-amperes if No. 10 AWG copper conductors are installed to increase the voltage at the load.

7. The electrician might consider installing a single circuit for branch circuits 12 and 9. Is this a good idea? Explain in detail. Do the changes required by this strategy over-shadow its advantages?

   _____

   _____

   _____

PART II: SWITCHES AND RECEPTACLES

Note: Refer to the *CEC* or the blueprints as necessary.

1. Indicate which of the following switches may be used to control the loads listed below.

   A.   AC/DC   10A–125V   5 A–250 V

   B.   AC only   10A–120V

   C.   AC only   15 A–120/277–347 V

   D.   AC/DC   20 A–125 VT   10 A–250 V

   a. A 120-volt incandescent lamp load (tungsten filament) consisting of ten 150-watt lamps. _____

   b. A 120-volt fluorescent lamp load (inductive) of 1500 volt-amperes. _____

   c. A 347-volt fluorescent lamp load of 625 volt-amperes. _____

   d. A 120-volt motor drawing 10 amperes. _____

   e. A 120-volt resistive load of 1250 watts. _____

   f. A 120-volt incandescent lamp load of 2000 watts. _____

   g. A 230-volt motor drawing 3 amperes. _____

2. Using the information in Figures 10-9 and 10-10, select by number the correct receptacle for each of the following outlets.

   a. 120/208-volt, single-phase, 20-ampere outlet. _____

   b. 208-volt, three-phase, 50-ampere outlet. _____

   c. 208-volt, single-phase, 30-ampere outlet. _____

3. Show all wiring for the following circuit. The circuit consists of two 3-way switches controlling two luminaires with the feed at one luminaire. Indicate conductor colours.

To source

3-way Switch

3-way Switch

© Cengage Learning®

4. Show all wiring for the following circuit. The circuit consists of two 3-way switches and one 4-way switch controlling a light. One of the three switches is to be selected as the four-way switch; then complete the wiring accordingly. Indicate conductor colours.

© Cengage Learning®

UNIT 11

Branch-Circuit Installation

OBJECTIVES

After completing the study of this unit, the student will be able to

- complete the installation of various raceway products, using the correct materials
- select the correct raceway size for a branch-circuit installation, using appropriate calculations
- select the correct size and type of box for the installation, and the conductors to be installed to the box
- specify the proper raceway support
- demonstrate the proper method of connecting copper and aluminum conductors

BRANCH-CIRCUIT INSTALLATION

In a commercial building, the major part of the electrical work is the installation of the branch-circuit wiring. The electrician must have the ability to select and install the correct materials to ensure a successful job.

The term "raceway," which is used in this unit as well as others, is defined by the *Canadian Electrical Code (CEC)* as any channel that is designed and used expressly for holding wires, cables, or bus bars.

The following paragraphs describe several types of materials that are classified as raceways, including rigid metal conduit, electrical metallic tubing, liquidtight flexible conduit, flexible conduit, rigid non-metallic conduit, and electrical non-metallic tubing.

RIGID AND FLEXIBLE METAL CONDUIT (*RULES 12–1000* TO *12–1014*)

Rigid metal conduit, Figure 11-1, is of heavy-wall construction to provide a maximum degree of physical protection to the conductors run through it. Rigid conduit is available in either steel or aluminum. The conduit can be threaded on the job, or nonthreaded fittings may be used where permitted by local codes, Figure 11-2.

Rigid conduit bends can be purchased, or they can be made using special bending tools. Bends

FIGURE 11-2 Rigid metal conduit fittings.

Photo courtesy Appleton Group LLC, Emerson Industrial Automation

in 16-mm, 21-mm, and 27-mm conduit can be made using hand benders or hickeys, Figure 11-3. Hydraulic benders must be used to make bends in larger sizes of conduit.

Installation of Rigid and Flexible Metal Conduit

Rigid metal conduit and electrical metallic tubing are installed according to the requirements of

Photo Love/Shutterstock.com

FIGURE 11–1 Rigid steel conduit.

Courtesy Ideal Industries

FIGURE 11-3 Conduit bender.

Rules 12–1000 to *12–1014*. The following points summarize these rules. All conduit runs should be level, straight, plumb, and neat with a good standard of workmanship, *Rule 2–112*.

The rigid types of metal conduit

- May be installed in concealed and exposed work;

- May be installed in or under concrete when of the type approved for this purpose;

- Must be threaded as specified in *Table 40, Rule 12–1006;* (This rule is to ensure a minimum thread length as well as to avoid a running thread that would prevent the joints from being securely tightened.)

- May contain up to four quarter bends (for a total of 360 degrees) in any run;

- Must be supported at least every 1.5 metres for 16-mm and 21-mm trade size, every 2.0 metres for 27-mm and 35-mm, and every 3.0 metres for 41-mm to 155-mm;

- May be installed in wet or dry locations if the conduit is of the type approved for this use;

- Must have the ends reamed to remove rough edges and a bushing installed on the end of the conduit;

- Must be run through drilled, bored, or punched holes in framing members, such as studs or joists, in order for conduit to be considered adequately supported; when installed on the surface, conduit may be supported with straps approved for the purpose.

◢▶ ELECTRICAL METALLIC TUBING (*RULES 12–1400* TO *12–1414*)

Electrical metallic tubing (EMT) is a thin-wall metal raceway that is not to be threaded, Figure 11-4. Several types of nonthreaded fittings and connectors are available for use with EMT. The specifications for the commercial building allow the use of EMT under 53-mm trade size for several types of installations. EMT must not be installed where subject to mechanical damage.

Courtesy of COLUMBIA-MBF

FIGURE 11-4 Electrical metallic tubing.

EMT Fittings

Since EMT is not to be threaded, tubing sections are joined together and connected to boxes and cabinets by fittings called *couplings* and *connectors*. Several styles of EMT fittings are available, including setscrew, compression, and indenter styles. Indenter-style fittings are not commonly used. Fittings are available in die-cast aluminum or steel.

Setscrew. When used with this type of fitting, the EMT is pushed into the coupling or connector and is secured in place by tightening the setscrews, Figure 11-5. This type of fitting is classified as concrete-tight.

Compression. EMT is secured in these fittings by tightening the compressing nuts with a wrench or pliers, Figure 11-6. These fittings are classified as raintight and concrete-tight types.

Photo courtesy of Appleton Grp LLC, Emerson Industrial Automation

FIGURE 11-5 EMT connector and coupling, setscrew type.

Photo courtesy of Appleton Grp LLC, Emerson Industrial Automation

FIGURE 11-6 EMT connector and coupling, compression type.

Indenter. A special tool is used to secure EMT in this style of fitting, which places an indentation in both the fitting and the tubing. It is a standard wiring practice to make two sets of indentations at each connection. This type of fitting is classified as concrete-tight.

The efficient installation of EMT requires the use of a bender similar to the one used for rigid conduit, Figure 11-7. This tool is commonly available in hand-operated models for EMT in sizes from 16 mm to 35 mm and in power-operated models for EMT in sizes greater than 27 mm.

Five kinds of bends can be made with the use of the bending tool. The offset bend, saddle bend, stub bend (90°), back-to-back bend, and angle bend/kick are shown in Figure 11-8. The manufacturer's instructions that accompany each bender indicate the method of making each type of bend.

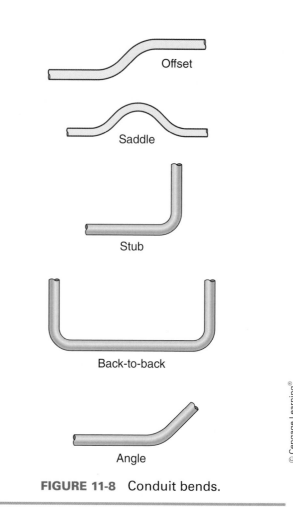

Offset

Saddle

Stub

Back-to-back

Angle

© Cengage Learning®

FIGURE 11-8 Conduit bends.

Notch mark for saddle bends

Mark for stub bends

Mark for back to back bends

Degree marks for offsets and saddle bends

Hook that holds conduit

Handle

Foot pedal

Courtesy of Sam Maltese

FIGURE 11-7 EMT bender.

ELECTRICAL NON-METALLIC TUBING (*RULES 12–1500 TO 12–1514*)

This type of raceway, Figure 11-9, is very popular where some degree of flexibility is required. It is referred to as ENT and is not to be confused with EMT. ENT is discussed later in this unit.

Copyright Thomas & Betts Corp. (www.tnb.com) Used with permission

FIGURE 11-9 Electrical non-metallic tubing.

Flexible metal conduit shall be not less than 16 mm except when connecting equipment, in which case 12 mm may be used for runs not exceeding 1.5 metres, *Rule 12–1004.*

FIGURE 11-10   Flexible metal conduit.

FLEXIBLE CONNECTIONS (*RULES 12–1000* TO *12–1014* AND *12–1300* TO *12–1308*)

The installation of certain equipment requires flexible connections, both to simplify the installation and to stop the transfer of vibrations. The two basic types of material used for these connections are flexible metal conduit, Figure 11-10, and liquidtight flexible metal conduit, Figure 11-11.

Flexible metal conduit shall not be less than 16-mm trade size, *Rule 12–1004,* except when connecting equipment, in which case 12-mm trade size may be used in lengths of 1.5 metres or less.

Rules 12–1000 to *12–1014* cover the use and installation of flexible metal conduit. This type of conduit is similar to armoured cable, except that the conductors are installed by the electrician. For armoured cable, the cable armour is wrapped around the conductors at the factory to form a complete cable assembly.

Some of the more common installations using flexible metal conduit are shown in Figure 11-12. Note that the flexibility required to make these installations is provided by flexible metal conduit. The figure calls attention to the requirement to install a separate bonding conductor in the raceway, *Rule 12–1306.*

Flexible metal conduit is commonly used to connect recessed luminaires (lighting fixtures). When used in this manner, electricians refer to it as a "fixture whip" or a "fixture drop." *Note:* Flexible metal conduit always requires a separate bonding conductor in the conduit.

The use and installation of liquidtight flexible metal and liquidtight flexible nonmetallic conduit are described in *Rules 12–1300* to *12–1308.* Liquidtight flexible metal conduit has a tighter fit to its spiral turns as compared to standard flexible metal conduit. Liquidtight metal conduit also has a thermoplastic outer jacket that is liquidtight. Liquidtight flexible conduit is commonly used as the flexible connection to a central air-conditioning unit located outdoors, Figure 11-13.

As shown in Figure 11-14, liquidtight flexible non-metallic conduit may be used:

- In exposed or concealed locations;
- Where flexibility is required;
- Where not subject to damage;
- Where the combination of ambient and conductor temperature does not exceed that for which the flexible conduit is approved;
- In lengths not over 1.5 metres for 12-mm conduit;

FIGURE 11-11   Liquidtight flexible conduit and fitting.

A Always install a separate bonding conductor with flexible metal conduit, *Rule 10–610*.

B When the method of raceway changes from rigid to flexible metal conduit, always ensure proper bonding connection by installing a bonding jumper inside raceway.

C A separate bonding conductor is required in accordance with *Rule 10–610*.

Bonding jumper

Flexible metal conduit

© Cengage Learning®

FIGURE 11-12 Installations using flexible metal conduit.

- Only with fittings identified for use with the flexible conduit;
- In sizes 16 mm to 53 mm; 12 mm is permissible for enclosing motor leads;
- For flexible connection to equipment, such as a motor.

FIGURE 11-13 A use of liquidtight flexible conduit. Install separate bonding conductor in liquidtight flexible conduit, *Rule 12–1306*. Air-conditioner must have disconnect within 3 metres of equipment, *Rule 28–604(5)*.

A separate equipment bonding conductor must be installed. This conductor is sized according to *Table 16A*. Figure 11-15 illustrates the application of this table.

There are three types of liquidtight flexible non-metallic conduit, which are marked as follows:

- "FNMC-A" for layered conduit
- "FNMC-B" for integral conduit
- "FNMC-C" for corrugated conduit

Both metal and nonmetal fittings listed for use with the various types of FNMC will be marked with the "A," "B," or "C" designations.

A Maximum length for 12 mm size is 1.5 metres. Use of liquidtight flexible conduit is restricted by *Rule 12–1302*.

B Liquidtight flexible metal conduit may not be used as a grounding means in any size, *Rule 12–1306*.

© Cengage Learning®

FIGURE 11-14 Liquidtight flexible metal conduit, *Rule 12–1300*.

FIGURE 11-15 Separate equipment bonding conductor in flexible metal conduit installation.

100 ampere overcurrent protection

Minimum No. 6 copper equipment bonding conductor according to Table 16A

No. 3 T90 conductors rated for 115 amperes (Table 2)

© Cengage Learning®

FIGURE 11-16 (A) Clean conduit and apply solvent cement. (B) Insert conduit into fitting and rotate a quarter turn. (C) Heat conduit until soft using approved heater or heat gun (D) for bending conduit. (E) Use a template for 90° bends. *Rule 12–1108 and Appendix B.*

Courtesy of Sandy F. Gerolimon

RIGID PVC CONDUIT (*RULE 12–1100* TO *12–1124*)

Rigid PVC conduit (PVC) is made of polyvinyl chloride, which is a plastic material. Solvent-welded fittings may be used for PVC connections and terminations, Figure 11-16.

PVC may be used:

- Concealed in walls, floors, and ceilings, provided no thermal insulation is present;
- In cinder fill;
- In wet and damp locations;
- For exposed work where not subject to physical damage unless identified for such use;
- For underground installations.

PVC may **not** be used:

- In hazardous locations, *Section 18;*
- For support of luminaires;

- Where exposed to physical damage;
- Where subject to ambient temperatures exceeding 75°C (167°F);
- When conductor ampacities are based on values that exceed those permitted for a 90°C (194°F) conductor;
- Where enclosed in thermal insulation.

On outdoor exposed runs of PVC, it is important that expansion joints (Figure 11-17) be installed to comply with the requirements set forth in *Rule 12–1118* and *Appendix B,* where the expansion characteristics of PVC are given. An equipment bonding conductor must be installed in PVC, and, if metal boxes are used, the bonding conductor must be connected to the metal box. For additional information regarding the installation of PVC, see Unit 7 of *Electrical Wiring: Residential* and *Rule 12–1100.*

Solvent weld

Solvent weld

Piston

Pipe strap should be anchored tightly to barrel

Pipe strap is mounted loosely to allow movement

FIGURE 11-17 Expansion fitting for PVC conduit. *Rule 12–1118* and *Appendix B*.

PVC Fittings and Boxes

A complete line of fittings, boxes, and accessories is available for PVC. Where PVC is connected to a box or an enclosure, an offset connector is recommended over a male adapter because the offset is stronger and provides a gentle radius for wire pulling, as shown in Figure 11-18, item (c).

Electrical Non-metallic Tubing (*Rules 12–1500* to *12–1514*)

Electrical non-metallic tubing (ENT) is a pliable, corrugated raceway that is made of polyvinyl chloride, a plastic material that is also used to make PVC. ENT is hand bendable and is available in coils and reels to facilitate a single length installation between pull points. ENT is designed for use within building walls, encased in concrete, or buried directly in the earth. It is not intended for outdoor use. The specification for the commercial building allows the use of ENT for feeders, branch circuits, and telephone wiring where a raceway size of 53-mm trade size or less is required. ENT is available in a variety of colours, which allows the colour coding of the various systems (e.g., electrical circuits could be blue, telephone circuits could be red, etc.).

ENT may be used:

- Concealed in masonry walls, see Table 11-1;
- In corrosive environments (check manufacturer's recommendations);
- In damp and wet locations, where fittings identified for the purpose are used;
- For direct burial, if approved for the purpose;
- Embedded in poured concrete, including slab-on-grade or below grade, where fittings identified for the purpose are used.

ENT may **not** be used:

- In hazardous locations;
- For luminaire or equipment support, *Rule 12–1512;*
- Subject to ambient temperature in excess of 75°C (167°F);
- When conductor ampacities are based on values in excess of those permitted for a 90°C (194°F) conductor;
- Where exposed, unless provided with protection from mechanical damage.

ENT Fittings

Two styles of listed mechanical fittings are available for ENT: one-piece snap-on and a clam-shell variety. Solvent cement PVC fittings are approved for use with ENT. The package label will indicate which of these fittings are concrete-tight without tape.

ENT Boxes and Accessories

Wall boxes and ceiling boxes are available with knockouts for 16-mm to 27-mm ENT. ENT can also be connected to any PVC box or access fitting. ENT mud boxes listed for fixture support are available, with knockouts for 16-mm to 27-mm trade sizes.

RACEWAY SIZING

The conduit size required for an installation depends upon three factors: (1) the number of conductors to be installed, (2) the cross-sectional area of the conductor, and (3) the permissible raceway fill. These factors are described in *Rule 12–1014*. After the student examines the plans and determines the number of conductors to be installed to a certain point, either of the following procedures can be used to find the conduit size.

If all of the conductors have the same insulation and are of the same wire size, the conduit size can be determined directly by referring to

A SERVICE/ENTRANCE/CAP

B METER HUB

C OFFSET CONNECTOR

D RIGID METAL TO PVC CONDUIT ADAPTOR/TYPE 'FA'

E O-RING EXPANSION JOINT CLOSED

F O-RING EXPANSION JOINT OPEN

G TYPE 'LB'

H TYPE 'LL'

I TYPE 'LR'

J TYPE 'T'

K TYPE 'C'

L TYPE 'E'

M TYPE 'TA'

N PVC COUPLING

O 2-HOLE STRAP

P EXPANSION JOINT

Courtesy Ipex Inc.

FIGURE 11-18 Various PVC fittings.

TABLE 11-1

Summary of conduit and tubing types.

	RIGID METAL CONDUIT	FLEXIBLE METALLIC CONDUIT	RIGID PVC CONDUIT	ELECTRICAL METALLIC TUBING	ELECTRICAL NON-METALLIC TUBING
CEC	Subsection 12–1000	Subsection 12–1000	Subsection 12–1100	Subsection 12–1400	Subsection 12–1500
Min.–Max. Size	16–155 mm	12–53 mm	16–155 mm	16–103 mm	16–27 mm
Min. Radii of Bends, *Rule 12–924*	See *CEC Table 7*		See *CEC Table 7*	See *CEC Table 7*	See *CEC Table 7*
Maximum Spacing of Supports	1.5 to 3 m depending on size of conduit, *Rule 12–1010*	300 mm from box/ outlet. 1.5 m apart, *Rule 12–1010(3)*	0.75 to 2.5 m depending on size of conduit, *Rule 12–1114*	1 m from box/ outlet. 1.5 m to 3 m apart depending on size of conduit. *Rules 12–1406* and *12–1010*	1 m from box/ outlet. 1 m apart.
Uses Allowed	All conditions. Under cinder fill if protected from corrosion, *Rule 12–936.*	Dry, accessible locations.	All conditions except in hazardous, *Rule 12–1100.*	As permitted, *Rule 12–1402.*	Concealed in masonry walls, encased in concrete, or direct burial, if approved.
Uses Not Allowed	See *Rule 12–936.*	Hazardous. Underground in concrete.	Hazardous. Thermal insulation, *Rule 12–1102.* High temperature, *Rule 12–1104.*	Where subject to damage. Burial. Hazardous. Wet. See *Rule 12–1402.*	Hazardous. Support of equipment. Exposed. Thermal insulation, *Rule 12–1508.*
Miscellaneous	Grounding. Threaded with taper die. See *Rule 12–1002* for damp and wet locations	Liquidtight with non-metallic jacket. See *Rule 12–1304* for fill of 12 mm (3/8-in.)	Expansion joints required. Nongrounding— bonding conductor required.	Not threaded.	Pliable. Corrugated. Nongrounding.

Tables 6A to 6K (Table 11-2). For example, three No. 8 T90 Nylon conductors may be installed to the air-conditioning unit in the drugstore in the commercial building. Refer to *Table 6K* to find the insulation type (T90 Nylon) and the conductor size (No. 8 AWG). It can be seen that 16-mm conduit or tubing is allowed.

Now assume that in addition to the three No. 8 T90 Nylon conductors, a No. 10 T90 Nylon equipment bonding conductor is to be installed. This type of installation will meet the requirements of *Rule 10–610* if a flexible connection is used to connect to the air-conditioning unit. The No. 10 equipment bonding conductor must be included in the determination of the required conduit size, as specified by *Rule 12–1014*.

Thus, for the case of three No. 8 T90 Nylon conductors and one No. 10 T90 Nylon equipment bonding conductor, refer to *Table 10A* (Table 11-4) to determine the conductor area. The total cross-sectional area of all of the conductors can now be calculated.

1 No. 10 T90 Nylon	13.63 mm²
3 No. 8 T90 Nylon @ 23.66 mm²	70.98 mm²
Total conductor cross-sectional area	84.61 mm²

Next, refer to *Table 8* (Table 11-3) in the category "Conductors … (not lead-sheathed)" and the subheading "4" conductors. This shows a maximum fill of 40%. Now refer to *Table 9B CEC* (Table 11-5) under the 40% column to find a value equal to or greater than 84.61. This takes us to the third number in the column, which is 134. Read across to the left to find the conduit size. For this situation, 21-mm trade size conduit is required.

▶ SPECIAL CONSIDERATIONS

Following the selection of the circuit conductors and the branch-circuit protection, a number of special considerations are generally factors for

TABLE 11-2

A section of *CEC Table 6K*, showing TWN75 and T90 nylon conductors' conduit fill.

TABLE 6K
MAXIMUM NUMBER OF THERMOPLASTIC CONDUCTORS TYPES TWN75 AND T90, OF ONE SIZE IN TRADE SIZES OF CONDUIT OR TUBING
(SEE RULE 12–910(5) AND APPENDIX B)

Conductor size, AWG, kcmil	Size of Conduit or Tubing													
	16	21	27	35	41	53	63	78	91	103	116	129	155	200
14	7	16	30	48	70	128	191	200	200	200	200	200	200	200
12	5	12	22	35	51	93	139	200	200	200	200	200	200	200
10	3	7	14	22	32	58	86	132	180	200	200	200	200	200
8	1	4	8	12	18	33	49	76	103	133	171	200	200	200
6	1	3	5	9	13	24	36	55	75	96	124	152	200	200
4	1	1	3	5	8	14	22	33	46	59	76	93	132	200
3	0	1	3	4	6	12	18	28	39	50	64	79	112	198
2	0	1	1	4	5	10	15	24	32	42	54	66	94	167
1	0	1	1	2	4	7	11	17	24	31	40	49	69	123
1/0	0	1	1	1	3	6	9	14	20	26	33	41	58	103
2/0	0	0	1	1	3	5	8	12	17	21	28	34	48	86
3/0	0	0	1	1	1	4	6	10	14	18	23	28	40	72
4/0	0	0	1	1	1	3	5	8	11	15	19	23	33	59
250	0	0	0	1	1	3	4	7	9	12	15	19	27	48
300	0	0	0	1	1	2	3	6	8	10	13	16	23	42
350	0	0	0	1	1	1	3	5	7	9	12	14	20	37
400	0	0	0	1	1	1	3	4	6	8	10	13	18	33
450	0	0	0	0	1	1	2	4	5	7	9	11	16	29
500	0	0	0	0	1	1	2	3	5	6	8	10	15	27

Courtesy of CSA Group

TABLE 11-3

CEC Table 8.

TABLE 8
MAXIMUM ALLOWABLE PERCENT CONDUIT AND TUBING FILL
(SEE RULE 38–032.)

	Maximum Conduit and Tubing Fill, %				
	Number of Conductors or Multi-Conductor Cables				
	1	2	3	4	Over 4
Conductors or multi-conductor cables (not lead-sheathed)	53	31	40	40	40
Lead-sheathed conductors or multi-conductor cables	55	30	40	38	35

Courtesy of CSA Group

TABLE 11-4

CEC Table 10A.

TABLE 10A
DIMENSIONS OF STRANDED CONDUCTORS CALCULATING CONDUIT AND TUBING FILL
(SEE *RULE 12–910* AND *APPENDIX B*)

Conductor Size AWG/kcmil	R90XLPE*, RW75XLPE*, RW90XLPE* 600 V		R90XLPE*, RW75XLPE*, RW90XLPE* 1000 V		R90XLPE†, RW75XLPE†, R90EP†, RW75EP†, RW90XLPE†, RW90EP‡ 600 V		TWN75 T90 NYLON		TW, TW75		TWU, TWU 75 RWU90XLPE*	
	Dia. (mm)	Area (mm²)	Dia. (mm)	Area (mm²)	Dia. (mm)	Area (mm²)	Dia. (mm)	Area (mm²)	Dia. (mm)	Area (mm²)	Dia. (mm)	Area (mm²)
14	3.36	8.89	4.12	13.36	4.12	13.36	2.80	6.18	3.36	8.89	4.88	18.70
12	3.84	11.61	4.60	16.55	4.60	16.75	3.28	8.47	3.84	11.61	5.36	22.56
10	4.47	15.67	5.23	21.45	5.23	21.45	4.17	13.63	4.47	15.67	5.97	27.99
8	5.99	28.17	5.99	28.17	6.75	35.77	5.49	23.66	5.99	28.17	7.76	47.29
6	6.95	37.98	7.71	46.73	8.47	56.39	6.45	32.71	7.71	46.73	8.72	59.72
4	8.17	52.46	8.93	62.67	9.69	73.79	8.23	53.23	8.93	62.67	9.95	77.76
3	8.88	61.99	9.64	73.05	10.40	85.01	8.94	62.83	9.64	73.05	10.67	89.42
2	9.70	73.85	10.46	85.88	11.22	98.82	9.76	74.77	10.46	85.88	11.48	103.5
1	11.23	99.10	12.49	122.6	13.51	143.4	11.33	100.9	12.49	122.6	13.25	137.9
1/0	12.27	118.3	13.53	143.9	14.55	166.4	12.37	120.3	13.53	143.9	14.28	160.2
2/0	13.44	141.9	14.70	169.8	15.72	194.2	13.54	144.0	14.70	169.8	15.45	187.5
3/0	14.74	170.6	16.00	201.0	17.02	227.5	14.84	172.9	16.00	201.0	16.76	220.6
4/0	16.21	206.4	17.47	239.7	18.49	268.5	16.31	209.0	17.47	239.7	18.28	262.4
250	17.90	251.8	19.17	288.5	21.21	353.2	18.04	255.7	19.43	296.4	20.20	320.5
300	19.30	292.6	20.56	332.1	22.60	401.2	19.44	296.9	20.82	340.5	21.54	364.4
350	20.53	331.0	21.79	372.9	23.83	446.0	20.67	335.6	22.05	381.9	22.81	408.6
400	21.79	373.0	23.05	417.3	25.09	494.5	21.93	377.8	23.31	426.8	24.07	455.0
450	22.91	412.2	24.17	458.8	26.21	539.5	23.05	417.3	24.43	468.7	25.19	498.4
500	23.95	450.5	25.21	499.2	27.25	583.2	24.09	455.8	25.47	509.5	26.24	540.8
600	26.74	561.7	27.24	582.9	30.04	708.8	—	—	28.26	627.3	29.02	661.4
700	28.55	640.0	29.05	662.6	31.85	796.5	—	—	30.07	710.0	30.82	746.0
750	29.41	679.3	29.91	702.6	32.71	840.3	—	—	30.93	751.3	31.69	788.7
800	30.25	718.7	30.75	742.6	33.55	884.0	—	—	31.77	792.7	32.53	831.1
900	31.85	796.6	32.35	821.8	35.15	970.2	—	—	33.37	874.5	34.13	914.9
1000	33.32	872.0	33.82	898.4	36.62	1053	—	—	34.84	953.4	35.60	995.4
1250	37.56	1108	38.32	1153	42.38	1411	—	—	39.08	1200	39.08	1199
1500	40.68	1300	41.44	1349	45.50	1626	—	—	42.20	1399	42.96	1449
1750	43.58	1492	44.34	1544	48.40	1840	—	—	45.10	1598	45.86	1652
2000	46.27	1681	47.03	1737	51.09	2050	—	—	47.79	1794	48.55	1851

* Unjacketed

† Jacketed

‡ Includes EPCV

Note: *Aluminum conductors may not be available in the same range of sizes as copper conductors.*

Courtesy of CSA Group

TABLE 11-5

CEC Table 9B.

TABLE 9B
CROSS-SECTIONAL AREAS OF FLEXIBLE METAL CONDUIT
(SEE *RULE 12–910*)

Nominal Conduit Size	Internal Diameter, mm	Cross-Sectional Area of Flexible Metal Conduit, (mm²)							
		100%	55%	53%	40%	38%	35%	31%	30%
12	9.52	71	39	38	28	27	25	22	21
16	15.88	198	109	105	79	75	69	61	59
21	20.62	334	184	177	134	127	117	104	100
27	25.4	507	279	269	203	193	177	157	152
35	31.75	792	435	420	317	301	277	245	238
41	38.1	1140	627	604	456	433	399	353	342
53	50.8	2027	1115	1074	811	770	709	628	608
63	63.5	3167	1742	1678	1267	1203	1108	982	950
78	76.2	4560	2508	2417	1824	1733	1596	1414	1368
91	88.9	6207	3414	3290	2483	2359	2173	1924	1862
103	101.6	8107	4459	4297	3243	3081	2838	2513	2432

Courtesy of CSA Group

each installation. The electrician is usually given the responsibility of planning the routing of the conduit to ensure that the outlets are connected properly. As a result, the electrician must determine the length and the number of conductors in each raceway. In addition, the electrician must derate the conductors' ampacity as required, make allowances for voltage drops, recognize the various receptacle types, and be able to install these receptacles correctly on the system.

BOX STYLES AND SIZING

The style of box required on a building project is usually established in the specifications. However, the sizing of the boxes is usually one of the decisions made by the electrician. Box sizing should be determined by *Tables 22* and *23* (Tables 11-6 and 11-7) and *Rule 12–3034*.

TABLE 11-6

CEC Table 22.

TABLE 22
SPACE FOR CONDUCTORS IN BOXES
(SEE *RULE 12–3034*)

Size of Conductor	Usable Space Required for Each Insulated Conductor
AWG	Millilitres
14	24.6
12	28.7
10	36.9
8	45.1
6	73.7

Courtesy of CSA Group

Switch Boxes (Device Boxes)

Switch boxes are 76 × 51 mm (2 × 3 in.) in size and are available with depths ranging from 38 to 89 mm ($1\frac{1}{2}$ in. to $3\frac{1}{2}$ in.), Figure 11-19. These boxes

TABLE 11-7

Table 23, CEC. See Rule 12–3034.

TABLE 23
NUMBER OF CONDUCTORS IN BOXES

Box Dimensions Inches Trade Size	Capacity Millilitres (Cubic Inches)	Maximum Number of Conductors Size AWG				
		14	12	10	8	6
Octagonal						
4 × 1½	245 (15)	10	8	6	5	3
4 × 2⅛	344 (21)	14	12	9	7	4
Square						
4 × 1½	344 (21)	14	12	9	7	4
4 × 2⅛	491 (30)	20	17	13	10	6
4¹¹⁄₁₆ × 1½	491 (30)	20	17	13	10	6
4¹¹⁄₁₆ × 2⅛	688 (42)	28	24	18	15	9
Round						
4 × ½	81 (5)	3	2	2	1	1
Device						
3 × 2 × 1½	131 (8)	5	4	3	2	1
3 × 2 × 2	163 (10)	6	5	4	3	2
3 × 2 × 2¼	163 (10)	6	5	4	3	2
3 × 2 × 2½	204 (12.5)	8	7	5	4	2
3 × 2 × 3	245 (15)	10	8	6	5	3
4 × 2 × 1½	147 (9)	6	5	4	3	2
4 × 2⅛ × 1⅞	229 (14)	9	8	6	5	3
4 × 2⅜ × 1⅞	262 (16)	10	9	7	5	3
Masonry						
3¾ × 2 × 2½	229 (14)/gang	9	8	6	5	3
3¾ × 2 × 3½	344 (21)/gang	14	12	9	7	4
4 × 2¼ × 2⅜	331 (20.25)/gang	13	11	9	7	4
4 × 2¼ × 3⅜	364 (22.25)/gang	14	12	9	8	4
Through Box						
3⅜ × 2	3.8/mm (6/inch)	4	3	2	2	1
Concrete Ring						
4	7.7/mm (12/inch)	8	6	5	4	2
FS						
1 Gang	229 (14)	9	8	6	5	3
1 Gang Tandem	557 (34)	22	19	15	12	7
2 Gang	426 (26)	17	14	11	9	5
3 Gang	671 (41)	27	23	18	14	9
4 Gang	917 (56)	37	32	24	20	12
FD						
1 Gang	368 (22.5)	15	12	10	8	5
2 Gang	671 (41)	27	23	18	14	9
3 Gang	983 (60)	40	34	26	21	13
4 Gang	1392 (85)	56	48	37	30	18

Courtesy of CSA Group

can be purchased with knock-outs for 16-mm or 21-mm conduit, or with cable clamps. Each side of the switch box has holes through which a suitable nail can be inserted for nailing the box to wood studs. The boxes can be ganged by removing the common sides of two or more boxes and connecting the boxes together. Plaster ears may be provided for use on plasterboard or for work on old installations. Device boxes are also available with the sides welded in place to provide a stronger construction.

Masonry Boxes

Masonry boxes are designed for use in masonry block or brick. These boxes do not require an extension cover; they will accommodate devices directly, Figure 11-20. Masonry boxes are available with

3 × 2 inch conduit device boxes in assorted depths are available with or without ears

Photo courtesy of Appleton Grp LLC, Emerson Industrial Automation

FIGURE 11-19 Device box for use with conduit.

depths of 64 mm ($2\frac{1}{2}$ in.) and 90 mm ($3\frac{1}{2}$ in.) and in through-the-wall depths of 90 mm ($3\frac{1}{2}$ in.), 140 mm ($5\frac{1}{2}$ in.), and 190 mm ($7\frac{1}{2}$ in.). Boxes of this type are available with knockouts up to 27-mm (1-in.) trade size.

Utility boxes, Figure 11-21, are generally used in exposed surface-mount installations and are available with 16-, 21-, and 27-mm knockouts. Since these boxes range in depth from 32 to 64 mm ($1\frac{1}{4}$ to $2\frac{1}{2}$ in.), they can accommodate a device without the use of an extension cover.

102-mm (4-in.) Square Boxes

A 102-mm (4-in.) square box, Figure 11-22, is used commonly for surface or concealed installations.

Available in 1-, 2-, 3-, 4-, and 5-gang sizes.

Photo courtesy of Appleton Grp LLC, Emerson Industrial Automation

FIGURE 11-20 Four-gang masonry device box.

Photo courtesy of Appleton Grp LLC, Emerson Industrial Automation

FIGURE 11-21 Utility box and receptacle cover.

Extension covers of various depths are available to accommodate devices in those situations where the box is surface mounted. This type of box is available with knockouts up to 27-mm trade size.

Octagonal Boxes

Boxes of this type are used primarily to install ceiling outlets. Octagonal boxes are available for mounting in concrete, surface mounting, or concealed installations, Figure 11-23. Extension covers are available, but are not always required. Octagonal boxes are commonly used in depths of 38 mm ($1\frac{1}{2}$ in.) and 54 mm ($2\frac{1}{8}$ in.). These boxes are available with knockouts up to 21-mm trade diameter.

119-mm ($4\frac{11}{16}''$) Boxes

These more spacious boxes are used where the larger size is required. Available with 16-, 21-, and 27-mm trade size knockouts, these boxes require an extension cover or a raised cover to permit the attachment of devices, Figure 11-24. These are primarily used for stove and dryer outlets and junction boxes.

Box Sizing

Rule 12–3034 indicates that outlet boxes, switch boxes, and device boxes must be large enough to provide ample room for the wires in them so that the wires do not have to be jammed in or crowded.

4-inch square drawn
outlet box with
eagle claw mounting
brackets, cable clamps,
and conduit knockouts

4-inch square
double offset device

4-inch square extension
ring in assorted knockout
combinations

Raised 4-inch square
cover available for toggle
switches and receptacles

4-inch square with 3/4-inch
raised cover for additional
4-1/2 cubic inches
of wiring space

© Cengage Learning®

FIGURE 11-22   102-mm (4-in.) square boxes and covers.

Photo courtesy of Appleton Grp LLC, Emerson
Industrial Automation

FIGURE 11-23   Octagonal box on telescopic
hanger and box extension.

Photo courtesy of Appleton Grp LLC, Emerson
Industrial Automation

FIGURE 11-24   119-mm ($4\frac{11}{16}''$) box.

If box contains NO fittings, devices, fixture studs, hickeys, switches, receptacles, or grounding conductors . . . ⟹ refer directly to *Table 23*.

If box contains ONE or MORE fittings, fixture studs, or hickeys . . . ⟹ count ONE conductor for each type.

For each strap containing ONE or MORE devices such as switches, receptacles, or pilot lights . . . ⟹ count TWO conductors for each strap.

For each pair of wire connectors . . . ⟹ count ONE conductor.

For ONE or MORE bare bonding conductors in the box . . . ⟹ do not count these conductors.

For conductors running through the box . . . ⟹ count ONE conductor for each conductor.

For conductors that originate outside of the box and terminate inside the box (except 18 and 16 AWG fixture conductors) . . . ⟹ count ONE conductor for each conductor.

If no part of the conductor leaves the box, for example, a "jumper" wire used to connect three devices on one strap . . . ⟹ DO NOT count these conductors.

When an insulated bonding conductor is installed (as for isolated ground receptacles, *Rule 10–904(8)(9)*) . . . ⟹ count ONE conductor for each insulated grounding conductor.

FIGURE 11-25 Checklist for determining proper box sizes.

When conductors are the same size, the proper box size can be determined by referring to *Table 23*. When conductors are of different sizes, refer to *Table 22*.

Tables 22 and *23* do not consider fittings or devices such as fixture studs, hickeys, switches, pilot lights, or receptacles which may be in the box. Refer to Figures 11-25 to 11-29 for information on determining the proper box size.

© Cengage Learning®

FIGURE 11-26 Transformer leads must be counted when selecting the proper size box, *Rule 12–3034(1)(b)*. In the example shown, the box is considered to contain four conductors.

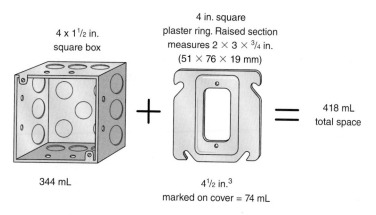

4 x 1½ in. square box

4 in. square plaster ring. Raised section measures 2 × 3 × ³⁄₄ in. (51 × 76 × 19 mm)

418 mL total space

344 mL

4½ in.³ marked on cover = 74 mL

© Cengage Learning®

FIGURE 11-27 Area of box and plaster ring.

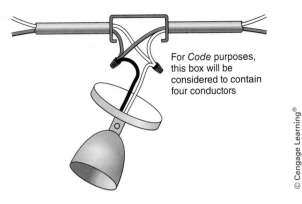

FIGURE 11-28 Fixture wires are not counted when determining correct box size. *Rule 12–3034 (1)(d)* specifies that fixture wires No. 18 and No. 16 AWG shall not be counted.

EXAMPLE

A box contains one fixture stud and two wire connectors. The number of conductors permitted in the box shall be two less than shown in the table. (Deduct one conductor for the fixture stud; deduct one conductor for the two wire connectors.) See Figure 11-25.

When the box contains different size wires, *Rule 12–3034,* do the following:

- Size the box based upon the total volume required for the conductors, according to *Table 22*.

One conductor

A conductor running through the box is counted as one conductor

There is no deduction for bare bonding conductors entering a box. If insulated (e.g., isolated bond conductor), then each conductor entering the box is counted as one conductor, *Rule 12–3034(1)(b)*

No deduction for bare conductor

Two conductors

Each conductor originating outside of the box and terminating inside the box is counted as one conductor

Conductors in which no part leaves the box are not to be counted

A deduction of two conductors shall be made for each strap containing one or more devices

Equivalent four conductors

FIGURE 11-29 Illustration of *Rule 12–3034* requirements.

- Then make the adjustments necessary for devices, wire connectors, and fixture studs.

- When conductors of different sizes are connected to device(s) on one yoke or strap, use the size of the largest conductor connected to the device(s) in computing the box fill.

- Count insulated bonding conductors the same as any other insulated conductor when computing the box fill.

Note: 1 cubic centimetre (cm³) = 1 millilitre (mL).

EXAMPLE

A box is to contain two devices: a duplex receptacle and a toggle switch. Two No. 12 conductors are connected to the receptacle, and two No. 14 conductors are connected to the toggle switch. There are two bonding conductors in the box, a No. 12 insulated and a No. 14 bare. There are two wire connectors in the box. The minimum box size would be computed as follows:

2 No. 12 conductors @ 28.7 cm³	57.4 cm³
2 No. 14 conductors @ 24.6 cm³	49.2 cm³
1 wire for connectors @ 28.7 cm³ (based on the size of the largest conductor in the box)	28.7 cm³
1 equipment bonding conductor @ 28.7 cm³	28.7 cm³
1 duplex receptacle @ 2 × 28.7 cm³ (based on the size of the connecting conductors)	57.4 cm³
1 toggle switch @ 2 × 24.6 cm³ (based on the size of the connecting conductors)	49.2 cm³
Minimum box volume	270.6 cm³ or mL

Therefore, select a box having a *minimum* volume of 270.6 mL. The volume may be marked on the box; otherwise, refer to the second column of *Table 23*, entitled "Capacity."

Two 76 × 51 × 51-mm (3 × 2 × 2-in.) boxes with a volume of 163 mL each would be adequate.

When sectional boxes are ganged together, the volume to be filled is the total volume of the assembled boxes. Fittings may be used with the sectional boxes, such as plaster rings, raised covers, and extension rings. When these fittings are marked with their volume, or have dimensions comparable to those boxes shown in *Table 23*, then their volume may be considered in determining the total volume to be filled. The following example illustrates how the volume of a plaster ring or of a raised cover, Figure 11-27, is added to the volume of the box to increase the total volume.

EXAMPLE

How many No. 12 conductors are permitted in this box and raised plaster ring, in Figure 11-27?

ANSWER See *Rule 12–3034* and *Tables 22* and *23* for the volume required per conductor. This box and cover will take 418 mL space ÷ 28.7 mL per No. 12 conductor = 14 No. 12 conductors maximum, less the deductions for devices and connectors, per *Rule 12–3034*.

Many devices, such as GFCI receptacles, dimmers, and timers, are much larger than a conventional receptacle or switch. To determine actual volume of a box where outlet devices are to be installed, subtract the volume of each outlet device from the box total. This will give a more accurate volume for wire fill in the box. It is always good practice to install a box that has ample room for the conductors, devices, and fittings rather than forcibly crowding the conductors into the box.

Insulated bonding conductors are the same size as the circuit conductors in cables having No. 14, No. 12, and No. 10 circuit conductors. Thus, box sizes can often be calculated using *Table 23* when the bonding conductor is insulated.

Another example of the procedure for selecting the box size is the installation of a four-way switch in the bakery. Four No. 12 conductors and one device are to be installed in the box.

Therefore, the box selected must be able to accommodate six conductors (two for the device). A 76 × 51 × 64-mm (3 × 2 × $2\frac{1}{2}$-in.) switch box is adequate.

When the box contains fittings such as fixture studs or hickeys, the number of conductors permitted in the box is reduced by one. For example, if a box contains one fixture stud, deduct one conductor from the allowable amount, as shown in *Table 23*. An example of this situation occurs in the bakery at the box where the conduit leading to the four-way switch is connected.

- There are two sets of travellers (one from each of the three-way switches) passing through the box on the way to the four-way switch (four conductors).

- There is a neutral, from the panelboard, that is spliced to a neutral that serves the luminaire attached to the box and to neutrals entering the two conduits leaving the box that go to other luminaires (four conductors).

- There is a switch return that is spliced in the box to a conductor that serves the luminaire that is attached to the box and to switch returns that enter the conduits leaving the box to serve other luminaires (four conductors).

- A fixture stud–stem assembly is used to connect the luminaire directly to the box (one conductor).

- This is a total of 13 conductors.

From *Table 23*, a 102 × 54-mm (4 × $2\frac{1}{8}$-in.) square or a 119 × 38-mm ($4\frac{11}{16}$ × $1\frac{1}{2}$-in.) square box would qualify. Since these boxes require plaster rings to accept finishing devices, a smaller size box could be used if the plaster ring volume added to the box volume were sufficient for the 13 conductors.

When necessary, conduit fittings, Figure 11-30, or pull boxes, Figure 11-31, are used. If pull boxes are used, they must be sized according to *Rule 12–3036,* as shown in Figure 11-32. The following *CEC* requirements apply to the installation of pull boxes and junction boxes.

FIGURE 11-30 Conduit fittings.

FIGURE 11-31 Threaded pull box.

ts = trade size

FIGURE 11-32 Straight pull box, *Rule 12–3036.*

- For straight pulls, the box must be at least eight times the diameter of the largest raceway when the raceways contain No. 4 or larger conductors, *Rule 12–3036(2)(a).*

- For angle (Figure 11-33) and U pulls (Figure 11-34), the distance *between* the raceways and the opposite wall of the box must

ts = trade size

For multiple entries on one side, minimum distance is 6 x largest trade size plus the trade size of all other entries on that side

© Cengage Learning®

FIGURE 11-33   Angle or U pulls, *Rule 12–3036.*

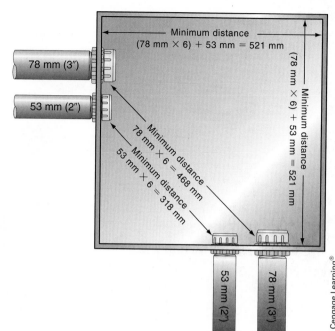

© Cengage Learning®

FIGURE 11-35   An example of angle pull when more than one raceway enters same wall of box.

be at least six times the diameter of the largest raceway. To this value, it is necessary to add the diameters of all other raceways entering in the same wall of the box, Figure 11-35.

© Cengage Learning®

FIGURE 11-34   An example of sizing pull and junction boxes for a U pull for two trade size 78 mm raceways.

This requirement applies when the raceways contain No. 4 or larger conductors. See *Rule 12–3036.*

- Boxes may be smaller than the preceding requirements when approved and marked with the size and number of conductors permitted.

- Boxes must have an approved cover.

- Boxes must be accessible.

▶ RACEWAY SUPPORT

Many devices are used to support conduit. The more popular devices are shown in Figures 11-36 and 11-37. The *CEC* section that deals with a particular type of raceway also gives the requirements for supporting that raceway. Figure 11-38 shows the support requirements for rigid metal conduit, electrical metallic tubing, flexible metal conduit, and liquid-tight flexible metal conduit.

Courtesy of Appleton Grp LLC

Courtesy of Appleton Grp LLC

Courtesy of Appleton Grp LLC

Courtesy of Eaton

Courtesy of Eaton

Photo is reprinted with permission and provided courtesy of ERICO International Corporation

FIGURE 11-36   Raceway support devices.

FIGURE 11-37   Non-metallic conduit and straps.

© Cengage Learning®

FIGURE 11-38   Supporting raceway. See *Rule 12–1010(1)* for securing and supporting raceways, boxes, etc., particularly for installation in suspended ceilings.

REVIEW

Note: Refer to the *CEC* or the blueprint package as necessary.

1. From the diagram that follows, determine the following dimensions.

   a. The minimum acceptable length dimension is _____ mm.

   b. The minimum acceptable width dimension is _____ mm.

2. From the diagram below, determine the following dimensions. All wires are No. 4 AWG and larger.

   a. Dimension a must be at least _____ mm.

   b. Dimension b must be at least _____ mm.

   c. Dimension c must be at least _____ mm.

© Cengage Learning®

3. Four No. 12 and four No. 10 conductors are to be installed in a conduit. What size conduit is required if

   a. the conductors are Type T90 Nylon?

   _____

   b. the conductors are Type RW90?

   _____

4. What is the smallest square box that will accommodate four No. 12 conductors and four No. 10 conductors if

   a. these conductors are pulled through without a splice?

   _____

   b. these conductors are all spliced in the box?

   _____

5. For each of the following raceways, how many supports are required to be installed? Both ends of each raceway terminate in outlet boxes.

   a. 15.2 m (50 ft) of 16-mm ($\frac{1}{2}$-in.) EMT

   _____

   b. 6.09 m (20 ft) of 16-mm ($\frac{1}{2}$-in.) EMT

   _____

6. What minimum volume is permissible for the outlet box shown below? Show your calculations.

7. Rigid metal conduit must be secured within _____ metres of a junction box.

8. Must additional support be provided for a conduit that is run through holes that have been bored in a series of studs? Explain your answer.

_____

_____

_____

_____

Appliance Circuits (Bakery)

OBJECTIVES

After completing the study of this unit, the student will be able to

- define "appliance"
- define "utilization equipment"
- understand the different wiring methods for connecting appliances
- determine branch-circuit ratings, conductor sizes, and overcurrent protection for appliances
- understand grounding requirements for appliances

APPLIANCES

Most of the rules that relate to the installation and connection of electrical appliances are found in *Section 26*. For motor-operated appliances, reference will also be made to *Section 28—Motors and Generators*. Where the appliance is equipped with hermetic refrigerant motor–compressor(s), such as refrigeration and air-conditioning equipment, *Subsection 28–700* applies.

"Appliance" could be defined as utilization equipment, generally other than industrial, of standardized sizes or types, that is installed or connected as a unit to perform such functions as air-conditioning, washing, food mixing, and cooking.

The *Canadian Electrical Code (CEC)* defines "utilization equipment" as equipment that utilizes electric energy for mechanical, chemical, heating, lighting, or similar purposes. The most common equipment types discussed in the *CEC* are the following:

- Permanently connected
- Cord connected
- Fixed

In commercial buildings, one of the most common tasks performed by the electrician is to make the electrical connections to appliances. The branch-circuit sizing may be specified by the consulting engineer in the specifications or on the electrical plans. In other cases, it is up to the electrician to do all of the calculations, and/or make decisions based upon the nameplate data of the appliance.

The bakery in the commercial building provides examples of several different methods that can be used to connect electrical appliances.

THE EXHAUST FAN (CONNECTED TO A LIGHTING BRANCH CIRCUIT)

The exhaust fan is a permanently connected motor-operated appliance. *Permanently connected* means "hard wired" with flexible conduit rather than being "cord connected." The exhaust fan is supplied by a 15-ampere branch circuit.

The rating of any one cord-and-plug-connected piece of utilization equipment shall not exceed 80% of the branch-circuit ampere rating.

This particular exhaust fan has a small, $\frac{1}{10}$-horsepower, 120-volt motor having a full-load ampere rating of 2.9 amperes. Because of its low current draw, the exhaust fan motor may be controlled by a standard single-pole switch, *Rule 28–500(3)(b)*, and supplied by the 15-ampere branch circuit.

THE BASICS OF MOTOR CIRCUITS

Disconnecting Means

The *CEC* requires that all motors be provided with a means to disconnect them from their electrical supply. These requirements are found in *Rule 28–600*. For most applications, the disconnect switch for a motor must be horsepower rated, *Rule 28–602*. There are some exceptions, the most common being the following:

- A manual across-the-line starter can serve as both the starter and disconnect if marked "suitable for motor disconnect."

- For three-phase stationary motors over 100 horsepower, a general-use disconnect switch or isolating switch is permitted when the switch is marked "Do not open under load," *Rule 28–602(3)(a)*.

- An attachment plug can serve as the disconnecting means for a portable motor if the attachment plug cap is rated not less than the ampacity of the conductors of the motor branch circuit.

Both the nameplate on all disconnect switches and the manufacturer's technical data furnish the horsepower rating, the voltage rating, and the ampere rating of the disconnect switch.

For those motors that are turned "off" and "on" by a controller (often referred to as a motor starter), a disconnecting means must be in sight of and within 9 metres of the controller and shall disconnect the controller, *Rule 28–604(3)(b)*, Figure 12-1A. That is why combination starters are so popular: they have the controller and the disconnecting means in one enclosure, as permitted

A

In sight of controller, and not more than 9 m from motor

B

In sight of controller, motor location, and driven machinery, and not more than 9 m from controller, motor location, and driven machinery

C

Another disconnect required

This disconnect is in sight of motor location and driven machinery

This disconnect is in sight of controller, but not in sight of the motor location, and driven machinery, or is more than 9 m from the motor location and driven machinery; then another disconnect is required

D

Lock-off feature

Disconnect need not be in sight of motor and driven machinery if the disconnect can be individually locked in the open (off) position. The locking provision must be of permanent type installed on the switch or circuit breaker. *Rule 28–604(4)(b)*

© Cengage Learning ®

FIGURE 12-1   Location of motor disconnect means.

in *Rule 28–602*, so the electrician does not have to mount a separate disconnect switch and a separate controller, Figure 12-1B. This results in labour savings.

According to *Rule 28–604(3)(a)*, Figure 12-1C, the disconnecting means must also be in sight of and within 9 metres of the motor location and the machinery driven by it. However, if the disconnect

FIGURE 12-2   Disconnect within 3 metres of air-conditioning equipment, *Rule 28–604(5)*.

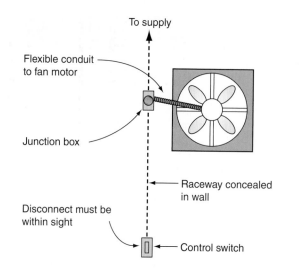

FIGURE 12-3   Exhaust fan installation.

is for air-conditioning equipment, it must be within sight of and within 3 metres of the equipment, *Rule 28–604(5)*, Figure 12-2. In the case of large units, the disconnect may be installed on or within the unit.

For a motor and the motor-disconnecting means to meet the *CEC* definition of being "in sight of" and "within 9 metres of" each other, both of the following conditions must be met:

- The disconnecting means must be visible from the motor, and the machinery driven by it.

- The disconnecting means must not be more than 9 metres from the motor.

Rule 28–604(4)(b) allows the motor disconnecting means to be located beyond the limit of 9 metres if it is capable of being locked in the open position, Figure 12-1D. To turn the exhaust fan on and off, an AC, general-use single-pole (toggle) switch is installed on the wall below the fan. The electrician will install an outlet box approximately 300 mm from the ceiling (see the electrical plan) from which he or she will run a short length of flexible metal conduit to the junction box on the fan. The fan exhausts the air from the bakery and is a through-the-wall installation, approximately 450 mm below the ceiling, Figure 12-3. *Rules 28–500(3)(b)* and *28–602(3)(e)* permit this switch to serve as both the controller and the required disconnecting means.

Motor Circuit Conductors

The conductors that supply a single motor shall be rated at not less than 125% of the full-load current rating of the motor, *Rule 28–106*. The branch-circuit conductors supplying the exhaust fan are No. 12 T90 Nylon conductors which have an ampacity well above the required 125% times the FLA of the motor. Refer to *CEC Table 2* for conductor ampacities.

Motor Overload Protection

Subsection 28–300 covers motor overload protection. The exhaust fan in the bakery draws 2.9 amperes, is less than 1 horsepower, and is not automatically started. According to *Rule 28–308(a)*, "any motor rated at 1 horsepower or less that is continuously attended while in operation and that is on a branch circuit having overcurrent protection rated or set at not more than 15 A does not require overload protection."

Unit 17 covers in detail the subject of overcurrent protection using fuses and circuit breakers. Time-delay fuses are an excellent choice for overload protection of motors, and can be sized close to the ampere rating of the motor. When the motor has integral overload protection or the motor controller has thermal overloads, time-delay fuses are selected to provide backup overload protection to the thermal overloads. See Table 12-1.

TABLE 12-1

Time-delay Fuse Protection From Thermal Overloads.

MOTOR NAMEPLATE RATING	OVERLOAD PROTECTION AS A PERCENTAGE OF MOTOR NAMEPLATE FULL-LOAD CURRENT RATING
Service factor not less than 1.15	125%
All other motors	115%

The momentary inrush current when a motor starts can be quite high, Figure 12-4. For example, a 4-ampere, full-load rated motor could have a starting current as high as 24 amperes. The fuse should not open needlessly because of this inrush current. In this situation, a 15-ampere ordinary fuse or breaker might be required to allow the motor to start. This would provide the branch-circuit overcurrent protection, but it would not provide overload protection for the appliance.

If the motor does not have built-in overload protection, sizing the branch-circuit overcurrent protection as stated above would allow the motor to burn out if it were overloaded or stalled.

Abuse, lack of oil, bad bearings, and jammed V-belts are a few of the conditions that can cause a motor to draw more than normal current. Too much current results in the generation of too much heat within the windings of the motor. For every 10°C (18°F) increase above the maximum temperature

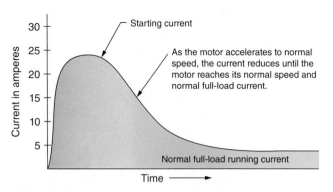

FIGURE 12-4 This diagram shows how the starting current of a typical electric motor reaches a very high value for a short time, then decreases as the motor reaches its normal full-load speed. Time-delay cartridge fuses are required to carry five times their ampere rating for at least 10 seconds, which is why this type of fuse is able to handle the momentary starting inrush current of the motor.

rating of the motor, the expected life of the motor is reduced by 50%. This is sometimes referred to as the "half-life rule."

The solution is to install a time-delay fuse that permits the motor to start, yet opens on an overload before the motor is damaged. Table 12-2 compares the load responses of ordinary fuses and time-delay fuses.

Table 12-3, Part I, shows how to size time-delay fuses for motors that are marked with a service factor of not less than 1.15 and for motors marked with a temperature rise of not over 40°C.

TABLE 12-2

Load Response of Fuses (Approximate Time in Seconds for Fuse to Blow).

LOAD (AMPERES)	4-AMPERE FUSE (TIME-DELAY)	4-AMPERE ORDINARY FUSE (NONTIME-DELAY)	8-AMPERE ORDINARY FUSE (NONTIME-DELAY)	15-AMPERE ORDINARY FUSE (NONTIME-DELAY)
5	Over 300 s	Over 300 s	Won't blow	Won't blow
6	250 s	5 s	Won't blow	Won't blow
8	60 s	1 s	Won't blow	Won't blow
10	38 s	Less than 1 s	Over 300 s	Won't blow
15	17 s	Less than 1 s	5 s	Won't blow
20	9 s	Less than 1 s	Less than 1 s	300 s
25	5 s	Less than 1 s	Less than 1 s	10 s
30	2 s	Less than 1 s	Less than 1 s	4 s

Table 12-3, Part II, shows how to size time-delay fuses for motors that have a service factor of less than 1.15 and for motors with a temperature rise of more than 40°C. For abnormal installations, time-delay fuses larger than those shown in Parts I and II of Table 12-3 may be required. These larger-rated fuses (or circuit breakers) will provide short-circuit protection only for the motor branch circuit.

Abnormal conditions might include

- Situations where the motor is started, stopped, jogged, inched, plugged, or reversed frequently;

- High inertia loads such as large fans and centrifugal machines, extractors, separators, pulverizers, and machines having large flywheels; and

- Motors having a high code letter and full-voltage start as well as some older motors without code letters.

The higher the code letter, the higher is the starting current. For example, two motors have exactly the same full-load ampere rating. One motor is marked Code J, while the other is marked Code D. The Code J motor will draw more momentary starting inrush current than the Code D. The code letter of a motor is found on the motor nameplate.

Refer to the *CEC* and to Bussmann's *Electrical Protection Handbook* for more data relating to motor overload and motor branch-circuit protection, disconnecting means, controller size, conductor

TABLE 12-3

Selection of Time-Delay Fuses for Motor Running Overload Protection Based on Motor Nameplate Ampere Rating.

PART I—MOTORS MARKED WITH NOT LESS THAN 1.15 SERVICE FACTOR

Motor Ampere Rating	Time-Delay Fuse Ampere Rating[*] (Max. 125%)	Motor Ampere Rating	Time-Delay Fuse Ampere Rating[*] (Max. 125%)
1.00 to 1.11	1 1/4	20.0 to 23.9	25
1.12 to 1.27	1 4/10	24.0 to 27.9	30
1.28 to 1.43	1 6/10	28.0 to 31.9	35
1.44 to 1.59	1 8/10	32.0 to 35.9	40
1.60 to 1.79	2	36.0 to 39.9	45
1.80 to 2.00	2 1/4	40.0 to 47.9	50
2.01 to 2.23	2 1/2	48.0 to 55.9	60
2.24 to 2.55	2 8/10	56.0 to 63.9	70
2.56 to 2.79	3 2/10	64.0 to 82.0	80
2.80 to 3.19	3 1/2	72.01 to 82.0	90
3.20 to 3.59	4	82.1 to 90	100
3.60 to 4.00	4 1/2	90.01 to 100	110
4.01 to 4.47	5	100.1 to 120	125
4.48 to 5.00	5 6/10	120.1 to 140	150
5.01 to 5.59	6 1/4	140.1 to 160	175
5.60 to 6.39	7	160.1 to 179	200
6.40 to 7.19	8	180 to 199	225
7.20 to 8.00	9	200 to 239	250
8.01 to 9.59	10	240 to 279	300
9.60 to 11.9	12	280 to 319	350
12.00 to 13.9	15	320 to 359[*]	400
14.00 to 16.0	17 1/2	360 to 399	450
16.01 to 19.9	20	400 to 479	500

(continues)

TABLE 12-3 *(Continued)*

PART II—ALL OTHER MOTORS (I.E., LESS THAN 1.15 SERVICE FACTOR)

All Other Motors Ampere Rating	Time-Delay Fuse Ampere Rating* (Max. 115%)	All Other Motors Ampere Rating	Time-Delay Fuse Ampere Rating* (Max. 115%)
1.00 to 1.08	1 1/8	21.8 to 25.0	25
1.09 to 1.21	1 1/4	26.1 to 30.0	30
1.22 to 1.39	1 4/10	30.5 to 34.7	35
1.40 to 1.56	1 6/10	34.8 to 39.1	40
1.57 to 1.73	1 8/10	39.2 to 43.4	45
1.74 to 1.95	2	43.5 to 50.0	50
1.96 to 2.17	2 1/4	52.2 to 60.0	60
2.18 to 2.43	2 1/2	60.9 to 69.5	70
2.44 to 2.78	2 8/10	69.6 to 78.2	80
2.79 to 3.04	3 2/10	78.3 to 86.9	90
3.05 to 3.47	3 1/2	87.0 to 95.6	100
3.48 to 3.91	4	95.7 to 108	110
3.92 to 4.34	4 1/2	109 to 125	125
4.35 to 4.86	5	131 to 150	150
4.87 to 5.43	5 6/10	153 to 173	175
5.44 to 6.08	6 1/4	174 to 195	200
6.09 to 6.95	7	196 to 217	225
6.96 to 7.82	8	218 to 250	250
7.83 to 8.69	9	261 to 300	300
8.70 to 10.0	10	305 to 347	350
10.5 to 12.0	12	348 to 391	400
13.1 to 15.0	15	392 to 434	450
15.3 to 17.3	17 1/2	435 to 480	500
17.4 to 20.0	20		

* Includes time-delay referred to in *Rule 14–200*.

Note: Disconnect switches must comply with *Rule 28–602*.

size, conduit size, and voltage drop (Cooper Bussmann, www.bussmann.com).

Time-delay fuses are designed to withstand motor-starting inrush currents and will not open needlessly. However, they will open if a sustained overload occurs. As illustrated in Unit 19, these fuses should be sized from 115% to 125% of the full-load ampere rating of the motor. For the exhaust fan motor that has a 2.9-ampere full-load current draw, select a time-delay fuse in the range of:

$$2.9 \times 1.15 = 3.34 \text{ amperes}$$

$$2.9 \times 1.25 = 3.63 \text{ amperes}$$

Referring to Table 12-3, find that a 3.5-ampere size is available. If the motor has built-in or other overload protection, the 3.5-ampere fuse will provide backup overload protection for the motor.

Motor Branch-Circuit Overcurrent Protection

It is beyond the scope of this text to cover every aspect of motor circuit design, since an entire text could be devoted to this subject, however, the basics of the typical motor installation will be presented.

- Motor branch-circuit conductors, controllers, and the motor must be provided with overcurrent protection, *Rules 28–200* and *14–012*.

- The overcurrent protection must have sufficient time-delay to permit the motor to be started, *Rule 28–200(e)*.

- *Rule 28–200(e)* and *Table 29* provide the maximum percentages permitted for fuses and circuit breakers, for different types of motors.

- When applying the percentages listed in *Table 29*, and the resulting ampere rating or setting does not correspond to standard equipment ratings, we must round down to the next lower standard rating.

- When applying the percentages listed in *Table 29*, and the resulting ampere rating or setting does not allow the motor to start, we are permitted to select a higher ampere rating. See Table 12-4 for maximum rating of overcurrent devices.

- Always check the overload relay table on a motor controller to see if the manufacturer has indicated a maximum size or type of overcurrent device, such as "Maximum Size Fuse 25 amperes." Do not exceed this ampere rating, even if the percentage values listed above result in a higher ampere rating.

The following example illustrates a typical squirrel cage motor, Code J. Refer to *Table 29* for other types of motors.

EXAMPLE

A motor, connected to a high inertia load, has an FLA rating of 27 amperes. First, determine the maximum ampere rating for time-delay fuses. Refer to *Table 29*.

27 × 1.75 = 47.25 amperes

The next lower standard rating fuse is 45 amperes, which is the maximum. If the 45-ampere fuse does not allow the motor to start, the fuse ampere rating may be increased, but cannot exceed 225%, *Rule 28–200(d)(ii)*.

27 × 2.25 = 60.75 amperes

The next lower standard ampere rating fuse is 60 amperes, which would be the absolute maximum size permitted to be installed.

TABLE 12-4

Maximum Rating of Overcurrent Devices.

RATINGS AS PERCENTAGE OF FLA	NORMAL MAXIMUM	ABSOLUTE MAXIMUM
Nontime-delay fuses not over 600 amperes	300%	400%
Time-delay fuses	175%	225%
Inverse time circuit breakers 100 amperes or less	250%	400%
Inverse time circuit breakers over 100 amperes	250%	300%
Fuses over 600 amperes	300%	300%
Instant trip circuit breakers	1300%	1300%

SEVERAL MOTORS ON A SINGLE BRANCH CIRCUIT

Two or more motors may be grouped on a single branch circuit provided that certain conditions, as laid out in *Rule 28–206*, are met. Generally, such circuits are limited to 15-A overcurrent protection or are for a single coordinated drive system or a single machine that has been designed for the application.

CONDUCTORS SUPPLYING SEVERAL MOTORS

Rule 28–108(1)(a) requires that the conductors supplying several motors on one circuit shall have an ampacity of not less than 125% of the largest motor plus the full-load current ratings of all other motors on that circuit.

For example, consider the bakery's multi-mixer and dough divider, which are supplied by a single branch circuit and have full-load ratings of 2.2 and 3.96 amperes:

1.25 × 3.96 amperes =	4.95 amperes
plus	2.20 amperes
Total	7.15 amperes

The branch-circuit conductors must have a minimum ampacity of 7.15 amperes. Specifications for this commercial building call for No. 12 minimum. Checking *Table 2*, T90 Nylon copper conductor has an ampacity of 30 amperes, more than adequate to serve the two appliances. When derating ampacities for more than three conductors in one raceway or for high temperatures, we would begin by using the 90°C column ampacity to determine the T90 Nylon conductor's ampacity before derating.

SEVERAL MOTORS ON ONE FEEDER

Rules 28–204(1) and *28–108* set forth the requirements for this situation. Each motor branch circuit has short-circuit and ground-fault overcurrent protection sized according to *Rule 28–200*. The individual motors have overload protection. The feeder fuses or circuit breakers will be sized by:

1. Determining the fuse or circuit breaker rating or setting per *Rule 28–200(g)* for the largest motor in the group,

2. Adding the value determined in Step 1 to the full-load current ratings for all other motors in the group.

EXAMPLE

To determine the maximum overcurrent protection using dual-element time-delay fuses for a branch circuit serving three motors, refer to *Table 12-4*. The FLA ratings of the motors are 27, 14, and 11 amperes.

First, determine the maximum time-delay fuse for the largest motor:

27 amperes × 1.75 = 47.25 amperes

Next, add the remaining ampere ratings to this value:

(47.25 + 14 + 11) amperes = 72.25 amperes

For these three motors the maximum ampere rating of dual-element time-delay fuses is 72.25 amperes. The next lower standard fuse of 70 amperes would be used.

Sometimes there is confusion when one or more of the motors are protected with instant trip breakers, sized at 1300% of the motor's full-load ampere rating. This would result in a very large (and possibly unsafe) feeder overcurrent device. *Rule 28–204(2)* limits this possibility by establishing 300% of feeder conductor ampacity for fuse sizing.

GROUNDING

Rule 10–408(1) states that exposed noncurrent-carrying metal parts of portable equipment must be bonded to ground.

- **Why?** *Rule 10–002* explains why equipment must be grounded and bonded.

- **Where?** *Rule 10–400* lists locations where equipment fastened in place or connected by permanent wiring methods (fixed) must be bonded to ground.

- **What?** *Rule 10–402* explains what equipment that is fastened in place or connected by permanent wiring methods (fixed) must be bonded to ground. *Rule 10–408* explains how portable equipment that is cord-and-plug connected must be bonded to ground.

- **How?** *Rule 10–618* explains how to bond equipment that is fastened in place and is connected by permanent wiring methods. *Rule 10–620(c)* explains how to bond to ground equipment that is cord-and-plug connected.

Rule 10–408(3), Appendix B, exempts tools and appliances that are double insulated and marked with the symbol ▣ from the bonding to ground requirement.

OVERCURRENT PROTECTION

Overcurrent protection for appliances is covered in *Sections 14* and *16* of the CEC. If the appliance is motor driven, then *Section 28* applies. In most cases, the overload protection is built into the appliance. The appliance must comply with Canadian Standards Association (CSA) standards or another accepted approval agency. If the appliance is double insulated, it does not need to be bonded to ground.

FIGURE 12-5   Cake mixer.

FIGURE 12-6   Rack conveyor.

TYPICAL BAKERY EQUIPMENT

A bakery can have many types and sizes of food preparation equipment, such as blenders, choppers, cutters, disposals, mixers, dough dividers, grinders, moulding and patting machines, peelers, slicers, wrappers, dishwashers, rack conveyors, water heaters, blower dryers, refrigerators, freezers, and hot-food cabinets; see Figures 12-5 and 12-6. All of this equipment is designed and manufactured by companies that specialize in food preparation equipment. These manufacturers furnish specifications that clearly state any technical data that is required in order to connect the appliance in a safe manner, such as voltage, current, wattage, phase (single- or three-phase), minimum required branch-circuit rating, minimum supply circuit conductor ampacity, and maximum overcurrent protective device rating. If the appliance is furnished with a cord-and-plug set, the CSA size and type will be specified.

The data for the mixers and dough divider are shown in Table 12-5.

A CSA 15–20R receptacle outlet, Figure 12-7, is provided at each appliance location. These are part of the electrical contract. Each of these appliances is purchased as a complete unit. After the electrician provides the proper receptacle outlets, the appliances are ready for use as soon as they are moved into place and plugged in.

TABLE 12-5

Ratings of Appliances.

TYPE OF APPLIANCE	VOLTAGE	AMPERES	PHASE	HP	HOW CONNECTED
Multi-mixer	208	3.96	3	$\frac{3}{4}$	CSA 15–20P 4-wire cord and plug on appliance
Multi-mixer	208	7.48	3	$1\frac{1}{2}$	Same as above
Dough divider	208	2.2	3	$\frac{1}{2}$	Same as above

FIGURE 12-7 CSA 15–20R receptacle and plate.

Courtesy of Hubbell Canada LP

THE DOUGHNUT MACHINE

An individual branch circuit provides power to the doughnut machine. This machine consists of

1. A 2000-watt heating element that heats the liquid used in frying, and

2. A driving motor that has a full-load rating of 2.2 amperes.

As with most food preparation equipment, the appliance is purchased as a complete, prewired unit. This particular appliance is equipped with a four-wire cord to be plugged into a receptacle outlet of the proper configuration.

Figure 12-8 is the control circuit diagram for the doughnut machine. The following components are indicated on the control circuit diagram:

FIGURE 12-8 Control wiring diagram for the doughnut machine.

© Cengage Learning®

S A manual switch used to start and stop the machine

T1 A thermostat with its sensing element in the frying tank (this thermostat keeps the oil at the correct temperature)

T2 Another thermostat with its sensing element in the drying tank (this thermostat controls the drive motor)

A A three-pole contactor controlling the heating element

B A three-pole motor controller operating the drive motor

M A three-phase motor

OL Overload units that provide overload protection for the motor (note that there is one thermal overload unit in each phase)

P Pilot light to indicate when power is "on"

Since this appliance is supplied by an individual circuit, its current draw is limited to 80% of the branch-circuit rating. The branch circuit supplying the doughnut machine must have sufficient ampacity to meet the minimum load requirements as indicated on the appliance nameplate. For example,

Heater load	=	2000	watts (VA)
Motor load			
2.2 amperes × 208 volts × 1.73	=	792	VA
Plus 25% of 792 VA	=	198	VA
Total	=	2990	VA

The maximum continuous load permitted on a 20-ampere, three-phase branch circuit is:

16 amperes × 208 volts × 1.73 = 5757.4 VA

The load of 2990 volt-amperes is well within the 5760 volt-amperes permitted loading of a 20-ampere branch circuit.

▶ THE BAKE OVEN

The bake oven installed in the bakery is an electrically heated commercial-type bake oven, Figure 12-9. The oven has a marked nameplate rating of 16 000 watts, 208 volts, three phase. This rated load includes all electrical heating elements, drive motors, timers, transformers, controls, operating coils, buzzers, lights, alarms, etc.

FIGURE 12-9 Bake oven.

Vereshchagin Dmitry/Shutterstock.com

The instructions furnished with the bake oven, as well as the nameplate on the oven, specify that the supply conductors shall have a minimum ampacity of 60 amperes. The line-connecting terminals in the control panel and in the branch-circuit panelboard are suitable for 75°C. This would match the temperature rating of Type TW75 wire. Type T90 Nylon wire has a temperature rating of 90°C, but for this application its ampacity is determined by referring to the 75°C column of *Table 2*.

According to the panel schedule, the bake oven is fed by a 60-ampere, three-phase, three-wire branch circuit using No. 6 T90 Nylon copper conductors. The metal raceway is considered acceptable for the equipment bond.

A three-pole, 60-ampere, 250-volt disconnect switch is mounted on the wall near the oven. From this disconnect switch, a conduit (flexible, rigid, or EMT as recommended by the oven manufacturer) is run to the control panel on the oven, Figure 12-10.

Conductor ampacity: See below

Disconnecting means

Overcurrent protection

Conductor ampacity:
from nameplate

Conduit size
Table 6

Appliance nameplate marking:
Rule 2–100
Generally will show:
• volts and amperes
         or
• volts and watts
• minimum supply circuit conductor
  ampacity
• maximum rating of the branch-circuit
  overcurrent protective device

© Cengage Learning®

FIGURE 12-10   Bake oven circuit.

The bake oven is furnished as a complete unit. Overcurrent protection is an integral part of the circuitry of the oven. The internal control circuit of the oven is 120 volts, which is supplied by an integral control transformer.

All of the previous discussion in this unit relating to circuit ampacity, conductor sizing, grounding, etc., also applies to the bake oven, dishwasher, and food-waste disposal. To avoid repetition, these *CEC* requirements will not be repeated. Additional information relating to appliances is found in Units 4 and 10.

For the bake oven:

1. Ampere rating = 44.5 amperes

2. Minimum conductor ampacity = $44.5 \times 1.25 = 55.6$ amperes

3. Conductor size No. 6 AWG T90 Nylon Copper

4. Branch-circuit overcurrent protection = 44.5 amperes $\times$ 1.25 = 55.6 amperes

It is permitted to install 60-ampere fuses in the 60-ampere disconnect switch.

REVIEW

Note: Refer to the *CEC* or the blueprint package as necessary.

1. What section of the *CEC* refers specifically to air-conditioning equipment?

   _____

2. What section of the *CEC* refers specifically to appliances?

   _____

3. A cord-and-plug connected appliance shall not exceed _____% of a branch-circuit ampere rating.

4. How many poles does a 15–30R receptacle have? _____

5. A disconnect switch for a motor must be _____ of the motor controller.

6. Where a branch-circuit disconnect is used as the motor disconnect, it must be located

   _____

7. What section of the *CEC* sets forth the requirements for the disconnecting means to be in sight of a motor and/or the controller for the motor?

   _____

8. Circle the correct answer: The definition of "in sight" is that the disconnecting means must be visible from the motor location (and) (or) be not more than 9 metres from the motor location.

9. Conductors that supply a motor shall be sized at not less than ____% of the motor full-load current rating.

10. Motor overload protection for a motor that is marked "S.F. 1.15" shall not exceed ____% of the motor full-load ampere rating.

11. What is the length of time that a time-delay fuse will carry five times its current rating?

_____

12. When motors run too hot under overloaded conditions, the life expectancy of the motor is cut in half for every ____°C that the motor is operating above the intended or marked temperature rating of the motor.

13. A 4-ampere, dual-element time-delay fuse protects a motor that momentarily draws 15 amperes when started. If the motor should fail to start, the fuse would protect the motor from burnout by opening the circuit in approximately ____ seconds. Refer to Table 12-2.

14. Motor #1 is marked Code B. Motor #2 is marked Code M. Both motors draw the same full-load current. Which motor has the larger momentary inrush current when started?

_____

15. The recommended ampere rating for a dual-element time-delay fuse for an 8-ampere motor is ____ amperes.

16. The conductors that supply a group of motors must be sized at a minimum of ____% of the largest motor, plus the ____ of the full-load ampere ratings of the other motors.

17. A motor has a full-load ampere rating of 16 amperes. The preferred sizing for dual-element time-delay fuses is (a) ____% of the motor full-load ampere rating. If, for some reason, the characteristic of the installation is that the motor will be started and stopped repeatedly, *Table D16* of the *CEC* indicates that these dual-element time-delay fuses could be sized at (b) ____% of the motor full-load ampere rating. In fact, if the repeated starting and stopping nuisance blows the fuse size indicated in *Table 29, Rule 28–200* allows the electrician to size the fuses at a maximum of (c) ____%. Fill in the blanks for (a), (b), and (c), and select the fuse and disconnect switch sizes for the above alternatives.

    a. _____

    b. _____

    c. _____

18. Three motors are connected to one feeder. Motor #1 draws 16 amperes. Motor #2 draws 11 amperes. Motor #3 draws 8 amperes. Calculate the minimum ampacity for the feeder conductors that supply these motors. The terminals in the switches and in the motor controllers are suitable for 75°C conductors. Type T90 Nylon conductors will be used.

_____

_____

_____

_____

19. What section of the *CEC* covers the subject of grounding?

_____

20. The doughnut machine has a _____-watt heating element plus a _____-ampere motor.

21. Calculate the current draw of an electric bake oven rated at 35 kilowatts, 208 volts, three-phase. _____

_____

22. What sizes and types of conductors supply the electric bake oven in the bakery?

_____

23. Circle the correct answer: When selecting overcurrent protection for a motor branch circuit, if no standard ampere rating given in *Table 13* equals the value required for the circuit, the next (higher) (lower) standard rating should be chosen.

24. When a motor controller's nameplate indicates that a 15-ampere maximum time-delay fuse is permitted for a given size overload relay, may a 20-ampere time-delay fuse be used? _____ May a 15-ampere circuit breaker be used? _____

Luminaires and Lamps

Part I: Luminaires

OBJECTIVES

After completing the study of Part I of this unit, the student will be able to

- install a lighting outlet
- install a luminaire
- identify different types of luminaires and state their application

After completing the study of Part II of this unit, the student will be able to

- list the types of lamps used in the commercial building
- define the technical terms relating to lamp selection and installation
- list the parts of three types of lamps
- list the operating characteristics of lamp types
- recognize the significance of lamp designations

DEFINITIONS

The terms "luminaire" and "lighting fixture" are used interchangeably. The Illuminating Engineering Society recommends the use of "luminaire" but the terms "lighting fixture" and "fixture" are commonly used. There is a danger in using the term "fixture" alone, for in different contexts it may have different meanings to the reader. In a heating application, for example, it is common to use the term "fixture" to refer to a baseboard heater. For the sake of clarity, and to follow good practice, in this text the word "luminaire" will be used except when addressing *Canadian Electrical Code (CEC)* requirements, in which case "lighting fixture" will be used.

The following definition from the *IES Handbook* applies to either of the terms. A luminaire (lighting fixture) is a complete lighting unit consisting of a lamp or lamps together with the parts designed to distribute the light, to position and protect the lamps, and to connect the lamps to the power supply.

A lighting outlet can be defined as an outlet intended for the direct connection of a lampholder, a lighting fixture, or a pendant cord terminating in a lampholder.

INSTALLATION

In this unit, we discuss the *CEC* requirements for installing luminaires in a commercial building. Surveys have proven that lighting is one of the largest consumers of energy (kilowatt-hours) in commercial buildings. It's important that you check with your local and provincial building codes to make sure that the types of luminaires and method of installation are in compliance with any and all lighting energy efficiency and conservation laws. These standards might require such things as certified airtight recessed luminaires, switching requirements, occupation sensors that turn off or turn down lighting levels when no motion is sensed and turn on or turn up lighting levels when motion is sensed, daylight controls, time clocks, limiting the use of luminaires that send light up into the atmosphere either by direct light or by reflected light, and similar devices that result in energy conservation of both indoor and outdoor nonessential (especially after-hours) lighting. It will be costly if, after you

do an installation, you find that the installation does not meet *CEC* requirements and you are required to bring it into compliance.

The IESNA is the Illuminating Engineering Society of North America, founded in 1906. Over all these years, the objective of this organization has been to "communicate information about all facets of good lighting practice to its members and to consumers" so that a lighting design meets national, provincial, and local lighting-lumens-per-square-foot lighting and energy efficiency requirements. This is accomplished through using high-efficiency, high power factor electronic ballasts and energy saving lamps, light-emitting diodes (LEDs), and compact fluorescent lamp (CFL) lamps, while at the same time designing a lighting installation that has aesthetics acceptable to the owner and the wishes of the architect.

The lighting layout for the commercial building follows general guidelines. The LED arena is fast moving, with new LED lamps (clusters, modules) and LED luminaires coming into the marketplace every day.

We have all heard the words, "Shut off the lights when you leave the room." Sometimes this works … sometimes not. *Electrical Wiring—Commercial* is used in Canada and in all 50 states in the United States. It is not the intent of this text to go into the detail of specific provincial and/or local energy requirements. Just be aware that requirements are being implemented relative to energy conservation, and you need to be aware of them.

Most fluorescent lamps with energy saving ballasts, CFLs, LEDs, and halogen lamps meet this high-efficacy requirement. Conventional incandescent lamps do not. As you study the lighting layout for the commercial building, you will note that incandescent lamps are not specified.

In a nutshell, today's lighting incorporates a combination of well-designed, high-efficiency (airtight) luminaires, high-efficiency lamps, high-efficiency ballasts, and various levels of control such as occupancy sensors, timers, and dimmers.

Occupancy sensors are available that install in the same way as conventional toggle switches. They provide automatic shut-off control when no motion is detected after a preset time, and as soon as motion is detected, they turn the lamps on. See Figure 13-1. Also available are timers, dimmer switches, and

Courtesy of Legrand/Pass & Seymour

FIGURE 13-1 An occupancy (motion) sensor.

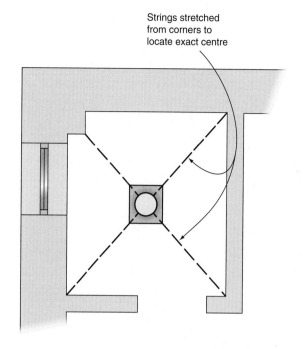

Strings stretched from corners to locate exact centre

© Cengage Learning®

FIGURE 13-2 Luminaire location.

astronomical controls that are calibrated onsite for the longitude and latitude. All of the above have a manual *on–off* override control.

The installation of luminaires is a frequent part of the work required for new building construction and for remodelling projects where customers are upgrading the illumination of their facilities. To execute work of this sort, the electrician must know how to install luminaires and, in some cases, select the luminaires.

The luminaires required for the commercial building are described in the specifications and indicated on the plans. The installation of luminaires, lighting outlets, and supports is covered in *Section 30 of the CEC.*

This section sets forth the basic requirements for the installation of the outlets and supports in what is commonly referred to as the rough-in. The rough-in must be completed before the ceiling material can be installed. The exact location of the luminaires is rarely dimensioned on the electrical plans, and in some remodelling situations there are no plans. In either case, the electrician should be able to rough in the outlet boxes and supports so that the luminaires will be correctly spaced in the area.

If a single luminaire is to be installed in an area, the centre of the area can be found by drawing diagonals between opposite corners, Figure 13-2. When more than one luminaire is required in an area, the procedure shown in Figure 13-3 should be followed.

Uniform light distribution is achieved by spacing the luminaires so that the distances between the luminaires and between the luminaires and the

walls follow these recommended spacing guides. The spacing ratios for specific luminaires are given in the data sheets and published by each manufacturer. This number, usually between 0.5 and 1.5, when multiplied by the mounting height, gives the maximum distance that may separate the luminaires to provide uniform illuminance on the work surface.

Supports

Both the lighting outlet and the luminaire must be supported from a structural member of the building. To provide this support, a large variety of clamps and clips are available, Figures 13-4 and 13-5. The selection of the type of support depends upon the way in which the building is constructed.

Surface-Mounted Luminaires

For surface-mounted and pendant-hung luminaires, the lighting outlets and supports must be roughed in so that the luminaire can be installed after the ceiling is finished. Support rods should be placed so that they extend about 25 mm below the finished ceiling. The support rod may be either a threaded

Luminaire layout for uniform lighting

Note: d should not exceed spacing ratio times mounting height. Mounting height is from luminaire (fixture) to work plane for direct, semidirect, and general diffuse luminaires and from ceiling to work plane for semi-indirect and indirect luminaires.

FIGURE 13-3 Luminaire spacing.

© Cengage Learning®

Installation instructions

FIGURE 13-4 Rod hangers for connection to flange.

© Cengage Learning®

rod or an unthreaded rod, Figure 13-6. If the luminaires are not available when the rough-in is necessary, luminaire construction information should be requested from the manufacturer. The manufacturer can provide drawings that will indicate the exact dimensions of the mounting holes in the back of the luminaire, Figure 13-7.

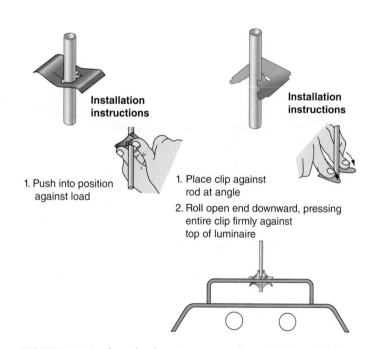

Installation instructions

9.5 mm
(3/8 in.) hole

Wood joist or wall-
concrete, etc.

Wood or concrete

FIGURE 13-5   Rod hanger supports for flat surfaces.

Installation
instructions

Installation
instructions

1. Push into position
against load

1. Place clip against
rod at angle

2. Roll open end downward, pressing
entire clip firmly against
top of luminaire

FIGURE 13-6   Luminaire support using unthreaded rod.

Recessed Luminaires

For recessed luminaires, the lighting outlet box will be located above the ceiling and must be accessible. It is connected to the luminaire by means of a metal raceway that is at least 450 mm long, but not more than 2 m long. Conductors suitable for the temperatures encountered are necessary.

This information is marked on the luminaire. Branch-circuit conductors may be run directly into the junction box on approved prewired luminaires *(Rule 30–910(8))*, Figure 13-8. Recessed luminaires are usually supported by rails installed on two sides of the rough-in opening. The rails can be heavy lather's channel or another substantial type of material.

© Cengage Learning®

FIGURE 13-7 Shop drawing indicating mounting holes for luminaire.

All recessed incandescent luminaires are required to have thermal protection and be so marked. IC when blanketed by thermal insulation, *Rule 30–906*.

At least 450 mm (18 in.) but not more than 2 m (6 ft) of suitable metal raceway with conductors having insulation suitable for temperature encountered. See *Rule 30–910(3)*.

Branch-circuit conductors with insulation suitable for the temperature requirements marked on the "listed" prewired recessed luminaire may be run directly to junction box, per *Rule 30–910(2)*.

Junction box at least 1 ft (300 mm) from luminaire, per *Rule 30–910(4)(b)*.

For 12 mm flex "fixture whip," maximum length is 1.5 m. *Rule 12–1004(a)*.

"Listed" prewired luminaire with junction box.

These boxes must be accessible per *Rule 30–910(4)(c)*.

"Listed" luminaire without junction box.

At least ½ in. (13 mm) clearance from combustible material except at point of support per *Rule 30–902*.

Keep insulation at least 76 mm (3 in.) from luminaire unless it is suitable for direct contact with insulation per *Rule 30–902*.

© Cengage Learning®

FIGURE 13-8 Clearance requirements for installing recessed luminaires.

LABELLING

Always carefully read the label(s) on the luminaires. The labels will list the types of lamps and the maximum wattage for the luminaires; a power factor for determining the volt-amperes from the lamp wattage may be listed as well. The labels may also indicate whether the luminaires are

- For wall mount only,
- For ceiling mount only,
- Required to have access above ceiling,
- Suitable for air handling use,
- For chain or hook suspension only,
- Suitable for operation in ambient temperatures not exceeding ___°C (___°F),

- Suitable for installation in poured concrete,
- For installation in poured concrete only,
- Suitable for use in suspended ceilings,
- Suitable for use in uninsulated ceilings,
- Suitable for use in insulated ceilings,
- Suitable for damp locations (such as bathrooms and under eaves),
- Suitable for wet locations,
- Suitable for use as a raceway.

The instructions on these labels and the bonding requirements of *Section 10* must be followed to ensure a safe installation.

The luminaire manufacturers' catalogues and literature are excellent sources of information on how to install luminaires correctly. The Canadian Standards Association (CSA) lists, tests, and certifies luminaires and publishes relevant information.

Requirements for Installing Recessed Luminaires

The electrician must very carefully follow the requirements given in *Rule 30–900* for the installation and construction of recessed luminaires. Of particular importance are the restrictions on conductor temperature ratings, luminaire clearances from combustible materials, and maximum lamp wattage when luminaires are installed in thermal insulation.

Recessed luminaires generate a considerable amount of heat within the enclosure and are a **definite fire hazard** if not wired and installed properly, Figures 13-8, 13-9, and 13-10. In addition, the excess heat will have an adverse effect on lamp and ballast life and performance.

If other than a Type IC luminaire is being installed, the electrician should work closely with the installer of the insulation to be sure that the clearances required by the *CEC* are followed.

According to *Rule 30–910(3)(4),* a tap conductor must be run from the luminaire terminal connection to an outlet box. For this installation, the following conditions must be met:

- The conductor insulation must be suitable for the temperatures encountered.

- The outlet box must be at least 30 cm from the luminaire.
- The tap conductor must be installed in a suitable metal raceway or cable.
- The raceway shall be at least 450 mm but not more than 2 m long.

The branch-circuit conductors are run to the junction box, where they are connected to conductors from the luminaire. These fixture wires have insulation suitable for the temperature encountered at the lampholder. Locating the junction box at least 30 cm from the luminaire ensures that the heat radiated from the luminaire cannot overheat the wires in the junction box. The conductors must run through at least 450 mm of the metal raceway (but not more than 2 m) between the luminaire and the junction box. Thus, any heat being conducted by the metal raceway will be dissipated considerably before reaching the junction box. Many recessed luminaires are factory equipped with a flexible metal raceway containing high-temperature wires that meet the requirements of *Rule 30–910(1).*

Some recessed luminaires have a box mounted on the side of the luminaire so that the branch-circuit conductors can be run directly into the box and then connected to the conductors entering the luminaire. Additional wiring is unnecessary with these prewired luminaires, Figure 13-11. It is important to note that *Rule 30–910(8)* states that branch-circuit wiring shall not be passed through an outlet box that is an integral part of an incandescent luminaire unless the luminaire is identified for through wiring. If a recessed luminaire is not prewired, the electrician must check the luminaire for a label indicating what conductor insulation temperature rating is required.

Recessed luminaires are inserted into the rough-in opening and fastened in place by various devices. One type of support and fastening method for recessed luminaires is shown in Figure 13-12. The flag hanger remains against the luminaire until the screw is turned. The flag then swings into position and hooks over the support rail as the screw is tightened.

Thermal Protection

Recessed luminaires have an inherent heat problem. Accordingly, they must be suitable for the intended

LUMINAIRE TYPE	COMMENT
Incandescent Recessed Luminaires Marked "Type IC" 	Usually of low wattage for installation in direct contact with insulation, *Rule 30–906*. May be blanketed with insulation. Recommended for homes, these luminaires operate under 90°C when covered with insulation. Separation from insulation is not required. Integral thermal protection is required in Type IC luminaires, which must be marked that thermal protection is provided. Check the manufacturer's markings for minimum clearances, maximum lamp wattage, and temperature rating for supply conductors. These luminaires may also be installed in noninsulated locations.
Incandescent Recessed Luminaires *Not* Marked "Type IC" 	For recessed incandescent luminaires *not* marked "Type IC," or for those luminaires marked for installing directly in poured concrete. Insulation must *not* be placed over the top of the luminaires. Insulation must be kept back at least 76 mm (3 in.) from the sides of the luminaire, *Rule 30–902*. Combustible material must be kept at least 13 mm from the luminaire except at the point of support for the luminaire. Watch for markings on the luminaire for maximum clearances, maximum lamp wattages, and temperature ratings for the supply conductors.
Suspended Ceilings 	Presently, CSA lists luminaires that are marked "Suitable for use in suspended ceilings." Watch for markings on these luminaires showing spacing requirements between luminaires, walls, and ceiling above the luminaire. Also, check the markings for lamp types and maximum wattages as well as temperature ratings required for the supply conductors. Types IC and non-IC luminaires may be installed in suspended ceilings.

(a and b) © Cengage Learning®

FIGURE 13-9 Types of recessed fixtures.

X = a distance of at least 76 mm
Insulation above the luminaire must not trap heat.
Insulation must be installed to permit free air
circulation, unless the luminaire is identified for
installation directly in thermal insulation. *Rule 30–902*

© Cengage Learning®

FIGURE 13-10 Clearances for recessed lighting luminaire installed near thermal insulation, *Rule 30–902.*

application, and must be properly installed. To protect against overheating, the CSA requires that some types of recessed luminaires be equipped with an integral thermal protector, Figure 13-13. These devices will cycle *on–off–on–off* repeatedly until the heat problem is reduced. They are factory-installed by the manufacturer of the luminaire.

Both incandescent and fluorescent recessed luminaires are marked with the temperature ratings required for the supply conductors if over 60°C. In the case of fluorescent luminaires, branch-circuit conductors within 75 mm of a ballast must have a temperature rating of at least 90°C, *Rule 30–308(3).*

All fluorescent ballasts—including replacement ballasts—installed indoors must have integral thermal protection, per *CSA Standard C22.2 No. 9.0.*

© Cengage Learning®

FIGURE 13-11 Installation permissible only with prewired recessed lighting luminaires, *Rule 30–910(7).*

Support rods or wires

Luminaire flag hanger

Ceiling tees

Luminaire support rails

Finished ceiling

© Cengage Learning®

FIGURE 13-12 Recessed luminaire support with flag hanger and support rails, *Rule 30–302(4).*

FIGURE 13-13 A thermal protector.

Opening in back of lighting luminaire lines up with outlet box in ceiling, *Rule 30–308*

FIGURE 13-14 Method of connecting fluorescent luminaire to outlet box, *Rule 30–308(1)*.

These thermally protected ballasts are called "Class P ballasts." This device will provide protection during normal operation, but it should not be expected to provide protection from the excessive heat that will be created by covering the luminaire with insulation. For this reason, fluorescent luminaires, just as incandescent luminaires, must have 13 mm clearance from combustible materials and 76 mm clearance from thermal insulation.

Because of the inherent risk of fires due to the heat problems associated with recessed luminaires, always read the markings on the luminaire and any instructions furnished with it.

Wiring

It was stated previously that it is very important to provide an exact rough-in for surface-mounted luminaires. *Rule 30–308* emphasizes this fact by requiring that the lighting outlet be accessible for inspection without removing the luminaire supports. The installation of the outlet meets the requirements of *Rule 30–308* if the lighting outlet is located so that the large opening in the back of the luminaire can be placed over it, Figure 13-14.

To meet the requirements of *Rule 30–910(2)*, branch-circuit conductors with a rating of 90°C may be used to connect luminaires. However, these conductors must be from the single branch circuit supplying the luminaires. All of the conductors of multi-wire branch circuits can be installed as long as these conductors are the grounded and ungrounded conductors of a single system, *Rule 30–310(1)*. For example, when a building has a three-phase, four-wire supply, the neutral and three hot wires, one on each phase, may be installed in a luminaire that has been approved as a raceway. This type of installation is suited to a long continuous row of luminaires, Figure 13-15.

DISCONNECTING MEANS

Rule 30–308 requires a means to be provided for disconnecting fluorescent luminaires that have double-ended lamps and ballast(s). This requirement is for the safety of those having to work on the luminaire, such as when changing out a ballast or socket. The disconnecting means is permitted to be inside or outside of the luminaire. Figure 13-16 shows a simple disconnecting means that meets the intent of *Rule 30–308*.

LOADING CALCULATIONS

The branch circuits are usually determined when the luminaire layout is completed. For incandescent luminaires, the VA allowance for each luminaire is based on the wattage rating of the luminaire. If an incandescent luminaire is rated at 300 watts, it must be included at 300 watts even though a smaller lamp is to be installed. For fluorescent and high-intensity discharge lamp luminaires, the VA allowance is based on the rating of the ballast. In the past it was a general practice to estimate this value, usually on the high side. With recent advances in ballast manufacture, and increased interest in reducing energy usage, the practice is to select a specific ballast type and base the allowance on the operation of that ballast. The contractor is required to install an "as good as" or "better than" ballast. Several types of ballasts are discussed later in the text. The design volt-amperes and watts for the luminaires selected for

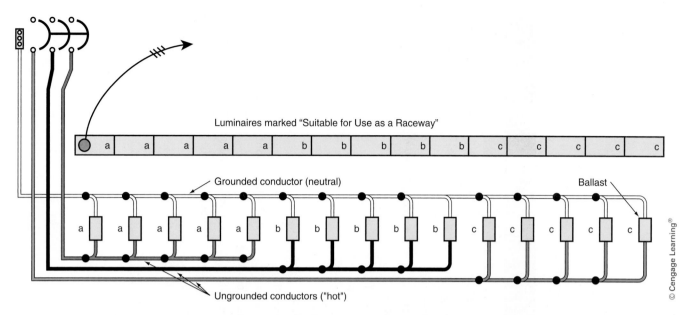

FIGURE 13-15 A single branch, multi-wire circuit consisting of one grounded and three ungrounded conductors of the same system supplying a continuous row of approved luminaires, *Rule 30–310(1)*. Luminaires can also be connected alternately: a-b-c-a-b-c-a-b-c-a-b-c-a-b-c.

FIGURE 13-16 A simple disconnecting means that allows anyone working on a luminaire to snap the device apart, killing power to the luminaire. Note that the disconnect breaks both the ungrounded and grounded conductors.

the commercial building are shown in Table 13-1. The following paragraphs discuss the styles of luminaires listed in this table.

The total wattage rating of all the luminaires on a branch circuit shall not exceed 80% of the branch overcurrent protection, *Rule 30–712(2)*.

Style A

The Style A luminaire is a popular fluorescent type that features a diffuser extending up the sides of the luminaire, Figure 13-17. These are often called wrap-around lenses. This type of luminaire provides good ceiling lighting, which is particularly important for low ceilings. The diffuser is usually available in either an acrylic or polystyrene material. Although a polystyrene diffuser is less expensive, it will yellow as it ages, and quickly becomes unattractive. Diffusers can be specified to be made with an acrylic material that does not yellow with age.

The major disadvantage of the Style A luminaire is the difficulty of locating replacements for yellowed or broken diffusers. The ballast chosen for this luminaire has an A sound rating and is of an energy-efficient type that, when operated with two F40T12/RS lamps, has a line current of 0.725 ampere at 120 volts (87 volt-amperes) and a power rating of 75 watts. This type of luminaire is used in the beauty salon.

Style B

The Style B luminaire, another popular style, has solid metal sides, Figure 13-18. The bright sides of

TABLE 13-1

Luminaire Schedule for the Commercial Building.

STYLE	NOMINAL SIZE	LAMP	BALLAST	LENS-LOUVRE	MOUNTING	DESCRIPTION	VA	WATTS
A	450 mm × 1200 mm	Two F40/30U	Energy Saving	Wraparound	Surface		87	75
B	600 mm × 600 mm	Two F40/30U	Energy Saving	Flat Opal	Surface		87	75
C	225 mm × 1275 mm	Two F40/30U	Energy Saving	Clear Acrylic	Surface	Enclosed, Gasketed	87	75
D	1200 mm strip	One F48CW	Std. Magnetic	Luminous Ceiling	Surface		74	32
E	225 mm dia.	Two 26W Quad T4	Electronic	Gold Alzak Reflector	Recessed		144	74
F	600 mm × 1200 mm	Four FO32/35K	Electronic	24 Cell Lens-Louvre	Recessed		132	106
G	300 mm × 2400 mm	Four FO32/35K	Electronic	24 Cell Lens-Louvre	Recessed		132	106
H	225 mm × 2400 mm	Two F40CWX	Energy Saving	Translucent Acrylic	Surface		87	75
I	300 mm × 1200 mm	Two F40CWX	Energy Saving	Prismatic Lens	Surface		87	75
J	350 mm dia.	One 150W, A221	N/A	Fresnel Lens	Recessed	IC Rated	150	150
K	400 mm square	One 70W HPS	Std. Magnetic	Vandal Resistant	Surface	With Photo Control	192	82
L	325 mm × 1200 mm	Two F40CWX	Energy Saving	None	Surface or Hung		87	75
M	175 mm dia.	12V 50W NFL	Transformer	Coilex Baffle	Recessed	Adjustable Spot	50	50
N	500 mm long	One 60W, A19	N/A	None	Surface	Exposed Lamp	60	60
O	1200 mm × 500 mm	Three F40CWX	Two Ballasts	Low-Brightness Lens	Recessed		143	129
P	600 mm × 500 mm	Two FB40/RS	One Ballast	Low-Brightness Lens	Recessed		87	75
Q	1200 mm × 500 mm	Three F40CWX	Two Ballasts	Small-Cell Parabolic	Recessed		143	129

Courtesy of Eaton's Cooper Lighting Business

FIGURE 13-17   Style A. A shallow, surface-mounted fluorescent luminaire, 1.2 m long and 450 mm wide with a wrap-around acrylic diffuser and equipped for two F40T12/RS rapid-start lamps.

Courtesy of Eaton's Cooper Lighting Business

FIGURE 13-19   Style C. A surface-mounted fluorescent luminaire 1.28 m long and 225 mm wide. Designed to prevent contamination of the area by having a totally enclosed and gasketed clear acrylic diffuser; equipped for two F40 rapid-start lamps.

Courtesy of Shining Electronic Co. Ltd.

FIGURE 13-18   Style B. A surface-mounted fluorescent luminaire, 600 mm square with solid sides, an opal glass diffuser, and equipped for two FB40 rapid-start lamps.

Style D

Luminous ceiling systems are used where a high level of diffuse light is required. The Style D system consists of fluorescent light strips (which may be ballasted for rapid-start, high-output, or very-high-output lamps) and a ceiling suspended 457 mm or more below the lamps, Figure 13-20. The diffuser may be of any translucent material, but usually consists of 0.6 m × 0.6 m or 0.6 m × 1.22 m panels that are easily removed for cleaning and lamp replacement. This lighting system is used in the drugstore.

Style E

The Style E luminaire is one of a type commonly referred to as a *downlight* or *recessed can*, Figure 13-21. It requires a ceiling opening of under 230 mm in diameter and has a height of 200 mm. This fluorescent version uses two 26-watt quad tube T4 lamps. A single unit will produce about 20 footcandles on a surface at 2.74 m distance. Each lamp has a current rating of 0.6 ampere at 120 volts (72 volt-amperes) and uses 37 watts. These are used in several locations in the interior of the commercial building. This lamp is not recommended for exterior applications because of its poor starting characteristics during cold weather.

the Style A luminaire, when used on a low ceiling, may be objectionable to people who must look at them for long periods of time. The Style B reduces this problem. The opal glass of the Style B luminaire provides a soft diffusion of the light, but any flat diffuser may be used. Glass is easily cleaned and does not exhibit the same aging effects shown by the plastic materials. This style is used in the sales area of the bakery.

Style C

The Style C luminaire is used where the possible contamination of the area is an important consideration, such as in the bakery, where food is prepared. Style C luminaires are suitable for use in bakeries, kitchens, slaughterhouses, meat markets, and food-packaging plants. The clear acrylic diffuser of this luminaire protects the area in the event of a broken lamp. At the same time, the interior of the luminaire is kept dry and free from dirt or dust, Figure 13-19. Raceways serving this luminaire may enter from the top or from either end.

Style F

The Style F is a 0.6 m × 1.2 m recessed fluorescent luminaire, Figure 13-22 (the luminaire is shown

FIGURE 13-20 Style D. A fluorescent luminaire strip 1.2 m long, equipped with an F48 high-output lamp. Installed as part of a luminous ceiling system with 610 mm × 610 mm panels with 12.7 mm × 12.7 mm cell size.

FIGURE 13-21 Style E luminaire.

FIGURE 13-22 Style F luminaire.

Courtesy of Eaton's Cooper Lighting Business

in this partial view with its distinctive end profile visible). This luminaire is equipped with four F032T8 lamps and an electronic ballast. The lamps and the ballast are a matched set, a system developed to maximize the ratio of light output to watts input. Compared to standard F40T12/RS lamps and a magnetic ballast, this combination provides about 160% as much light per watt. According to the manufacturer's data, each lamp in this luminaire will produce 2900 lumens initially, achieving a lamp–ballast efficacy of 110 lumens per watt. This type of luminaire is available with many styles of lenses and louvres.

With T-bar ceiling luminaires, chain supports attached to the construction above the T-bar ceiling should be provided. This is usually achieved by the use of S-hooks and jack chain.

Style G

The Style G luminaire is identical to the Style F luminaire except that it is 0.3 m × 2.4 m. A single ballast is used with four F032T8 lamps. This is an identical luminaire to the Style F except for the dimensions.

Style H

The Style H luminaire shown in Figure 13-23 is designed to light corridors, the narrow areas between storage shelves, and other long, narrow spaces. Style H luminaires are available in slightly different forms from various manufacturers and are relatively inexpensive. This luminaire has been chosen for use in the second-floor corridor. It is 225 mm wide and 2400 mm long, and has two F40 lamps placed in tandem (end to end). A single two-lamp ballast serves 2.4 m of luminaire.

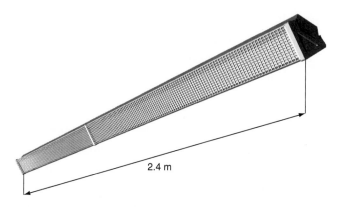

FIGURE 13-25 Style J luminaire.

FIGURE 13-23 Style H. A surface-mounted luminaire designed for corridor lighting with 9.5 mm cube, V-shaped diffuser 2.4 m long; equipped for two F40CW rapid-start lamps in tandem.

Style J

The Style J luminaire, Figure 13-25, is a recessed incandescent downlight luminaire that can be installed in the opening left by removing or omitting a single 0.3-m square ceiling tile. This luminaire is a Type IC, which has been approved to be covered with insulation. It has been specified to be equipped with a fresnel lens for wide distribution of the light and a 150-watt A221 lamp. This is used in protected exterior locations and in the doctor's office.

Style I

The Style I luminaire is similar to Style H but is 1.2 m long with the two lamps side by side. Style I luminaires will be used in the merchandising areas of the drugstore, where higher levels of illumination are desirable, Figure 13-24. They are 300 mm × 1200 mm fluorescent troffers c/w prismatic lens. Style I luminaires will be supplied with deluxe cool white lamps to improve the colour rendering in the display areas of the store.

Style K

The Style K luminaire is used on vertical exterior walls, where it provides a light pattern that covers a large area. The Style K luminaire uses a high-pressure sodium, high-intensity discharge (HID) lamp, Figure 13-26. This luminaire provides reliable security lighting around a building. The Style K luminaire is completely weatherproof and

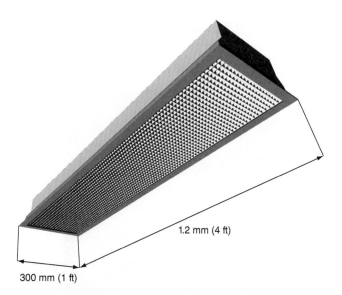

FIGURE 13-24 Style I. This luminaire is 300 mm wide and 1.2 m long and equipped for two F40CW rapid-start lamps.

FIGURE 13-26 Style K luminaire.

FIGURE 13-27 Style L. An industrial-type fluorescent luminaire, 1.2 m long with a steel white enameled reflector; equipped for two F40 lamps.

is equipped with a photoelectric cell to turn the light off during the day.

Style L

The Style L luminaire is of open construction, as shown in Figure 13-27. This type is generally used in storage areas and other locations where it is not necessary to shield the lamps. It is often called an industrial fixture. The Style L luminaires are to be suspended from the ceiling on chains. In this type of installation, armoured cable or flexible metal conduit will run from the luminaire to an outlet box on the ceiling. This type of luminaire is used in the basement storage areas of both the bakery and the drugstore.

Style M

The Style M luminaire is used to focus light on a specific object, Figure 13-28. This luminaire has a diameter of 175 mm. The lamp can be swivelled

FIGURE 13-28 Style M luminaire.

through 358° laterally and 40° from the zenith. This luminaire utilizes an MR16 lamp in sizes up to 50 watts and in a variety of light distribution patterns. A low-voltage transformer is provided with the luminaire. At a distance of 1.22 m, the 50-watt narrow flood lamp produces 26 footcandles. A dimmer switch used to operate these luminaires (such as that specified in the beauty salon) must be approved for use with low-voltage illumination systems. The dimmer is connected in the supply to the transformers. An ordinary dimmer will, at best, provide sporadic operation and is not approved for this application.

Style N

The Style N luminaire (Figure 13-29) is especially designed for installations where the lamp will be exposed to viewing. This luminaire can be installed on both sides of a mirror to provide excellent illumination of the face but avoid the glare that can be a problem when conventional incandescent lamps are installed in a similar fashion. The lamp is a 0.51-m linear incandescent with a 60-watt rating.

Style O

The Style O luminaire is 1.2 m long and 0.5 m wide. It is equipped with three F40CWX lamps and two ballasts. One of the ballasts supplies two lamps, and

FIGURE 13-29 Style N luminaire.

the other, one lamp. This arrangement provides for three levels of illumination in the room by using the single lamp in each luminaire, by using a pair of lamps in each luminaire, or by using all three lamps. In each of these arrangements, the illumination is uniformly distributed throughout the room. The luminaire is fitted with a white louvre and a lens, which provides a high-quality light in addition to having a very pleasing effect on the appearance of the room. This luminaire is used in the staff and reception areas of the insurance office.

Style P

The Style P luminaire is similar to the Style O luminaire but is only 600 mm long and has a single ballast supplying two FB40/RS lamps. The FB40 is a lamp that was 1.2 m long but has been bent into a U-shape for use in luminaires that are only 600 mm long. This arrangement allows the use of the standard F40 two-lamp ballast. Two of these luminaires are used in the staff area of the insurance office to fill in areas of the room where the Style O was too large.

Style Q

The Style Q luminaire is similar to the Style O except for the louvres. The lens specified in this room is a low-brightness lens especially designed for use in computer rooms. A high percentage of the light is directed downward to the horizontal surfaces. A minimal amount of light is produced on the computer monitor's vertical surface.

◗ LOCATION OF LUMINAIRES IN CLOTHES CLOSETS

Clothing, boxes, and other material normally stored in clothes closets are a potential fire hazard. These items may ignite on contact with the hot surface of an exposed light bulb. The bulb, in turn, may shatter and spray hot sparks and hot glass onto other combustible materials.

Rule 30–204 covers the special requirements for installing luminaires in clothes closets. It is significant to note that these rules cover *all* clothes closets—residential, commercial, and industrial.

Part II: Lamps for Lighting

For most construction projects, the electrical contractor is required to purchase and install lamps in the luminaires (lighting fixtures). Thomas Edison provided the talent and perseverance that led to the development of the incandescent lamp in 1879 and the fluorescent lamp in 1896. Peter Cooper Hewitt produced the mercury lamp in 1901. All three of these lamp types have been refined and greatly improved since they were first developed.

Section 30 contains the provisions for the wiring and installation of lighting fixtures, lampholders, lamps, receptacles, and rosettes.

Several types of lamps are used in the commercial building. In the *CEC,* these lamps are referred to as either incandescent or electric discharge lamps, *Rule 30–100.* In the industry, the incandescent lamp is also referred to as a filament lamp because the light is

produced by a heated wire filament. The electric discharge lamps include a variety of types but all require a ballast. The most common types of electric discharge lamps are fluorescent, mercury, metal halide, high-pressure sodium, and low-pressure sodium. Mercury, metal halide, and high-pressure sodium lamps are also classed as HID lamps.

◗ LIGHTING TERMINOLOGY

Electricians should be familiar with the following lighting terminology.

- *Candela* (cd): The luminous intensity of a source, when expressed in candelas, is the candlepower (cp) rating of the source.

- *Lumen* (lm): The amount of light received in a unit of time on a unit area at a unit distance from a point source of one candela, Figure 13-30. The surface area of a sphere is 12.57 times the square of its radius; therefore, a one-candela source produces 12.57 lumens. When the measurement is in imperial units, the unit area is 1 quare foot and the unit distance is 1 foot. If the units are in SI, then the unit area is 1 square metre and the unit distance is 1 metre.

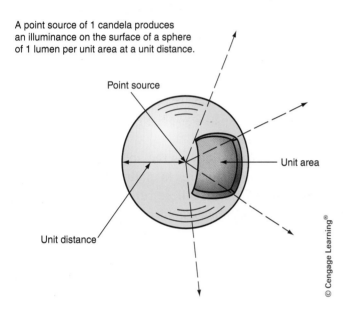

A point source of 1 candela produces an illuminance on the surface of a sphere of 1 lumen per unit area at a unit distance.

Point source

Unit area

Unit distance

© Cengage Learning®

FIGURE 13-30 Diagram showing definition of candela.

- *Illuminance*: The measure of illuminance on a surface is the lumen-per-unit area expressed in footcandles (fc) or lux (lx). The recommended illuminance levels vary greatly, depending on the task to be performed and the ambient lighting conditions. For example, while 54 lux (5 footcandles) is often accepted as adequate illumination for a dance hall, 2152 lx (200 fc) may be necessary on a drafting table for detailed work.

- *Lumen per watt* (lm/W): This is defined as the measure of the effectiveness (efficacy) of a light source in producing light from electrical energy. A 100-watt incandescent lamp producing 1670 lumens has an effective value of 16.7 lumens per watt.

- *Kelvin* (K): The kelvin (sometimes incorrectly called *degree Kelvin*) is measured from absolute zero; it is equivalent to a degree value in the Celsius scale plus 273.16. The colour temperature of lamps is given in kelvin. The lower the number, the warmer the light (more red/orange/yellow content); see Figure 13-31. The higher the number, the cooler the light (more blue content). Light of 2000–3000 K is considered to be "warm," and greater than 4000 K is considered to be "cool." Different manufacturers have different names for their lights: for example, white, warm white, warm white deluxe, warm white plus, cool white, cool white plus, cool white deluxe, natural white, and daylight. It all

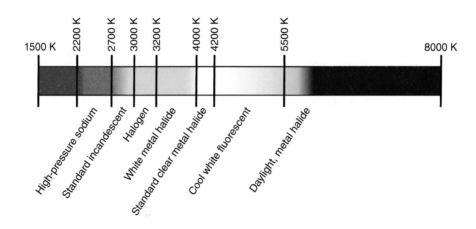

FIGURE 13-31 Kelvin temperature chart.

© Cengage Learning®

TABLE 13-2

Characteristics of Electric Lamps.

	FILAMENT	FLUORESCENT	MERCURY	METAL HALIDE	HPS	LPS
Lumen per watt	6 to 23	25 to 100	30 to 65	65 to 120	75 to 140	130 to 180
Wattage range	40 to 33k	4 to 215	40 to 1000	175 to 1500	35 to 1000	35 to 180
Life (hours)	750 to 8k	9k to 20k	16k to 24k	5k to 15k	20k to 24k	18k
Colour temperature	2400 to 3100	2700 to 7500	3000 to 6000	3000 to 5000	2000	1700
Colour rendering index	90 to 100	50 to 110	25 to 55	60 to 70	20 to 25	0
Potential for good colour rendition	High	Highest	Fair	Good	Colour discrimination	No colour discrimination
Lamp cost	Low	Moderate	Moderate	High	High	Moderate
Operational cost	High	Good	Moderate	Moderate	Low	Low

The values above are generic for general service lamps. A survey of lamp manufacturers' catalogues should be made before specifying or purchasing any lamp.

boils down to the Kelvin rating of the lamp. The fluorescent lamps specified for the commercial building are 3500 K. This is a good all-around choice.

- *Colour rendering index (CRI)*: This value is often given for lamps so that the user can have an idea of the colour rendering probability. The CRI uses filament light as a base for 100 and the warm white fluorescent for 50. The CRI can be used only to compare lamps that have the same colour temperature. The only sure way to determine if a lamp will provide good colour rendition is to see the material in the lamplight.

Refer to Table 13-2 for a comparison of the various types of electric lamps.

▶ LAMP ENERGY EFFICIENCY

In 1992, the federal government put in place the *Energy Efficiency Act* (EEACT) legislation to ensure the use of more energy efficient lamps and thereby reduce Canada's contribution to global greenhouse gas emissions. The current amendment to the energy efficiency regulations establishes energy efficiency standards (lumens per watt) for certain types of lamps imported into Canada or traded

interprovincially. Failing to abide by these regulations could result in very large fines.

Products restricted under this legislation include:

- All R and PAR shaped medium base, 115 to 130 V incandescent reflector lamps of 40 to 205 watts;
- Full wattage 4-foot medium bi-pin (T8, T10, T12) fluorescent lamps;
- Full wattage 2-foot U-shaped (T12, T8) fluorescent lamps;
- Full wattage 8-foot high-output fluorescent lamps;
- Full wattage 8-foot slimline fluorescent lamps.

Certain types of R and PAR incandescent lamps—such as halogen, coloured, and special purpose lamps—are exempt. The ER and BR shaped lamps are also acceptable because they meet the minimum lumens/watt required by the Act. The restricted full wattage fluorescent lamps have been replaced with reduced wattage products of the same physical size that will function with existing ballasts. General service fluorescent lamps that are exempted include lamps with a colour rendering index of 82 or greater, lamps for use in cold temperature applications, grow lamps, and reprographic and UV radiation lamps, as well as coloured, impact resistant, and reflectorized types.

INCANDESCENT LAMPS

The incandescent lamp has the lowest efficacy of the types listed in Table 13-2. However, incandescent lamps are very popular and account for more than 50% of the lamps sold in North America. This popularity is due largely to the low cost of incandescent lamps and luminaires.

FIGURE 13-32 Incandescent lamp.

Construction

The light-producing element in the incandescent lamp, Figure 13-32, is a tungsten wire called the *filament* supported in a glass envelope or bulb. The air is evacuated from the bulb and replaced with an inert gas such as argon. The lead-in wires connect the filament to the base, which supports the lamp and provides the connection means to the power source. The lamp's base may be any one of the styles shown in Figure 13-33.

Characteristics

Incandescent lamps are classified according to voltage rating, wattage, size, and shape.

Voltage Rating. Incandescent lamps are available with many different voltage ratings. When installing lamps, the electrician should be sure that a lamp with the correct rating is selected, since a small difference between the rating and the actual voltage has a great effect on the lamp life and lumen output, Figure 13-34.

Wattage. Lamps are usually selected according to their wattage rating. This rating is an indication of the consumption of electrical energy but is not a true

FIGURE 13-33 Incandescent lamp bases.

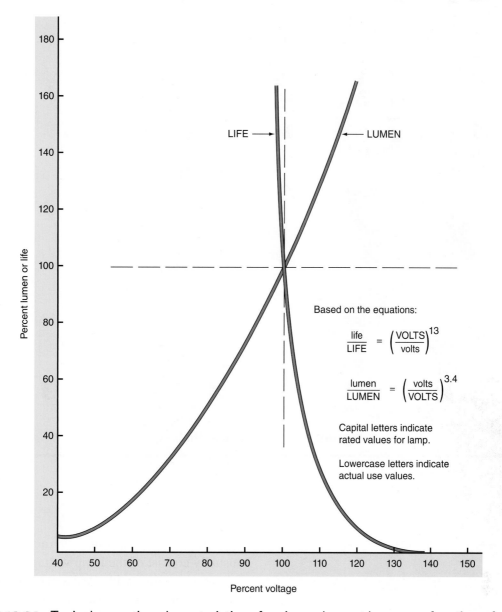

FIGURE 13-34   Typical operating characteristics of an incandescent lamp as a function of voltage.

measure of light output. For example, at the rated voltage, a 60-watt lamp produces 840 lumens, and a 300-watt lamp produces 6000 lumens; therefore, one 300-watt lamp produces more light than seven 60-watt lamps.

Size.   Figure 13-35 illustrates the common lamp configurations and their letter designations.

Shape.   Lamp size is usually indicated in eighths of an inch and is the diameter of the lamp at the widest place. Thus, the lamp designation A19 has an arbitrary shape and is 60.3 mm ($\frac{19}{8}$ or $2\frac{3}{8}$ in.) in diameter, Figure 13-36.

Operation

The light-producing filament in an incandescent lamp is a resistance load that is heated to a high temperature by the electric current going through the lamp. This filament is usually made of tungsten, which has a melting point of 3655 K. At this temperature, the tungsten filament produces light with an efficacy of 53 lumens per watt. However, to increase the life of the lamp, the operating temperature is lowered, which also means a lower efficacy. For example, if a 500-watt lamp filament is heated to a temperature of 3000 K, the resulting efficacy is 21 lumens per watt.

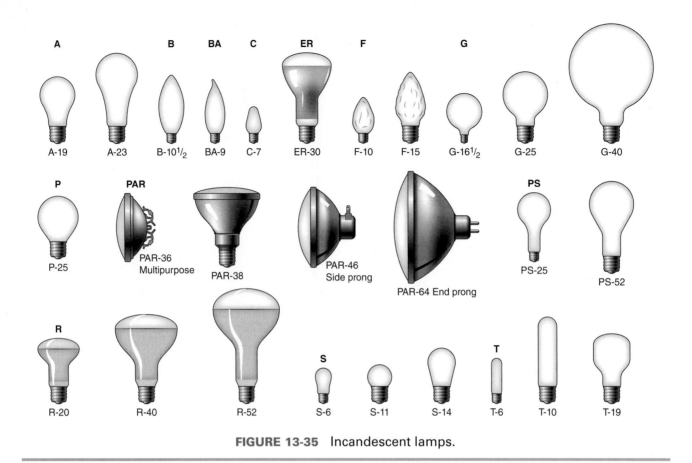

FIGURE 13-35 Incandescent lamps.

Catalogue Designations

Catalogue designations for incandescent lamps usually consist of the lamp wattage followed by the shape and ending with the diameter and other special designations as appropriate. Common examples are:

60A19	60 watt, arbitrary shape, $\frac{19}{8}$ inches diameter
75PAR38	75 watt, parabolic reflector, $\frac{38}{8}$ inches diameter
100R40/FL	100 watt, reflector flood, $\frac{40}{8}$ inches diameter

LOW-VOLTAGE INCANDESCENT LAMPS

In recent years, low-voltage (usually 12-volt) incandescent lamps have become very popular for accent lighting. Many of these lamps are tungsten halogen lamps. They have a very small source size, as shown

60.3 mm (19/8 in)

© Cengage Learning®

FIGURE 13-36 An A19 lamp.

FIGURE 13-37 Assortment of MR (multifaceted reflector) halogen lamps.

in Figure 13-37, which allows very precise control of the light beam. A popular size is the MR16, which has a reflector diameter of just 51 mm (2 in.). These lamps provide a whiter light than do regular incandescent lamps. When dimming these tungsten halogen lamps, a special dimmer is required because of the transformer that is installed to reduce the voltage. The dimmer is installed in the line voltage circuit supplying the transformer. Dimmed lamps will darken if they are not occasionally operated at full voltage.

Catalogue Designations

Except for some special cases, catalogue designations for low-voltage incandescent lamps are similar to other incandescent lamps, for example:

MR16 mirrored reflector, $\frac{16}{8}$ inches diameter.

▶ FLUORESCENT LAMPS

Luminaires using fluorescent lamps are considered to be electric-discharge lighting. Fluorescent lighting has the advantages of a high efficacy and long life.

Construction

A fluorescent lamp consists of a glass bulb with an electrode and a base at each end, Figure 13-38. The inside of the bulb is coated with a phosphor (a fluorescing material), the air is evacuated, and an inert gas plus a small quantity of mercury is released into the bulb. The base styles for fluorescent lamps are shown in Figure 13-39.

Characteristics

Fluorescent lamps are classified according to type, length or wattage, shape, and colour. See Figure 13-40.

Type.   The lamps may be preheat, rapid start, or instant start depending upon the ballast circuit.

Length or Wattage.   Depending on the lamp type, either the length or the wattage is designated. For example, both the F40 preheat and the F48 instant start are 40-watt lamps, 1.2 m (48 in.) long. The bases of these two lamps are different, however.

Shape.   The fluorescent lamp usually has a straight tubular shape. Exceptions are the circline lamp,

© Cengage Learning®

Bulb
Usually a straight glass tube; may be circular or U-shape.

Phosphor
Coating inside bulb transforms ultraviolet radiation into visible light. Colour produced depends on composition of phosphor.

Cathode
Hot cathode at each end of lamp is coated with material that emits electrons; usually made of coiled tungsten wire.

Exhaust tube
Air is exhausted through this tube during manufacture and inert gas introduced into bulb.

Mercury
A minute quantity of liquid mercury is placed in bulb to furnish mercury vapour.

Stem press
The lead-in wires have an airtight seal here and are made of Dumet wire to ensure about the same coefficient of expansion as the glass.

Lead-in wires
These connect to the base pins and carry the current to and from the cathodes and the mercury arc.

Base
Several different types are used to connect the lamp to the electric circuit and to support the lamp in the lampholder.

Gas
Gas is usually argon or a mixture of inert gases at low pressure. Krypton is sometimes used.

© Cengage Learning®

FIGURE 13-38   Basic parts of a typical hot cathode fluorescent lamp.

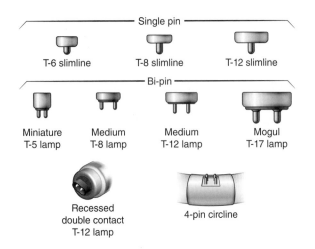

Single pin

T-6 slimline   T-8 slimline   T-12 slimline

Bi-pin

Miniature   Medium   Medium   Mogul
T-5 lamp   T-8 lamp   T-12 lamp   T-17 lamp

Recessed double contact
T-12 lamp

4-pin circline

FIGURE 13-39   Bases for fluorescent lamps.

which forms a complete circle; the U-shaped lamp, which is an F40T12 lamp having a 180° bend in the centre to fit a 600 mm (2 ft) long luminaire; and the PL lamp, which has two parallel tubes with a short connecting bridge at the ends opposite the base.

Colour.   The colour of a fluorescent lamp depends upon the phosphor mixture used to coat the inside of the lamp. The colour is indicated in the lamp designation. For example, an F40T12CW is a cool white

lamp and accents blue colours wherever it is used. Additional colour designations are warm white (WW), daylight (D), white (W), cool white deluxe (CWX), and warm white deluxe (WWX). Decorative colours such as blue (B), pink (PK), and green (G) are also available.

Catalogue Designations

Catalogue designations for fluorescent lamps begin with an "F," except for compact fluorescent and a few other cases. The following numbers indicate the wattage for rapid-start lamps and the length for most of the other types. This is followed by shape, diameter, colour, and special designations.

F40T12/CW	fluorescent, 40 watts, tubular shape, $\frac{12}{8}$ inches diameter, cool white
F48T12/CW/HO	fluorescent, 48 inches in length, tubular shape, $\frac{12}{8}$ inches diameter, cool white, high output
PL18/27	compact fluorescent, 18 watts, 2700 K colour temperature

FIGURE 13-40   Fluorescent lamps.

Operation

If a substance is exposed to ultraviolet light or X-rays and emits visible light as a result, then the substance is said to be fluorescing. The inside of the fluorescent lamp is coated with a phosphor material that serves as the light-emitting substance. When sufficient voltage is applied to the lamp electrodes, electrons are released. Some of these electrons travel between the electrodes to establish an electric discharge or arc through the mercury vapour in the lamp. As the electrons strike the mercury atoms, ultraviolet radiation is emitted by the atoms. This radiation is converted into visible light by the phosphor coating on the tube, Figure 13-41.

As the mercury atoms are ionized, the resistance of the gas is lowered. The resulting increase in current ionizes more atoms. If allowed to continue, this process will cause the lamp to destroy itself. As a result, the arc current must be limited. The standard method of limiting the arc current

FIGURE 13-41   How light is produced in a typical hot cathode fluorescent lamp.

is to connect a reactance (ballast) in series with the lamp.

Ballasts *(Rule 30–700)*

Although inductive, capacitive, or resistive means can be used to ballast fluorescent lamps, the most

Photo courtesy of Philips Lighting Electronics

FIGURE 13-42   Fluorescent ballast.

practical ballast is an assembly of a core and coil, a capacitor, and a thermal protector installed in a metal case, Figure 13-42. Once the assembled parts are placed in the case, it is filled with a potting compound to improve the heat dissipation and reduce ballast noise. Ballasts are available for the three basic operating circuits that are discussed next.

Preheat Circuit. The first fluorescent lamps developed were of the preheat type and required a starter in the circuit. This type of lamp is now obsolete and is seldom found except in smaller sizes that may be used for items such as desk lamps. The starter serves as a switch and closes the circuit until the cathodes are hot enough. The starter then opens and the lamp lights. The cathode temperature is maintained by the heat of the arc after the starter opens. Note in Figure 13-43 that the ballast is in series with the lamp and acts as a choke to limit the current through the lamp.

Rapid-Start Circuit. In the rapid-start circuit, the cathodes are heated continuously by a separate winding in the ballast, Figure 13-44, with the result that almost instantaneous starting is possible. This type of fluorescent lamp requires the installation of a continuous grounded metal strip within 25 mm of the lamp. The metal wiring channel or the reflector of the luminaire can serve as this grounded strip. The standard rapid-start circuit operates with a lamp current of 430 mA. Two variations of the basic circuit are available: the high-output (HO) circuit operates with a lamp current of 800 mA, and the very high-output circuit has 1500 mA of current. Although high-current lamps are not as efficacious as the standard lamp, they do provide a greater concentration of light, thus reducing the required number of luminaires.

Instant-Start Circuit. The lamp cathodes in the instant-start circuit are not preheated. Sufficient voltage is applied across the cathodes to create an instantaneous arc, Figure 13-45. As in the preheat circuit, the cathodes are heated during lamp operation by the arc. The instant-start lamps require

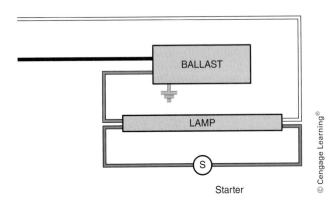

FIGURE 13-43   Basic preheat circuit.

FIGURE 13-44   Basic rapid-start circuit.

FIGURE 13-45 Basic instant-start (cold cathode) circuit.

single-pin bases, Figure 13-46, and are generally called *slimline lamps*. Some bi-pin base instant-start fluorescent lamps are available, such as the 40-watt F40T12/CW/IS lamp. For this style of lamp, the pins are shorted together so that the lamp will not operate if it is mistakenly installed in a rapid-start circuit.

Special Circuits

Most fluorescent lamps are operated by one of the circuits just covered: the preheat, rapid-start, or instant-start circuits. Variations of these circuits, however, are available for special applications.

Dimming Circuit.   The light output of a fluorescent lamp can be adjusted by maintaining a constant voltage on the cathodes and controlling the current passing through the lamp. Devices such as thyratrons, silicon-controlled rectifiers, and autotransformers can provide this type of control. The manufacturer of the ballast should be consulted about the installation instructions for dimming circuits.

Flashing Circuit.   The burning life of a fluorescent lamp is greatly reduced if the lamp is turned on and off frequently. Special ballasts are available that maintain a constant voltage on the cathodes and interrupt the arc current to provide flashing.

High-Frequency Circuit.   Fluorescent lamps operate more efficiently at frequencies above 60 hertz. The gain in efficacy varies according to the lamp size and type. However, the gain in efficacy and the lower ballast cost generally are offset by the initial cost and maintenance of the equipment necessary to generate the higher frequency.

Direct-Current Circuit.   Fluorescent lamps can be operated on a direct-current power system if the proper ballasts are used. A ballast for this type of

A
Single-pin base,
spring loaded end

B
Single-pin base,
stationary end

FIGURE 13-46 Single-pin base for instant-start fluorescent lamp.

Only this fuse is affected.

Branch-circuit fuse is
not affected; circuit
remains energized.

All remaining fixtures
continue to operate normally.

Faulty ballast is isolated
from circuit.

© Cengage Learning®

FIGURE 13-47   Each ballast is individually protected by a fuse. In this system, a faulty ballast is isolated from the circuit.

system contains a current-limited resistor that provides an inductive kick to start the lamp.

Special Ballast Designation

Special ballasts are required for installations in cold areas such as out of doors. Generally, these ballasts are necessary for installations in temperatures lower than 10°C (50°F). These ballasts have a higher open-circuit voltage and are marked with the minimum temperature at which they will operate properly.

Class P.   The National Fire Protection Association reports that the second most frequent cause of electrical fires in the United States is the overheating of fluorescent ballasts. To lessen this hazard, Underwriters Laboratories Inc. has established a standard for a thermally protected ballast that is designated as a Class P ballast. This type of ballast has an internal protective device that is sensitive to the ballast temperature. This device opens the circuit to the ballast if the average ballast case temperature exceeds 90°C when operated in a 25°C ambient temperature. After the ballast cools, the protective device is automatically reset. As a result, a fluorescent lamp with a Class P ballast is subject to intermittent off–on cycling when the ballast overheats.

It is possible for the internal thermal protector of a Class P ballast to fail in the "welded shut" mode or in the "open" mode. Since the welded shut mode

can result in overheating and possible fire, it is recommended that the ballast be protected with in-line fuses as indicated in Figure 13-47. External fuses can be added for each ballast so that a faulty ballast can be isolated to prevent the shutdown of the entire circuit because of a single failure. The ballast manufacturer normally provides information on the fuse type and its ampere rating.

The specifications for the commercial building require that all ballasts shall be individually fused; the fuse size and type are selected according to the ballast manufacturer's recommendations, Figure 13-48.

CSA Standard C22.2 No. 9.0 requires that all fluorescent ballasts installed indoors, both for new installations and replacements, must have built-in thermal protection.

Sound Rating.   All magnetic ballasts emit a hum that is caused by magnetic vibrations in the ballast

© Cengage Learning®

FIGURE 13-48   In-line fuseholder for ballast protection.

core. Ballasts are given a sound rating (from A to F) to indicate the severity of the hum. The quietest ballast has a rating of A. The need for a quiet ballast is determined by the ambient noise level of the location where the ballast is to be installed. For example, the additional cost of the A ballast is justified when it is to be installed in a doctor's waiting room. In the bakery work area, however, a ballast with a C rating is acceptable; in a factory, the noise of an F ballast probably will not be noticed.

Power Factor.   The ballast limits the current through the lamp by providing a coil with a high reactance in series with the lamp. An inductive circuit of this type has an uncorrected power factor of from 40% to 60%; however, the power factor can be corrected to within 5% of unity by the addition of a capacitor. In any installation where there are to be a large number of ballasts, it is advisable to install ballasts with a high power factor.

Compact Fluorescent Lamps

Compact fluorescent lamps usually consist of a twin tube arrangement that has a connecting bridge at the end of the tubes, Figure 13-49. Lamps are available that have two sets of tubes; these are called double twin tube or quad tube lamps, Figure 13-50.

This type of lamp has a rated life ten times that of an incandescent lamp and provides about three times the light per watt of power. A typical socket, along with a low-power factor ballast, is shown in Figure 13-51. Some lamps have a medium screw base, with the ballast in the base. This type of lamp can directly replace an incandescent lamp. The bases for these lamps often have a retractable pin for the base connection, Figure 13-52, which allows the positioning of the lamp.

FIGURE 13-50   Ten-watt, double twin tube or quad tube compact fluorescent lamp.

Courtesy of Natalie Barrington

FIGURE 13-49   Five-watt, twin tube, compact fluorescent lamp.

Courtesy of Natalie Barrington

FIGURE 13-51   Socket and ballast for compact fluorescent lamp.

© Cengage Learning®

Yuriy Boyko/Shutterstock.com

FIGURE 13-52 Medium base screw-in socket.

Class E Ballast

Manufacturers of certain fluorescent ballasts are required to meet a Ballast Efficiency Factor (BEF). Initially, this applied only to F40T12 one- and two-lamp ballasts and to two-lamp ballasts for F96T12 and F96T12HO lamps. The BEF for the two-lamp F40T12, as used in the commercial building, requires that the lamp operate at a minimum 84.8 lumens per watt efficacy.

◤ HIGH-INTENSITY DISCHARGE (HID) LAMPS

Two of the lamps in the high-intensity discharge category—mercury and metal halide—are similar in that they use mercury as an element in the light-producing process. The other HID lamp—high-pressure sodium—uses sodium in the light-producing process. It produces a golden-white light

in the 1900–2700 terahertz range. Low-pressure sodium produces a yellow-orange light around 1700 THz. Mercury vapour produces a blue-green light in the 3000–7000 THz range.

In all three lamps, the light is produced in an arc tube that is enclosed in an outer glass bulb which serves to protect the arc tube from the elements and to block ultraviolet radiation produced by the arc. A high-intensity discharge lamp may continue to give light after the bulb is broken, but it should be promptly removed from service. When the outer bulb is broken, people can be exposed to harmful ultraviolet radiation.

Mercury Lamps

Many people consider mercury lamps to be obsolete. They have the lowest efficacy of the HID family, which ranges from 30 lumens per watt for the smaller-wattage lamps to 65 lumens per watt for the larger-wattage lamps. Some of the positive features of mercury lamps are that they have a long life, with many lamps still functioning at 24 000 hours. With a clear bulb, they give a greenish light that makes them popular for landscape lighting.

Catalogue Designations

Catalogue designations vary considerably for HID lamps, depending on the manufacturer. In general, the designation for a mercury lamp will begin with an "H"; metal halide lamps have an "M"; high-pressure sodium lamps have either an "L" or a "C." Designations for two common types are:

MH250/C/U	metal halide, 250 watts, phosphor coated, base up (Philips)
MVR250/C/U	same as above (General Electric)
M250/C/U	same as above (Sylvania)
H38MP-100/DX	mercury, Type 38 ballast, 100 watts, deluxe white (Philips)
HR100DX38/A23	same as above (General Electric)
H38AV-100/DX	same as above (Sylvania)

HID lamp shapes are shown in Figure 13-53.

FIGURE 13-53   HID lamps.

Metal Halide Lamps

The metal halide lamp has the disadvantages of a relatively short life and a rapid drop-off in light output as the lamp ages. Rated lamp life varies from 5000 to 15 000 hours, during which the light output can be expected to drop by 30% or more. These lamps are considered to have good colour-rendering characteristics and are often used in retail clothing and furniture stores. The lamp has a high efficacy rating, which ranges from 65 to 120 lumens per watt. Only a few styles are available with ratings below 175 watts. Operating position (horizontal or vertical) is critical with many of these lamps and should be checked before a lamp is installed.

High-Pressure Sodium (HPS) Lamps

This type of lamp is ideal for applications in warehouses, parking lots, and similar places where colour recognition is necessary but high-quality colour rendition is not required. The light output is rather orange in colour. The lamp has a life rating equal to or better than that of any other HID lamp and has very stable light output over the life of the lamp. The efficacy is very good, ranging as high as 130 lumens per watt.

Low-Pressure Sodium Lamps

This lamp has the highest efficacy of any of the lamps, ranging from 140 to above 180 lumens per watt. The light is monochromatic, containing energy in only a very narrow band of yellow. This lamp is usually used only in parking and storage areas where no colour recognition is required. The lamp has a good life rating of 18 000 hours, during which it maintains a very constant light output. The lamp is physically longer than HID lamps but generally is shorter than fluorescent lamps.

Light-Emitting Diodes (LEDs)

"Reduce the electric bill! Save energy! Reduce energy consumption! Reduce the air-conditioning load!" It has been said that the incandescent lamp is from the dinosaur age, having been around since Thomas Edison applied for a patent on May 4, 1880.

Moving us from the dinosaur age to the space age is a new concept for lighting with low power consumption—LED lighting—and electricians had better get ready for this new type of light source in luminaires. LED lighting is not used yet for general lighting of a room, but that will come as the light output of LEDs increases.

As this text is being written, LED technology is rapidly changing to provide better colour rendition, lower energy costs, and more lumens per watt. Luminaires with LED lamps are coming onto the market faster than electrical distributors can print their catalogues.

LEDs are solid-state devices and have been around since 1962. You could call it "solid-state lighting." When connected to a direct-current source, the electrons in the LED smash together, creating light. Think of an LED as a tiny light bulb. Figure 13-54 shows a typical LED in part (a), and a cluster of LEDs in one small enclosure in part (b).

The size and number of LEDs determine the amount of lumen output of the device.

LEDs are all around us. They are commonly recognized by the tiny white, red, yellow, green, purple, orange, and blue lights found in the digital displays in TVs, radios, DVD and CD players, remotes, computers, printers, fax machines, telephones, answering machines, Christmas light strings, night lights, "locator" switches, traffic lights, digital clocks, meters, testers, automobile tail lights, strobe lights, occupation sensors, and other electronic devices, equipment, and appliances.

Individual LEDs are rather small. Using LEDs for lighting is a rather recent concept because the lumens per watt in these devices are on the increase. Putting a cluster (called an array) of LEDs together (i.e., 5, 20, 30, 60, 120) produces a lot of light. The result can be an LED lamp, usable in a luminaire in the same way as a typical medium Edison-base incandescent lamp. Figure 13-55 shows three different styles of LED lamps.

LED lighting is beginning to appear in exit, recessed, surface, and under cabinet luminaires, desk lamps, wall luminaires, downlights, stoplights in automobiles, and floor egress path lighting on airplanes, to name a few applications. LEDs in flashlights have been around for a while.

Courtesy of Lumileds

A

Courtesy Philips Lumileds Lighting Company, LLC

B

FIGURE 13-54 (A) A typical single light-emitting diode (LED); and (B) a cluster of LEDs in one device.

© Cengage Learning®

FIGURE 13-55   Three light bulbs (lamps) powered by a number of individual LEDs. These lamps screw into a conventional medium Edison-base lampholder.

A few lamp manufacturers have come out with many different types of lamps for accent, task, conventional shape, flood, spots, and so on. They are available with the standard Edison and candelabra bases to replace existing incandescent lamps. The light is white, but with some LEDs, other colours are available.

Today's LEDs last 60 000 to 100 000 hours. They produce virtually no heat, have no filament to burn out, can withstand vibration and rough usage, contain no mercury, operate better when cool, lose life and lumen output at extremely high temperatures, and can operate at −40°F (−40°C). Their lumen output slowly declines over time, but the industry seems to be settling on a lumen output rating of 70% after 50 000 hours of use.

LEDs use a tiny amount of electricity. Virtually all of the power consumption converts to light, whereas for an incandescent lamp, 5% of the power consumption converts to light and the rest converts to heat. When the purpose is to have illumination, heat is a waste.

Some LEDs may be dimmed, but others may not. Check with the manufacturer of the luminaire to verify the dimming capability. LEDs start instantly with no flickering.

LEDs put out directional light, as compared to the conventional incandescent lamp that shoots light out in almost all directions. A recent study compared an LED's predicted life of 60 000 hours (almost seven years of continuous burning, or 21 years at eight hours per day of usage) to a standard 60-watt

incandescent lamp that has a rated life of 1000 hours. Over the 60 000-hour life:

- 60 standard incandescent lamps would be used compared to one LED lamp.

- The standard incandescent lamps would use 3600 kWh, resulting in an electricity cost of $360. The LED lamp would use 120 kWh, resulting in an electricity cost of $12.

- The total cost (lamps and cost of electricity) for the incandescent lamps was $400 and for the LED lamp it was $47.

Today, the lumen output of LED lamps is similar to that of compact fluorescent lamps (CFL), producing approximately 50 lumens per watt. Philips recently announced an LED producing in excess of 200 lumens per watt. This is expected to increase further as technology improves in LED production. Compare the LED lumen output to a typical incandescent lamp that has a lumen output of 14–18 lumens per watt. That's a significant increase in lumens per watt!

Some LED lamps are direct replacements for the conventional 40-watt fluorescent lamp. These lamps both work on magnetic and electronic ballasts. When compared to a conventional 40-watt fluorescent lamp, the LED replacement lamp typically has a 10-year life as opposed to a two- or three-year life. Their power consumption is 20% less and their lumen output is comparable or slightly greater, and they can operate at 32°F (0°C). These are currently available in warm white, cool white, daylight, neutral, and bright white.

REVIEW

▶ PART I: LUMINAIRES

Note: Refer to the *CEC* or the blueprint package as necessary.

1. A luminaire weighing 3.2 kg shall not be supported by the screw-shell of the lamp-shade. True or false? Explain.

   _____

   _____

2. Flexible raceway or cable installed to a recessed luminaire must be not more than
   _____ m (_____ ft) or less than _____ m (_____ ft) long.

3. The neutral and phase wires of a three-phase, four-wire system may be installed in a continuous row of luminaires, provided the conductors have a _____°C rating if they are within 76 mm of the ballast.

4. A 2.4-m and a 1.2-m luminaire are to be installed end to end. Using the dimensions given in Figure 13-7, provide the following measurements from a wall 600 mm from the end of the 2.45-m luminaires.

   a. The first support rod hole is _____ m from the wall.

   b. The first outlet box opening is _____ m from the wall.

   c. The second support rod hole is _____ m from the wall.

   d. The third support rod hole is _____ m from the wall.

   e. The final support rod hole is _____ m from the wall.

5. The basic loading allowance for the beauty salon is _____ watts.

6. Book shelving, such as in a library, is best illuminated by using a luminaire similar to Style _____.

7. What are the requirements for installing a recessed luminaire that does not have an approved box furnished with the luminaire? _____

   _____

   _____

   _____

▶ PART II: LAMPS FOR LIGHTING

Note: Refer to the *CEC* or the blueprint package as necessary.

1. Describe the lamps specified by the following designations:

   a. 150 R_____

   b. 200 A_____

    c. 150 PAR _____

    d. F48T12/CW _____

    e. H37-5KC/DX _____

2. What is the size and shape of an A23 lamp?

    _____

3. The operating current of a 40-watt standard rapid-start lamp is _____ mA.

4. Match the following items:

Lumen per square metre	_____	a. Intensity
Lumen per watt	_____	b. Footcandle
One candela	_____	c. Efficacy
Starter	_____	d. Instant start
Cold cathode	_____	e. Rapid start
Constant heated cathode	_____	f. Preheat

5. The _____ is the light-producing element in the incandescent lamp.

6. The _____ is the light-producing element in the mercury lamp.

7. The _____ is the light-producing element in the fluorescent lamp.

8. The main purposes of a fluorescent ballast are to _____

    _____

    _____

9. When calculating the lighting load for a circuit, _____ volt-amperes are allowed for a Style A fluorescent luminaire.

10. Describe a Class P ballast. _____

    _____

    _____

11. Explain a method used to isolate a faulty ballast to prevent the shutdown of the entire circuit because of a single failure. _____

    _____

    _____

Reading Electrical Drawings—
Insurance Office/Beauty Salon

OBJECTIVES

After completing the study of this unit, the student will be able to

- list (tabulate) the materials required to install an electrical system
- make branch-circuit calculations
- lay out the raceway system
- cite *CEC* references

BLUEPRINTS

A journeyman electrician should be able to look at an electrical drawing (Figure 14-1) and prepare a list of the materials required to install the wiring system. A great deal of time can be lost if the proper materials are not on the job when needed, and so it is essential that the electrician prepare in advance for the installation so that the correct variety of material is available in sufficient quantities to complete the job.

LOADING SCHEDULE

The loading schedules for the insurance office and the beauty salon are included at this point to assist the student in reviewing the project. See Tables 14-1 and 14-2. In the insurance office there are 56 receptacles in the multioutlet assembly, and 15 elsewhere in the occupancy that qualify.

APPLIANCE CIRCUITS

Water Heater

A beauty salon (Figure 14-2) uses a large amount of hot water. To accommodate this need, the specifications indicate that a circuit in the beauty salon is to supply an electric water heater. The water heater is not furnished but is to be connected by the electrical contractor, Figure 14-3. The water heater is a full demand heater and is rated for 3 800 watts at 208 volts single phase. In this case, it is assumed that the water heater is not marked with the maximum overcurrent protection. The water heater is connected to an individual branch circuit.

The *branch-circuit rating* must be at least 125% of the marked rating. The water heater in the beauty salon is considered to be "continuously loaded." See *Rule 8–104(3)*.

Current rating: 3 800/208 = 18.3 amperes

Minimum branch-circuit rating

$$18.3 \times 1.25 = 22.9 \text{ amperes}$$

Conductor size is selected from *Table 2*. No. 10 Type TWN 75 copper conductors have an ampacity of 35 amperes.

The *branch-circuit overcurrent protection* for single, nonmotor-operated appliances might be marked on the nameplate of the appliance. If not marked, then refer to *Table 13*. In this case, a 30-ampere conductor requires a 30-ampere fuse as the maximum overcurrent protection, *Rule 14–104(2)*.

Washer-Dryer Combination

One of the uses of the hot water is to supply a washer–dryer combination used to launder the many towels and other items commonly used in a beauty

FIGURE 14-1 Electrical drawing for an insurance office. *Note:* For the complete blueprint, refer to Drawing E4 in the blueprint package included with the text.

TABLE 14-1

Insurance Office Service Calculation.

ITEM	COUNT	VA/UNIT	DEMAND LOAD	INSTALLED LOAD	DEMAND FACTOR		USE	
					Feeder	Service	Feeder	Service
MINIMUM BASIC LOAD								
(CEC Rule 8–210)	122 m²	50 W/m²	6 100 W		1	0.9	6 100	5 490
BASIC INSTALLED LOADS								
Style E luminaires	10	144		1 440			1 440	1 440
Style O luminaires	16	143		2 288			2 288	2 288
Style P luminaires	2	87		174			174	174
Style Q luminaires	4	143		572			572	572
General receptacles	12	120		1 440			1 440	1 440
Basic installed loads							5 914	5 914
BASIC LOAD	(Use the greater of the demand load or installed loads, *CEC Rule 8–106(2)*.)						5 914	5 914
SPECIAL LOADS								
Copier outlet	1	1 500		1 500			1 500	1 500
Computer outlets	9	500		4 500			4 500	4 500
Roof receptacle	1	1 440		1 440			1 440	1 440
Multioutlet assembly	56	180		10 080			10 080	10 080
Floor receptacles	4	120		480			480	480
Motors	**Volts**	**FLA**	**Phase**					
Cooling unit								
• Compressor	208	20.2	3	7 269	1.25		9 086	7 269
• Evaporator motor	208	3.2	1	666			666	666
• Condenser motor	208	3.2	1	666			666	666
Total load							34 518	32 515

$$I = \frac{P}{E \times 1.73} = \frac{34\ 518}{208 \times 1.73} = 95.9 \text{ A}$$

Assuming all loads are continuous, using 75°C rated equipment (*Rule 4–006*) and RW90-XLPE copper conductors from *Table 2*, 75°C column, *CEC*, the ampere rating of the circuit would be:

$$\frac{95.9}{0.8} = 119.8 \text{ A}$$

The minimum size feeder would be four No. 1 RW90-XLPE copper conductors in a 41-mm conduit, supplied by a 200-A disconnect c/w 125-A fuses. Conductor sized based on 75°C termination temperature.

salon. The unit is rated for 4 000 volt-amperes at 120/208 volts single phase. The branch circuit supplying this unit must have a rating of at least 125% of the appliance load.

- The branch-circuit demand ampacity is 20 amperes (4 000/208 = 19.23).

- The overcurrent protective device must be rated at least 25 amperes (load in amperes multiplied by 1.25, then raised to the next standard size overcurrent device).

- The conductor ampacity must be at least 25 amperes to allow the use of an overcurrent device with a rating of 25 amperes, *Rule 8–104(6)(a)*.

For appliances that contain one or more motors and other loads, such as heating elements, reference should be made to the manufacturer's nameplate, which must state the minimum supply circuit conductor ampacity and the maximum rating of the circuit overcurrent protective device.

TABLE 14-2

Beauty Salon Service Calculation.

ITEM	COUNT	VA/UNIT	DEMAND LOAD	INSTALLED LOAD	DEMAND FACTOR Feeder	DEMAND FACTOR Service	USE Feeder	USE Service
MINIMUM BASIC LOAD								
(CEC Rule 8–210)	45 m²	30 W/m²	1 350 W		1	0.9	1 350	1 215
BASIC INSTALLED LOADS								
Style A luminaires	5	87		435			435	435
Style E luminaires	5	144		720			720	720
Style M luminaires	9	50		450			450	450
General receptacles	6	120		720			720	720
Basic installed loads							2 325	2 325
BASIC LOAD	(Use the greater of the demand load or installed loads, *CEC Rule 8–106(2)*.)						2 325	2 325
SPECIAL LOADS								
Receptacle outlets	3	1 500		4 500			4 500	4 500
Roof receptacle	1	1 440		1 440			1 440	1 440
Washer/dryer	1	4 000		4 000			4 000	4 000
Water heater	1	3 800		3 800			3 800	3 800
Motors	**Volts**	**FLA**	**Phase**					
Cooling unit								
• Compressor	208	14.1	3	5 074	1.25		6 342	5 074
• Condenser/evaporator	208	3.3	1	686			686	686
Total load							23 093	21 825

$$I = \frac{P}{E \times 1.73} = \frac{23\ 093}{208 \times 1.73} = 64.2\ A$$

Assuming all loads are continuous, using 75°C rated equipment (*Rule 4–006*) and RW90 copper conductors from *Table 2*, 75°C column, *CEC*, the ampere rating of the circuit would be:

$$\frac{64.2}{0.8} = 80.25\ A$$

The minimum size feeder would be four No. 4 RW90-XLPE copper conductors in a 35-mm conduit, supplied by a 100-A disconnect c/w 90-A fuses.

FIGURE 14-2   Electrical drawing for a beauty salon and doctor's office. *Note:* For the complete blueprint, refer to Drawing E4 in the blueprint package included with the text.

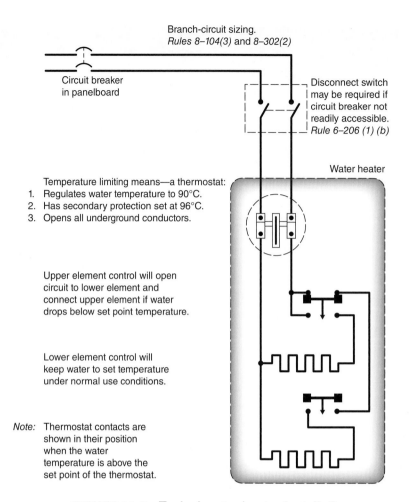

Branch-circuit sizing.
Rules 8–104(3) and 8–302(2)

Circuit breaker
in panelboard

Disconnect switch
may be required if
circuit breaker not
readily accessible.
Rule 6–206 (1) (b)

Water heater

Temperature limiting means—a thermostat:
1. Regulates water temperature to 90°C.
2. Has secondary protection set at 96°C.
3. Opens all underground conductors.

Upper element control will open
circuit to lower element and
connect upper element if water
drops below set point temperature.

Lower element control will
keep water to set temperature
under normal use conditions.

Note: Thermostat contacts are
shown in their position
when the water
temperature is above the
set point of the thermostat.

FIGURE 14-3   Typical water heater installation.

When stackable (combination) washer–dryers are installed, all of the power required for the unit comes from a single receptacle. For the combination washer–dryer in the beauty salon, a standard dryer receptacle (14-30R) is used and will be protected by a 30-amp breaker. The minimum size conductor permitted to supply the receptacle would be No. 10 AWG copper. If the conductors for the receptacle were run in the same raceway as conductors to other loads, they would have to be derated in accordance with *Table 5C.*

REVIEW

Note: Refer to the *Canadian Electrical Code (CEC)* or the blueprint package as necessary.

1. How many of each style of luminaire are located in the insurance office?

   a. Style E luminaire _____

   b. Style O luminaire _____

   c. Style P luminaire _____

   d. Style Q luminaire _____

2. How many lamps are required?

   a. F40/CW _____

   b. FB40/CW _____

   c. 26-watt quad T4 _____

3. Indicate the number of switches used in the insurance office.

   a. Single-pole switches _____

   b. Three-way switches _____

4. Three types of receptacles are specified for use in the insurance office. Identify the three types and the required number of each type.

   a. Type _____   Count _____

   b. Type _____   Count _____

   c. Type _____   Count _____

5. Branch circuit No. 14 serves the receptacle outlet for the copy machine.

   a. How many 3-m lengths of conduit will be needed to install this circuit if the conduit can be placed in the slab? _____ lengths

   b. How many 3-m lengths of conduit will be needed to install this circuit if the conduit is run overhead? _____ lengths

   Note: Although colour coding is not required by the *CEC*, the use of coloured conductors is highly recommended when it is necessary to keep track of specific circuits.

   c. How many metres and which colours of conductors should be provided for this run if it is installed in the slab?

   Colours _____

   Length _____

Complete the following loading schedule by referring to the *CEC*, the blueprint package, and the specifications.

6. The water heater (can) (cannot) be connected to the electrical power by a receptacle outlet. (Circle the correct answer.) Explain why and cite the *CEC* reference.

   _____

   _____

7. The rating of the circuit breaker or fuse that should be installed for the water heater circuit is _____ amperes. Explain and cite the *CEC* reference.

   _____

   _____

8. A separate disconnect switch (is) (is not) required for the water heater. (Circle the correct answer.) Explain and cite the *CEC* reference.

   _____

   _____

9. Make a list of the materials required to install the raceway system for the beauty salon. In actual practice, many of the luminaires will be equipped with a pull-through

junction box. For this exercise, assume that all connections will require a junction box. Indicate in the blanks provided the quantities, sizes, and styles.

Wall boxes for switches and receptacles: _____

_____

_____

_____

Ceiling boxes for lighting and junctions: _____

_____

_____

_____

Receptacles: _____

_____

_____

_____

_____

Switches: _____

_____

_____

Conduit: _____

_____

_____

10. Draw the configurations for the different styles of receptacles used in the beauty salon.

11. If an electric water heater is not marked with the maximum overcurrent protection rating, the water heater's ampere rating is to be sized not greater than _____% of the maximum overcurrent protection.

12. When an appliance is marked with a maximum branch-circuit rating that is not equal to a standard rating given in *Table 13*, then a standard rating that is next (lower) (higher) than the marked rating should be selected. (Circle the correct answer.)

UNIT 15

Alternate Systems

OBJECTIVES

After completing the study of this unit, the student will be able to

- install surface metal raceway
- install multioutlet assemblies
- install communication circuits
- install floor outlets

A number of alternate systems are found in almost every commercial building. Although these systems usually are a minor part of the total electrical work to be done, they are essential systems and it is recommended that the electrician be familiar with their installation requirements.

FIGURE 15-2 Surface metal raceway supports.

© Cengage Learning®

SURFACE RACEWAYS

Surface raceways, either metal or non-metallic, are generally installed as extensions to an existing electrical raceway system or used where it is impossible to conceal conduits, such as in desks, counters, cabinets, and modular partitions. This installation of surface metal raceways is governed by *Rules 12–1600* through *12–1614*. The number and size of the conductors to be installed in surface raceways is limited by the design of the raceway. Catalogue data from the raceway manufacturer will specify the permitted number and size of the conductors for specific raceways, *Rule 12–1606*. Conductors to be installed in raceways may be spliced at junction boxes or within the raceway if the cover of the raceway is removable. See *Rule 12–1610*.

Surface raceways are available in various sizes, Figure 15-1, and a wide variety of special fittings, such as the supports in Figure 15-2, make it possible to use surface metal raceway in almost any dry location. Two examples of the use of surface raceways are shown in Figures 15-3 and 15-4.

MULTIOUTLET ASSEMBLIES

The definition of a multioutlet assembly is given in *Section 0* of the *Canadian Electrical Code (CEC)*. Multioutlet assemblies, Figure 15-5, are similar to surface raceways and are designed to hold both conductors and devices. These assemblies offer a high degree of flexibility in an installation and are particularly suited to heavy-use areas where many outlets are required, or where there is a likelihood of changes in the installation requirements.

The plans for the insurance office specify the use of a multioutlet assembly that will accommodate both power and communications cables, Figure 15-6. This installation will allow the tenant in the insurance office to alter and expand the office facilities as the need arises.

Loading Allowance

The load allowance for a multioutlet assembly is specified by *Rule 8–304(4)*:

- Each 1.5 m of assembly counts as one outlet when normal loading conditions exist, or
- Each 300 mm of assembly counts as one outlet when heavy loading conditions exist.

The usage in the insurance office is expected to be intermittent and made up of only small appliances; thus it would qualify for the allowable minimum of one outlet for each 1.5 metres. However, the contractor is required to install a duplex receptacle every 450 mm for a total of 56 receptacles in the office space. The number of receptacles is

FIGURE 15-1 Surface non-metallic raceways.

© Cengage Learning®

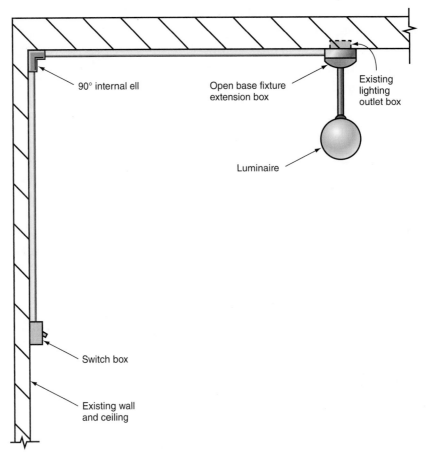

© Cengage Learning®

FIGURE 15-3 Use of surface raceway to install switch on existing lighting installation.

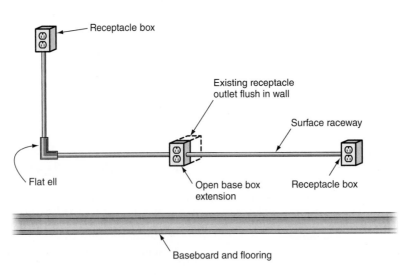

© Cengage Learning®

FIGURE 15-4 Use of surface metal raceway to install additional receptacle outlets.

limited to 12 outlets per circuit, *Rule 8–304(1)*. This load is connected to seven branch circuits, which satisfies the continuous loading maximum as set out in *Rule 8–304*. This allows for some growth, and additional circuits can easily be installed if required.

When determining the feeder size, the requirement for the receptacle load is calculated by

FIGURE 15-5 Multioutlet assemblies.

The connection of the identified conductor or the neutral (if fed from a three-phase, four-wire system) to the receptacle must utilize tails and wire connectors. This permits a receptacle to be removed without the neutral or identified conductor being broken. The neutral or identified conductor provides the return for the other two phases.

An electrician could be working on a circuit A receptacle, which is off (dead), while circuits B and C remain on (alive). These circuits are fed from a multiwire branch circuit. Since the neutral has not been broken, there is less possibility of the electrician becoming the return path to ground. **Any shock current greater than 6 milliamperes is dangerous.**

the application of the demand factors set out in *Table 14*.

Receptacle Wiring

The blueprints indicate that the receptacles to be mounted in the multioutlet assembly must be spaced 450 mm apart. The receptacles may be connected in either of the arrangements shown in Figure 15-7. That is, all of the receptacles on a phase can be connected in a continuous row, or the receptacles can be connected on alternate phases.

■ COMMUNICATIONS SYSTEMS

The installation of the telephone system in the commercial building will consist of two separate installations. The *electrical contractor* will install an empty conduit system according to the specifications for the commercial building and in the locations indicated on the plans. In addition, the installation will meet the rules, regulations, and requirements of the communications company that will serve the building. Once the conduit system is complete, the electrical contractor, telephone installing company, or telephone company will install a complete telephone system.

FIGURE 15-6 Multioutlet assembly for power and communications systems.

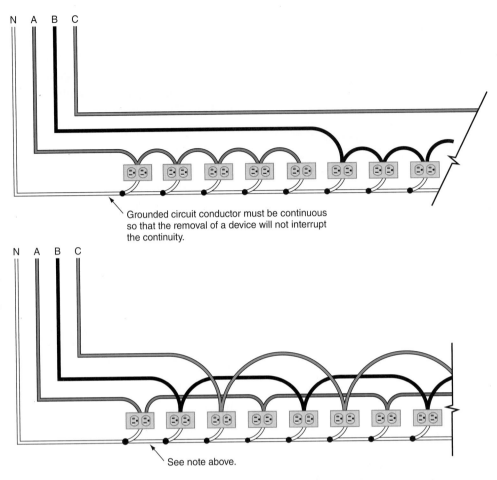

Grounded circuit conductor must be continuous so that the removal of a device will not interrupt the continuity.

See note above.

FIGURE 15-7   Connection of receptacles.

Power Requirements

An allowance of 2 500 volt-amperes is made for the installation of the telephone equipment to provide the power required to operate the special switching equipment for a large number of telephones. Because of the importance of the telephone as a means of communication, the receptacle outlets for this equipment are connected to the emergency power system.

The Telephone System

Telephone service requirements vary widely according to the type of business and the extent of the communication convenience desired. In many situations, the telephone lines may be installed in exposed locations. However, improvements in building construction techniques have made it more important to provide facilities for installing concealed telephone wiring. The use of new wall materials and wall

insulation, the reduction or the complete omission of trim around windows and doors, the omission of baseboards, and the increased use of metal trim make it more difficult to attach exposed wires. For these reasons (and many others), the wiring is more conspicuous if the installation is exposed. In addition, the unprotected wiring is more subject to damage and interference.

To solve the problems of exposed wiring, conduits are installed and the proper outlet boxes and junction boxes are placed in position during the construction. The material and construction costs are low when compared to the benefits gained from this type of installation.

Since it is generally more difficult to conceal the telephone wiring to the floors above the first floor after the building construction is completed, it is recommended that telephone conduits be provided to these floors. The materials used in the installation and the method of installing the conduits for the telephone lines are the same as those used for light and power wiring, with the following modifications:

- To avoid small openings and limited space, junction boxes are used rather than the standard conduit fittings such as ells and tees.

- Since multipair conductors are used extensively in telephone installations, the size of the conduit should be at least 21-mm trade size. The current and potential ratings for an installation of this type are not governed by the same rules as conventional wiring for light and power.

- The number of bends and offsets is kept to a minimum; when possible, bends and offsets are made using greater minimum radii or larger sweeps. The allowable minimum radius is 152 mm.

- A fishwire is installed in each conduit for use in pulling in the cables.

- When basement telephone wiring is to be exposed, the conduits are dropped into the basement and terminated with a bushing (junction boxes are not required). No more than 50 mm of conduit should project beyond the joists or ceiling level in the basement.

- The inside conduit drops *must* be grounded in accordance with *Section 10* and *Rule 60–302*.

- The service conduit carrying telephone cables from the exterior of a building to the interior must be permanently and effectively grounded, *Rule 60–302*.

- If the cable has a metal shield or sheath, the covering shall be bonded to ground, *Rules 60–306* and *60–700*.

Installing the Telephone Outlets

The blueprints for the commercial building indicate that each of the occupancies requires telephone service. As shown previously, the insurance office uses a multi-outlet assembly. Telephone outlets are installed in the balance of the occupancies. These outlets are supplied by electrical metallic tubing (EMT), which runs to the basement, Figure 15-8, where the cable then runs exposed to the main terminal connections. Telephone company personnel will install the cable and connect the equipment.

Data Network

The insurance office has a complete data network installed between the office computers and the computer room. Industry standard is that data networks must perform to a minimum Category 5e standard. Often the customer may require a higher standard depending on the type of work performed.

The data cables will share the same compartment in the surface raceway as the telephone cables. Data jacks are to be installed in the surface raceway at intervals not to exceed 1.2 metres. It is common practice to use a data jack of a different colour than that used for the telephone jacks in order to avoid confusion when plugging in equipment.

All of the data cables terminate in a floor-mounted data rack. This rack is used to mount equipment such as patches, panels, network switches, routers, uninterruptible power supplies, and other data-related equipment.

An EMT raceway is supplied from the computer room to the location of the telephone equipment in the basement. An Internet service provider will most likely install its equipment in the vicinity of the phone equipment. The raceway is then used to install copper or optical cable to link the equipment in the computer room to the Internet.

It is extremely important that all metal components of a data network system be bonded to ground. Failure to bond equipment such as raceways can cause electromagnetic interference (EMI) and radio frequency interference (RFI). These interferences can cause data errors and overall poor performance of the network. Care must also be taken to ensure that copper data cables are separated from 120 V power and lighting cable by a minimum of 50 mm, *Rule 60–308(1)*. If power cable crossovers are necessary, they should cross at 90 degrees.

▶ FLOOR OUTLETS

In an area the size of the insurance office, it may be necessary to place equipment and desks where wall outlets are not available. Floor outlets may be installed to provide the necessary electrical supply to such equipment. Two methods can be used to provide floor outlets:

1. Installing underfloor raceway, or

2. Installing floor boxes.

4 × 1½ in. square box
with plaster ring

Trade size 21 mm
EMT or ENT

Drywall

Block

Brick

Partition
wall

3¾ × 2 × 3½ in. masonry box

EMT or ENT roughed in
before concrete is poured

Bushings

Fishwire

50 mm (2″)
maximum

Foundation wall

© Cengage Learning®

FIGURE 15-8   Raceway installation for telephone system.

Underfloor Raceway (Subsections 12–1700 to 12–1718)

In general, underfloor raceway is installed to provide both power and communications outlets using a dual duct system similar to the one shown in Figure 15-9. The junction box is constructed so that the power and communications systems are always separated from each other. Service fittings are available for the outlets, Figure 15-10.

Floor Boxes

Floor boxes, either metallic or non-metallic, can be installed using any approved raceway such as rigid conduit, rigid non-metallic conduit, or EMT. Some

Duct end plug

Support coupler

Support coupler used as support

Marker screw assembly

Cabinet connector

Distribution system standard insert duct

Junction box

Box opening plug

"Y" take-off fitting

Vertical elbow

Feeder system standard blank duct

Conduit adapter

FIGURE 15-9 Underfloor raceway.

For high-potential service

Dimensions: 105 mm long; 105 mm wide; 75 mm high

FIGURE 15-10 Service fittings.

boxes must be installed to the correct height by adjusting the levelling screws before the concrete is poured, Figure 15-11. One type of non-metallic box can be installed without adjustment and cut to the desired height after the concrete is poured, Figures 15-12 and 15-13.

COMPUTER ROOM CIRCUITS

Although a room in the insurance office is especially designed to be a computer room, no special requirements are applicable. Three actions were taken to

Cover cap

Brass cover

Neoprene gasket

Metal plate

Duplex grounded receptacle

Gasket

Adjusting ring

Inner ring

Box body

Height: 54 mm

Surface diameter: 108 mm

Exploded view of floor box

© Cengage Learning®

FIGURE 15-11 A floor box with levelling screws.

comply with good practice: special receptacles are specified, a special grounding system is provided, and a separate neutral is provided with each of the 120-volt, single-phase circuits.

The special receptacles were discussed in Unit 11; they have an isolated grounding connection and provide surge protection. The grounding terminal of each receptacle is connected to an insulated grounding conductor installed in the conduit with the circuit conductors. The conduit is grounded according to *Section 10.* Since three conduits are involved, the three grounding conductors are connected to a special isolated grounding terminal in the panelboard; see *Rule 10–904(9).*

From here, a grounding conductor is installed with the feeder circuit and is connected to a grounding electrode at the switchgear. This grounding system reduces, if not eliminates, the EMI that often is present in the conventional grounding system. The surge protection feature of the receptacle provides protection from lightning and other powerful electrical surges that may occur in the electrical system. These surges are often severe enough to cause a loss

of computer memory and, in some cases, damage to the computer. Surge protectors are available in separate units, which can be used to provide protection on any type of sensitive equipment, such as radios and televisions.

The separate neutrals or identified conductors are installed to greatly reduce the possibility of conductor overheating. The common practice is to use one neutral to serve three single-phase circuits of a four-wire wye system. Loads such as computers produce odd harmonics—the third, the fifth, the seventh, and so on. For example, the third harmonic is 3×60 Hz $= 180$ Hz. Since the loads are not all resistance, the harmonics from different phases do not cancel out as ordinary current does. Under these conditions, the common neutral current could be the sum of the individual circuits and damage could occur to the wiring system. Doubling the size of the neutral or installing zero-sequence filters, *Rule 26–268,* will help to alleviate the effects of the additive harmonic RMS current on the neutral of a three-phase, four-wire system.

1 Level the floor box using a non-metallic slotted screw (Max. ht. adjustment 1¾ in.)

2 Secure mounting pads to form, rebar, etc.

3 Make connection into outlets by solvent cementing conduit into proper outlet in box.

4 Slip on throwaway plastic cap and seal unused outlets with solvent-cemented plug.

5 After the pour, install appropriate plates/barrier.

6 Install appropriate cover. Secure with screws provided.

FIGURE 15-12   Installation of multifunction non-metallic floor box.

© Cengage Learning®

1 Fasten the box to the form or set on level surface.

2 Make connections into outlets by solvent-cementing conduit into the proper outlet in box.

3 Slip on temporary plastic cover and seal unused outlets with solvent-cemented plugs.

4 Remove temporary plastic cover and determine thickness of flooring to be used. Scribe a line around box at this distance from the floor.

5 Using a handsaw, cut off box at scribed line.

6 Install leveling ring to underside of cover so that four circular posts extend through slots in outer ring.

7 Apply PVC cement to leveling ring and to top inside edge of box.

8 Press outer ring cover assembly into floor box for perfect flush fit.

NOTE: Box may extend any distance above finish concrete. Cut box off to exact height and trim out with hand-saw. The "Leveling Ring" guarantees a LEVEL top every time, even if the box is knocked at an angle during the pouring process.

4-15/16 in. (125 mm)

6 in. (152 mm)

3/4 in. (21 mm)

1 in. (27 mm)

4-1/8 in. (105 mm)

© Cengage Learning®

FIGURE 15-13 Installation and trim-out of non-metallic floor box.

REVIEW

Note: Refer to the *CEC* or the blueprint package as necessary.

1. Surface raceway may be extended through dry partitions if the _____.

2. In surface raceway, the conductors (including taps and splices) shall fit not more than _____% of the area.

3. Where power and communications circuits are run in combination raceway, the different systems shall be run in _____ _____.

4. In areas of light usage, _____ outlet(s) is/are allowed for each _____ metres of multi-outlet assembly.

5. How many floor outlets are to be installed in the insurance office? _____

6. To install the multioutlet assembly in the insurance office, _____ end cap(s) and _____ internal ell(s) are required.

7. How many telephone conduits, for all occupants, are stubbed into the basement? _____

Reading Electrical and Architectural Drawings (Blueprints)—Drugstore

OBJECTIVES

After completing the study of this unit, the student will be able to

- identify symbols associated with the reading of electrical drawings
- determine the requirements of the electrical contract
- calculate plan dimensions

◗ THE DRUGSTORE BLUEPRINTS

Branch circuits are an important part of the electrical system in a building. Unit 3 presents the essential information on branch circuits, including how to determine, according to the *Canadian Electrical Code (CEC),* the number of branch circuits required and their correct size. The student should pay close attention to the drugstore branch-circuit panelboard schedule on Drawing E6. This schedule gives the branch-circuit numbers and a brief description of the loads served, in addition to other information. These items provide important links between the panelboard schedule and the electrical drawings. See Figure 16-1.

A feature of the drugstore wiring is the low-voltage remote-control system offering flexibility of control that is not available in the traditional control system. See Unit 20 for a complete discussion. The switches used in this system operate on 24 volts, and the power wiring, at 120 volts, goes directly to the electrical load. This reduces branch-circuit length and voltage drop. A switching schedule gives details on the system operation, and a wiring diagram provides valuable information to the installer.

The illumination system in each area of the drugstore is somewhat different. In the merchandise area, nine luminaires are installed in a continuous row. It is necessary to install electrical power to only one point of a continuous row of luminaires. From this point, the conductors are installed in the wiring channel of the luminaire. A luminous ceiling in the pharmacy area consists of rows of strip fluorescents and a ceiling that will transmit light. The installation of the ceiling, in many jurisdictions, is the responsibility of the electrician. For this system to be efficient, the surfaces above the ceiling must be highly reflective (white).

Drugstore Feeder Loading Schedule

The various fixtures and other loads are calculated in Table 16-1 to show the total demand watts on the feeders and the service. This will require a 100-ampere switch with 90-ampere fuses in the main switchgear. The No. 4 RW90-XLPE copper conductor will feed the 100-ampere panel located in the drugstore.

FIGURE 16-1   Electrical drawing for a drugstore. *Note:* For complete blueprint, refer to Drawing E3 in the blueprint package provided with the text.

TABLE 16-1

Drugstore Service Calculation.

ITEM	COUNT	VA/UNIT	DEMAND LOAD	INSTALLED LOAD	DEMAND FACTOR		USE	
					Feeder	Service	Feeder	Service
MINIMUM BASIC LOAD								
(CEC Rule 8–210)	225 m²	30 W/m²	6 750		1	1	6 750	6 750
INSTALLED BASIC LOADS								
Style D luminaires	15	74		1 110			1 110	1 110
Style E luminaires	4	144		576			576	576
Style I luminaires	27	87		2 349			2 349	2 349
Style N luminaires	2	60		120			120	120
Style L luminaires	9	87		783			783	783
Track lighting	15	110		1 650			1 650	1 650
Sign	1	1 200		1 200			1 200	1 200
General receptacles	21	120		2 520			2 520	2 520
Basic installed loads							6 903	6 903
BASIC LOAD	(Use the greater of the demand or installed loads, *Rule 8–106(2)*.)						6 903	6 903
SPECIAL LOADS								
Show window rec.	3	500		1 500			1 500	1 500
Roof receptacle	1	1 440		1 440			1 440	1 440
Motors	**Volts**	**FLA**	**Phase**					
Cooling unit								
• Compressor	208	20.2	3	7 269	1.25		9 086	7 269
• Evaporator motor	208	3.2	1	666			666	666
• Condenser motor	208	3.2	1	666			666	666
Total load				8 601			23 666	21 849

$$I = \frac{P}{E \times 1.73} = \frac{23\ 666}{208 \times 1.73} = 65.76\ \text{A}$$

Assuming continuous loads, using 75°C rated equipment (*Rule 4–006*), and RW90-XLPE copper conductors (*Table 2*, 75°C column), the ampere rating of the circuit would be:

$$\frac{65.76}{0.8} = 82.2\ \text{A}$$

The feeder would be four No. 4 AWG RW90-XLPE copper conductors in a 35-mm raceway, supplied by a 100-ampere disconnect c/w 90-ampere fuses. Conductor is sized based on 75°C termination temperature.

REVIEW

1. Identify the following symbols. Refer to the Electrical Symbol Schedule on Drawing E1 of the blueprint package.

a. _____

b. _____

c. $\oint$ _____

d. $\bigcirc$ _____

e. ◀ _____

f. ⌷° _____

g. ⊗ _____

In the following questions, fill in the blanks.

2. The elevation of the finished first floor is _____ m.

3. The drugstore ceiling material is _____.

4. The north wall of the drugstore is constructed of _____ and is _____ mm (_____ in.) plus finish material thick.

5. The south wall of the drugstore is constructed of _____ and is _____ mm (_____ in.) plus finish material thick.

6. The ceiling height of the drugstore is _____.

7. The outside dimensions of the building are _____ by _____.

8. The drugstore basement has an area of _____ m².

9. How many luminaires are controlled by switch RC$_a$? _____.

10. How many receptacles are installed on branch circuit #2? _____.

Overcurrent Protection: Fuses and Circuit Breakers

OBJECTIVES

After completing the study of this unit, the student will be able to

- list and identify the types, classes, and ratings of fuses and circuit breakers

- describe the operation of fuses and circuit breakers

- develop an understanding of switch sizes, ratings, and requirements

- define interrupting rating, short-circuit currents, I^2t, I_p, RMS, and current limitation

- apply the *CEC* to the selection and installation of overcurrent protective devices

- use time–current characteristics curves and peak let-through charts

Overcurrent protection is one of the most important components of an electrical system. The overcurrent device opens an electrical circuit whenever an overload or short circuit occurs. Overcurrent devices in an electrical circuit may be compared to pressure relief valves on a boiler. If dangerously high pressures develop within a boiler, the pressure relief valve opens to relieve the high pressure. The overcurrent device in an electrical system also acts as a "safety valve."

The *Canadian Electrical Code (CEC)* requirements for overcurrent protection are contained in *Section 14*. *Rule 14–010* states, in part, that devices must be provided to automatically open the electrical circuit if the current reaches a value that will produce a dangerous temperature in the equipment or conductor. Additionally, in the event of a ground fault all normally ungrounded conductors must be de-energized, subject to *Rule 14–102*. Also refer to *Rule 14–012* for voltage and interrupting current rating requirements.

Two types of overcurrent protective devices are commonly used: fuses and circuit breakers. The Canadian Standards Association (CSA) and the National Electrical Manufacturers Association (NEMA) establish standards for the ratings, types, classifications, and testing procedures for fuses and circuit breakers.

The standard ampere ratings for fuses and/or nonadjustable circuit breakers are listed in *Table 13* of the *CEC* for ampacities up to and including 600 amperes. Standard ampacities in excess of 600 amperes are 700, 800, 1000, 1200, 1600, 2000, 2500, 3000, 4000, 5000, and 6000. Although fuses are readily available in all of these ampacities, circuit breakers may not be available in all sizes, depending on the manufacturer. In addition, there are standard ratings for fuses of 1, 3, 6, and 10 amperes.

The *CEC* describes circumstances in which it is permissible to increase the rating or setting of the overcurrent device. For example, *Rule 14–104(a)* states that if the ampacity of a conductor is less than 600 amperes and does not match a standard ampere rating fuse or breaker, then it is allowable to use the rating found in *Table 13*. Therefore, for a conductor that has an ampacity of 115 amperes, the *CEC* permits the overcurrent protection to be sized at 125 amperes.

When protecting conductors where the overcurrent protection is above 600 amperes, it is vital to ensure that the overcurrent device is equal to or lower than the ampacity of the conductor. For example, three 500 kcmil copper conductors are installed in parallel. Each has an ampacity of 380 amperes. Therefore, the total ampacity is $380 \times 3 = 1140$ amperes.

It would be a *CEC* violation to protect these conductors with a 1200-ampere fuse. The correct thing to do is to round down to an ampere rating less than 1140 amperes. Fuse manufacturers provide 1100 ampere rated fuses.

Fused switches are available in ratings of 30, 60, 100, 200, 400, 600, 800, 1200, 1600, 2000, 2500, 3000, 5000, and 6000 amperes in both 250-and 600-volt versions. They are for use with copper conductors only unless marked that the terminals are suitable for use with aluminum conductors. The ratings of the switches are based on 60°C wire (No. 14 through No. 1) and 75°C for wires No. 1/0 and larger, unless otherwise marked. Switches also may be equipped with ground-fault sensing devices. These will be clearly marked as such.

Switches may have labels that indicate their intended application, such as "Continuous load current not to exceed 80% of the rating of the fuses employed in other than motor circuits."

- Switches intended for isolating use only are marked "For isolation use only—Do not open under load."

- Switches that are suitable for use as service switches are marked "Suitable for use as service equipment."

- When a switch is marked "Motor-circuit switch," it is only for use in motor circuits.

- Switches of higher quality may have markings such as "Suitable for use on a circuit capable of delivering not more than 100 000 RMS symmetrical amperes, 600 volts maximum: Use Class T fuses having an interrupting rating of no less than the maximum available short-circuit current of the circuit."

- Enclosed switches with horsepower ratings in addition to ampere ratings are suitable for use in motor circuits as well as for general use.

- Some switches have a dual-horsepower rating. The larger horsepower rating is applicable when using dual-element time-delay fuses.

Dual-element time-delay fuses are discussed later in this unit.

- Fusible bolted pressure contact switches are tested for use at 100% of their current rating at 600 volts alternating-current (AC), and are marked for use on systems having available fault currents of 100 000, 150 000, and 200 000 RMS symmetrical amperes.

In the case of externally adjustable trip circuit breakers, the rating is considered to be the breaker's maximum trip setting. The exceptions to this are:

1. If the breaker has a removable and sealable cover over the adjusting screws, or

2. If it is located behind locked doors accessible only to qualified personnel, or

3. If it is located behind bolted equipment enclosure doors.

In these cases, the adjustable setting is considered to be the breaker's ampere rating. This is an important consideration when selecting proper size phase and neutral conductors, equipment grounding conductors, overload relays in motor controllers, and other situations where sizing is based upon the rating or setting of the overcurrent protective device.

▶ FUSES AND CIRCUIT BREAKERS

For general applications, the voltage rating, the continuous current rating, the interrupting rating, and the speed of response are factors that must be considered when selecting the proper fuses and circuit breakers.

Voltage Rating

According to *Rule 14–012*, the voltage rating of a fuse or circuit breaker must be sufficient for the voltage of the circuit in which the fuse or circuit breaker is to be used.

Continuous Current Rating

The continuous current rating of a fuse or circuit breaker should be equal to the ampacity of the conductor being protected. For example, referencing *Table 2*, we find that a No. 8 type T90 Nylon

conductor has an ampacity of 55 amperes. The proper overcurrent protection for this No. 8 type T90 Nylon conductor would be a 50-ampere fuse or 50-ampere circuit breaker. See *Table 13*.

There are several situations that permit the overcurrent protective device rating to exceed the ampacity of the conductor. For example, the rating of a fuse or circuit breaker may be greater than the current rating of the conductors that supply an electric motor or a refrigeration compressor.

Protection of Conductors

- *When the overcurrent device is rated 600 amperes or less.* When the ampacity of a conductor does not match the ampere rating of a standard fuse, or does not match the ampere rating of a standard nonadjustable circuit breaker, the use of the next higher standard ampere-rated fuse or circuit breaker is permitted, *Rule 14–104*. Refer to Figure 17-1.

- *When the overcurrent device is rated above 600 amperes.* When the ampere rating of a fuse or of a circuit breaker that does not have an overload trip adjustment exceeds 600 amperes, the conductor ampacity must be equal to or greater than the rating of the fuse or circuit breaker. The *CEC* allows the use of fuses and circuit breakers that have ratings less than the standard sizes. Adjustable trip circuit breakers were discussed above. See *Rule 14–104* and refer to Figure 17-2.

FIGURE 17-1   *Rule 14–104* allows the use of a 600-ampere fuse or circuit breaker.

FIGURE 17-2   *Rule 14–104* requires the use of a 1500-ampere fuse or circuit breaker.

SAFETY ALERT

The interrupting rating of a circuit breaker or fuse is *that* circuit breaker's or fuse's ability to safely interrupt fault currents not exceeding its interrupting rating. This marked interrupting rating has nothing to do with the *withstand ratings* of the downstream components, such as other circuit breakers, panelboards, switchboards, bus bars and their bracing, conductors, contactors, controllers, relays, ground-fault circuit interrupters, and so on.

Interrupting Rating
(*Rule 14–012(a)*)

The interrupting rating of an overcurrent device is a measure of the ability of a fuse or circuit breaker to safely open an electrical circuit under fault conditions, such as overload currents, short-circuit currents, and ground-fault currents. In other words, the interrupting rating of an overcurrent device is the maximum short-circuit current that it can interrupt safely at its rated voltage.

It can be very hazardous to consider the interrupting ratings of a circuit breaker while ignoring the withstand ratings of the downstream connected components.

The withstand rating of all electrical components is based upon **how much current will flow** and **how long the current will flow**. Units 18 and 19 discuss the important issues of "current limitation" and "peak let-through current."

Overload currents have the following characteristics:

- They are greater than the normal current flow.

- They are contained within the normal conducting current path.

- If allowed to continue, they will cause overheating of the equipment, the conductors, and the insulation of the conductors.

Short-circuit and ground-fault currents have the following characteristics:

- They flow outside of the normal current path.

- They may be less than, equal to, or greater than the normal current flow.

Short-circuit and ground-fault currents, which flow outside of the normal current paths, can cause conductors to overheat. In addition, mechanical damage to equipment can occur as a result of arcing and the magnetic forces of the large current flow. Some short-circuit and ground-fault currents may be no larger than the normal load current, but they can be thousands of times larger.

The terms "interrupting rating" and "interrupting capacity" are interchangeable in the electrical industry, but there is a significant difference. Interrupting rating is the value of the test circuit capability. Interrupting capacity is the actual current that the contacts of a circuit breaker "see" when opening under fault conditions.

For example, the test circuit in Figure 17-3 is calibrated to deliver 14 000 amperes of fault current. The standard test circuit allows 1.22 m of conductor to connect between the test bus and the breaker's terminals.

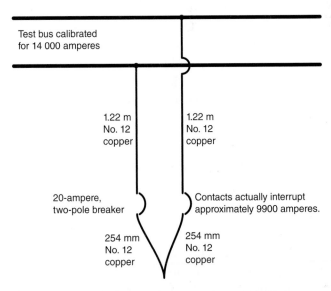

Typical laboratory test circuit for a moulded-case circuit breaker. The label on this breaker is marked 14 000 amperes interrupting rating. Its actual interrupting capacity is approximately 9900 amperes. The connecting conductors have an ampacity that is the same as the breaker's ampere rating.

FIGURE 17-3   Test circuit for a moulded-case circuit breaker.

The standard test further allows 254 mm of conductor per pole to be used as the shorting wire. Therefore, when the impedance of the connecting conductors is taken into consideration, the actual fault current that the contacts of the breaker "see" as they open is approximately 9900 amperes. The label on this circuit breaker is marked "14 000 amperes interrupting rating," yet its true interrupting capacity is 9900 amperes.

As described above, it is important to ensure that the ratings on equipment conform to the requirements of *Rule 14–012*.

For those interested in the complete details of how test circuits for breakers are connected, reference should be made to *CSA Standards C22.2 No. 4 M89* and *C22.2 No. 5 M91*.

Speed of Response

The time required for the fusible element of a fuse to open varies inversely with the magnitude of the current that flows through the fuse. In other words, as the current increases, the time required for the fuse to open decreases. The time–current characteristic of a fuse depends upon its rating and type. A circuit breaker also has a time–current characteristic. For the circuit breaker, however, there is a point at which the opening time cannot be reduced further due to the inertia of the moving parts within the breaker. The time–current characteristic of a fuse or circuit breaker should be selected to match the connected load of the circuit to be protected. Time–current characteristic curves are available from the manufacturers of fuses and circuit breakers.

▶ TYPES OF FUSES

Dual-Element, Time-Delay Fuse

The dual-element time-delay fuse, Figure 17-4, provides a time delay in the low-overload range to eliminate unnecessary opening of the circuit because of harmless overloads. However, this type of fuse is extremely responsive in opening on short circuits. This fuse has two fusible elements connected in series. Depending upon the magnitude of the current flow, one element or the other will open. The thermal cutout element is designed to have a delayed opening for currents up to approximately 500%

FIGURE 17-4   Cutaway view of a Fusetron dual-element time-delay fuse. On overloads, the spring-loaded trigger assembly opens. On short-circuits or heavy ground faults, the fuse elements in the short-circuit section open. The fuse elements are generally made of copper.

of the fuse rating. The short-circuit element opens immediately when a large short circuit or overload occurs. That is, the element opens at current values of approximately 500% or more of the fuse rating.

The thermal element is designed to open at approximately 140°C (284°F) as well as on damaging overloads. Thus, the thermal element will open whenever a loose connection or a poor contact in the fuseholder causes heat to develop. As a result, a true dual-element fuse also offers thermal protection to the equipment in which it is installed.

Dual-element fuses are suitable for use on motor circuits and other circuits having high-inrush characteristics. This type of fuse can be used as well for mains, feeders, subfeeders, and branch circuits. Dual-element fuses may be used to provide backup protection for circuit breakers, bus ducts, and other circuit components that lack an adequate interrupting rating, bracing, or withstand rating (covered later in this unit).

Using Fuses for Motor Overload Protection

Dual-element, time-delay fuses used on single-motor branch circuits are generally sized not to exceed 125% of the full-load running current of the motor, in accordance with *Rules 28–302(2)* and *28–306* for motor overload protection. Sizing these fuses slightly higher than the motor's overload setting will provide "backup" overload protection for the motor. The motor now has double overload protection.

Experience has shown that sizing dual-element time-delay fuses in the range of 115% to 125% of the motor's full-load current rating meets the *CEC* requirements for branch-circuit, short-circuit, and ground-fault protection, as well as providing "backup" motor overload protection.

Rules 28–300 to *28–318* set forth motor overload protection requirements. There are many questions to ask when searching for the proper level of motor overload protection. These include: What are the motor's starting characteristics? Is the motor manually or automatically started? Is the motor continuous duty or intermittent duty? Does the motor have integral built-in protection? Is the motor impedance protected? Is the motor larger than one horsepower? Is the motor fed by a general-purpose branch circuit? Is the motor cord-and-plug connected? Will the motor automatically restart if the overload trips? When these questions are answered, refer to *Rules 28–300* to *28–318* for the level of overload protection needed.

The procedure for selecting the overload protection is shown in Figure 17-5. The motor type is selected in the first column. The motor nameplate full-load current rating is increased according to the value in the second column.

EXAMPLE

What is the ampere rating of a dual-element, time-delay fuse that is to be installed to provide branch-circuit protection as well as overload protection for a motor with a full-load current rating of 16 amperes?

$$16 \times 1.25 = 20 \text{ amperes}$$

	MAXIMUM OVERLOAD SIZE, RULE 28–306
Service factor not less than 1.15	125%
All other motors	115%

FIGURE 17-5   Maximum overload size for motors.

Using Fuses for Motor Branch-Circuit Overcurrent Protection

Table 29 shows that for typical motor branch-circuit, overcurrent protection, the maximum size permitted for dual-element fuses is 175% of the full-load current of the motor. An exception to the 175% value is given in *Rule 28–200(e)(ii)*, where permission is given to go as high as 225% of the motor full-load current if the lower value is not sufficient to allow the motor to start.

EXAMPLE

Overcurrent protection for the motor branch circuit in the previous example:

$$16 \times 1.75 = 28 \text{ amperes (maximum)}$$

The next lower standard rating is 25 amperes.

If for some reason the 25-ampere fuse cannot handle the starting current, *Rule 28–200(e)(ii)* allows the selection of a dual-element time-delay fuse not to exceed

$$16 \times 2.25 = 36 \text{ amperes (maximum)}$$

The next lower standard rating is 35 amperes.

Dual-Element, Time-Delay, Current-Limiting Fuses

The dual-element, time-delay, current-limiting fuse, Figure 17-6, operates in the same manner as the standard dual-element, time-delay fuse, but has a faster response in the short-circuit range and thus is more current limiting. Sizing this type of fuse for motor circuits, where time-delay is necessary for starting the motor, is done according to the procedure discussed in the previous paragraph.

Courtesy of Eaton

FIGURE 17-6   Cutaway view of a low-peak, dual-element, time-delay, current-limiting fuse. On overloads, the spring-loaded trigger assembly opens. On large short circuits or ground faults, the fuse elements in the short-circuit section open. The fuse elements are generally made of silver.

Fuse opened on overload or high temperature condition

Fuse opened on short circuit

FIGURE 17-7A   Typical "P" or "D" fuse operation.

FUSE TERMINOLOGY

Interrupting Rating: The maximum current that a fuse can safely interrupt, usually as a result of a short circuit.

Class H Fuse: The standard fuse that has an interrupting rating of 10kA. Often referred to as a "Code fuse."

HRC: "High rupturing capacity" fuse with a minimum interrupting rating of 100kA. Most HRC fuses have an interrupting rating of 200kA.

HRCI or Form I: Can be used for both short-circuit and overload protection. Usually a Class J or Class R fuse.

HRCII or Form II: Provides short-circuit protection only. Usually a Class C fuse.

Supplemental Fuse: A fuse that is used to provide supplemental protection for an individual piece of equipment. Usually a small or midget fuse.

Dual Element: Often confused with the term "time delay." Dual element describes the construction of the fuse and may be a time-delay or nontime-delay fuse.

Time Delay: These fuses meet the CSA-specified time–current characteristics for the fuse type. These fuses will carry:

HRC and Code fuses:	500% of the fuse's continuous current rating for a minimum of 10 seconds.
Plug fuses:	200% of the fuse's continuous current rating for a minimum of 12 seconds
Supplemental fuses:	For fuses of 3 amperes and less, 200% of the fuse's continuous current rating for a minimum of 5 seconds
	For fuses of greater than 3 amperes, 200% of the fuse's continuous current rating for a minimum of 12 seconds

Fast Acting: A common term for a nontime-delay HRC fuse

Type "D": A time-delay fuse with low melting point characteristics. This can be a plug or cartridge type fuse.

Type "P": A nontime-delay fuse with low melting point characteristics. This can be a plug or cartridge type fuse.

FIGURE 17-7B   Fuse terminology.

The short-circuit element in the current-limiting fuse can be silver or copper surrounded by a quartz sand arc-quenching filler. Silver-link fuses are more current-limiting than copper-link fuses.

Some of the common terms referring to fuses are described in Figure 17-7B.

Current-Limiting Fuses (Nontime Delay)

The straight current-limiting fuse, Figure 17-7C, has an extremely fast response to both overloads and short circuits. When compared to other types of fuses, this one has the lowest energy let-through values. Current-limiting fuses are used to provide better protection to mains, feeders, subfeeders, circuit breakers, bus duct, switchboards, and other circuit components that lack adequate interrupting rating, bracing, or withstand rating.

FIGURE 17-7C   Cutaway view of a Limitron, current-limiting, fast-acting, single-element fuse.

	PERCENT OF FULL-LOAD CURRENT		
TYPE OF MOTOR	**MAXIMUM FUSE RATING**		**MAXIMUM SETTING TIME-LIMIT TYPE CIRCUIT BREAKER**
	TIME-DELAY* FUSES	**NONTIME DELAY**	
Alternating Current			
Single phase, all types	175	300	250
Squirrel-cage and synchronous:			
Full-voltage, resistor and reactor starting	175	300	250
Auto-transformer and star delta starting:			
Not more than 30 A	175	250	200
More than 30 A	175	200	200
Wound rotor	150	150	150
Direct Current	150	150	150

Table 29 (See Rules 28–200, 28–206, 28–208, and 28–308 and Table D16.)

RATING OR SETTING OF OVERCURRENT DEVICES FOR THE PROTECTION OF MOTOR BRANCH CIRCUITS

Courtesy of CSA Group

* Includes time-delay "D" fuses referred to in *Rule 14–200.*

NOTES: (1) Synchronous motors of the low-torque low-speed type (usually 450 rpm, or lower) such as are used to drive reciprocating compressors, pumps, and so on, and which start up unloaded, do not require a fuse rating or circuit-breaker setting in excess of 200% of full-load current.

(2) For the use of instantaneous trip (magnetic only) circuit interrupters in motor branch circuits see *Rule 28–210.*

FIGURE 17-8   *CEC Table 29.*

Current-limiting fuse elements can be made of silver or copper surrounded by a quartz sand arc-quenching filler. Silver-link fuses are more current limiting than copper-link fuses.

Current-limiting fuses may be selected and loaded at 100% of their rating for noninductive circuit loads. However, it is a good design practice not to exceed 80% of the rating for feeder and circuit loading.

A standard current-limiting fuse does not have the spring-loaded overload assembly found in dual-element fuses. To use straight current-limiting fuses for motor circuits, refer to *Table 29* (Figure 17-8) under the column for nontime-delay fuses.

To be classified as current limiting, the overcurrent device must reduce heavy (high magnitude) fault currents, to a value less than the fault current that would have flowed into the circuit had there been no fuse or breaker.

When used on motor circuits or other circuits having high current-inrush characteristics, the current-limiting nontime-delay fuses must be sized at a much higher rating than the actual load. That is, for a motor with a full-load current rating of 10 amperes, a 30- or 40-ampere current-limiting fuse may be required to start the motor. In this case,

the fuse is considered to be the motor branch-circuit short-circuit protection only.

CARTRIDGE FUSES

A cartridge fuse must be marked to show its

- ampere rating;
- voltage rating;
- interrupting rating when over 10 000 amperes;
- current-limiting type, if applicable;
- trade name or name of manufacturer.

The CSA requires that the fuse class be indicated on the fuse (e.g., Class R or Class C1).

Plug fuses and all fuses carrying the CSA Class listings are tested on AC circuits and are marked for AC use. When fuses are to be used on direct-current (DC) systems, the electrician should consult the fuse manufacturer since it may be necessary to reduce the fuse voltage rating and the interrupting rating to ensure safe operation.

The variables in the physical appearance of fuses include length, ferrule diameter, and blade size (length, width, and thickness), as well as other distinctive features. For these reasons, it is difficult to insert a fuse of a given ampere rating into a fuseholder rated for less amperage than the fuse. The differences in fuse construction also make it difficult to insert a fuse of a given voltage into a fuseholder with a higher voltage rating. Figure 17-9 indicates several examples of the method of ensuring that fuses and fuseholders are not mismatched.

Fuseholders for current-limiting fuses should be designed so that they cannot accept fuses that are noncurrent limiting, Figure 17-10.

A 250 volt fuse shall be constructed so that it cannot be inserted into a 600 volt fuseholder.

B A 60 ampere fuse is constructed so that it cannot be inserted into a 30 ampere fuseholder.

FIGURE 17-9 Examples of fuse construction ensuring that fuses and fuseholders are not mismatched.

A Fuse A is a Class H, non-current-limiting fuse. This fuse does not have the notch required to match the rejection pin in the fuse clip of the Class R fuseblock (C).

B Fuse B is a Class R fuse (either Class RK1 or RK5), which is a current-limiting fuse. This fuse has the required notch on one blade to match the rejection pin in the fuse clip of the Class R fuseblock (C).

FIGURE 17-10 Examples of *Rule 14–210* requirements.

In general, fuses may be used at system voltages that are less than the voltage rating of the fuse. For example, a 600-volt fuse combined with a 600-volt switch may be used on 575-volt, 277/480-volt, 120/208-volt, 240-volt, 50-volt, and 32-volt systems. Fuses that have silver links perform the best, followed by fuses with copper links, followed by fuses with zinc links.

P- and D-Type Fuses

P- and D-type fuses are standard fuses that have low melting point characteristics. P-type fuses are nontime-delay; D-type fuses are time-delay. They have been designed primarily to offer better protection for residential applications. This type of fuse provides overload and short-circuit protection and has low melting point characteristics. This means that the fuse opens in an ambient temperature of 200°C within 24 hours with no current flowing. These fuses are not current limiting.

Fuse Standards

Fuses that are made and sold in Canada are required to meet CSA standards. As a result of the North American Free Trade Agreement, fuse standards in Canada have been harmonized with those of the United States and Mexico.

Standard Code Fuses

In Canada, standard (code) fuses have included the current Class H and Class K. Most low-cost renewable and non-renewable fuses are Class H. If you have a fusible disconnect in your house, it will probably contain Class H fuses. Class H fuses have two voltage ratings (250 V and 600 V) and are available in current ratings from 1/10 A to 600 A. They have an interrupting rating of 10 000 A. If a fuse does not have its class marked on the label, it will be a Class H fuse. (See Figure 17-11.)

Class K Fuses

Class K fuses have the same physical dimensions as Class H fuses. They have higher interrupting ratings than Class H (K5 has a 50 000 A interrupting rating) and are available in the same voltage and current

FIGURE 17-11   Class H cartridge fuse. Illustration shows renewable-type fuse in which the blown link may be replaced.

fuses provide protection for a minimum of 200 000 amperes RMS symmetrical and are able to clear a high short-circuit fault current before it reaches its maximum available peak value. The fault current is reduced to zero in a specified time period. This gives these fuses current-limiting characteristics. Within the HRC type of fuse there are various classes and types of fuses, each with specific characteristics that make the fuse suitable for specific applications. The two large categories are HRCI (formerly called HRC-I or Form I) and HRCII (formerly called HRC-II or Form II).

HRCI fuses provide both overload and overcurrent (short-circuit) protection.

HRCII fuses *only* provide overcurrent (short-circuit) protection.

Another term often referred to on fuses is *time delay.* If a fuse is marked "time delay" it must carry 500% of its ampere rating for 10 seconds minimum. In view of the fact that some manufacturers have had difficulty in meeting the 10-second delay requirement, the CSA standard has been revised to allow for an 8-second delay on 250-volt 15- to 30-ampere rated fuses only. Some manufacturers

rating as Class H fuses. Class K is an older UL class fuse that has been supplanted by the newer Class R. Both Class H and Class K fuses are available in time-delay and nontime-delay versions, Figure 17-12.

HRC Fuses (High Rupturing Capacity)

These fuses were developed to protect against fault currents higher than 10 000 amperes. These

A
One-time, single-element non-current-limiting fuses.
1/8–6 amperes Class K5 50 000 amperes interrupting rating. 250–600 volt ratings. 70–600 amperes.
Class H 10 000 amperes interrupting rating.
250 and 600 volt ratings.

B
Limitron, single-element current-limiting, fast-acting fuses.
1–600 amperes.
Class RK1 200 000 amperes interrupting rating.
250 and 600 volts ratings.

C
Fusetron, dual-element time-delay current-limiting fuses. 1/10–600 amperes.
Class RK5 200 000 amperes interrupting rating.
250 and 600 volt ratings.

D
Low-peak, dual-element time-delay current-limiting fuses. 1/10–600 amperes.
Class RK1 300 000 amperes interrupting rating.
250 and 600 volt ratings.

FIGURE 17-12   Class H, K5, RK1, and RK5 fuses.

mark their fuses with phrases such as "dual element," "delay fuse," "time lag," or "super lag," but these manufacturers' terms do not mean that the fuses are a time-delay type.

HRCI-R Fuses (Class R)

A higher quality nonrenewable one-time fuse is also available, called an HRCI-R fuse. This class of fuses is divided into Class RK1 or Class RK5 in the United States. CSA only recognizes the higher Class RK5 level of protection, therefore caution should be exercised when replacing HRCI-R fuses. They are available in two types of overload performance characteristics; time-delay and non-time-delay. They have a 200 000-ampere interrupting rating. The peak let-through current (I_p) and the total clearing energy (I^2t) values are specified for this class of fuse. It is easy to identify this better grade of fuse because the HRCI-R mark and its interrupting rating are marked on the label.

The ferrule-type HRCI-R fuse has a rating range of $\frac{1}{10}$ to 60 amperes and can be distinguished by the annular ring on one end of the case, Figure 17-13(A). The knife blade-type HRCI-R fuse has a rating range of 70 to 600 amperes and has a slot in the blade on one end, Figure 17-13(B). When a fuseholder is designed to accept an HRCI-R fuse, it will be impossible to install a standard Class H or Code fuse (these fuses do not have the annular ring or slot of the HRCI-R fuse). However, the HRCI-R fuse can be installed in older-style fuse clips on existing installations. As a result, the Class R fuse may be called a *one-way rejection fuse*.

Electrical equipment manufacturers will provide the necessary rejection-type fuseholders in their equipment, which is then tested with an HRCI-R fuse at short-circuit current values of 200 000 amperes.

A Class R fuse will have a 200 000 ampere interrupting rating only if installed in a similarly rated fuseholder. Such fuseholders have the rejection feature so that only HRCI-R fuses can be installed. If Class R fuses are installed in fuse-holders without the rejection feature, the assembly will have an interrupting rating of only 10 000 amperes.

HRCI-J (Class J). HRCI-J fuses are current-limiting and are so marked, Figure 17-14. They are approved by CSA with an interrupting rating of 200 000 RMS symmetrical amperes. HRCI-J fuses are physically smaller than standard Code fuses. Therefore, when a fuseholder is installed to accept the HRCI-J fuse, it will be impossible to install a Code fuse (Class H) in the fuseholder, *Rule 14–210*.

FIGURE 17-13 Class R cartridge fuses (may be RK1 or RK5).

Courtesy of Eaton

A Class J Limitron current-limiting, fast-acting fuses; 200 000-ampere interrupting rating.

B Class J low-peak, current-limiting time-delay fuses. 1–600 amperes, 300 000 amperes interrupting rating, 600-volt rating.

Courtesy of Eaton

FIGURE 17-14 Class J current-limiting fuses.

Both fast-acting, current-limiting HRCI-J fuses and time-delay, current-limiting HRCI-J fuses are available in ratings ranging from 1 to 600 amperes at 600 volts AC.

HRCII-C (Class C).   HRCII-C fuses are designed to meet British Standard dimensions. They are different from HRCI-J and HRCI-R fuses in that they are intended to be bolted directly to bus bars or installed in dead-front fuseholders. The current-limiting levels are the same as those of HRCI-R fuses, with limits not to exceed maximum values of I^2t and I_p. Form II overload characteristic is neither time-delay nor nontime-delay, but is specified to provide short-circuit protection for motor circuits that already have overload protection, *Rule 14–212(c)*.

These fuses have ratings from 2 to 600 amperes at 600 volts or less, with a 200 000 A.I.R. See Figure 17-14A.

HRCI-L (Class L).   HRCI-L fuses, Figure 17-15 parts A and B, are listed by the CSA in sizes ranging from 601 to 6 000 amperes at 600 volts AC. These fuses have specified maximum values of I^2t and

Offset blade 2–200 amperes

Centre blade 80–200 amperes

Centre blade 250–600 amperes

Short blade 250–400 amperes

FIGURE 17-14A   HRCII-C low-voltage fuses.

A

Limitron current-limiting, time-delay Class L fuse; 200 000-ampere interrupting rating. Links are generally made of copper. Will hold 500% of rated current for a minimum of 10 seconds. Good for use on high inrush circuits (such as motors, transformers), but is the least current-limiting of Class L fuses.

B

Low-peak, current-limiting, time-delay multi-element fuse. 601-6000 amperes Class L–300 000 amperes interrupting rating. 600-volt rating. Links made of silver. Will hold 500% of rated current for a minimum of 4 seconds. Good for use on high inrush circuits (motors, transformers, and other inductive loads). The best choice for overall protection. Generally sized at 150–225% for motors and other high inrush loads.

Courtesy of Eaton

FIGURE 17-15   Class L fuses. All Class L fuses are rated 600 volts. Listed is 601–6000 ampere ratings. The smallest switch for Class L fuses is 800 amperes. Class L fuses that have fuse elements rated at 600 amperes and less are available. These special ampere-rated fuses are physically the same size as the 800-ampere size.

I_p. They are current-limiting fuses and have an interrupting rating of 200 000 RMS symmetrical amperes (A.I.R.). These bolt-type fuses are used in bolted pressure contact switches and are not interchangeable with any other fuse type. HRCI-L fuses are available in both a fast-acting, current-limiting type and a time-delay, current-limiting type. Both types of HRCI-L fuses meet CSA requirements.

HRCI-T (Class T). HRCI-T fuses, Figure 17-16, are current-limiting fuses and are so marked. These fuses are the same as the UL listed HRCI-T fuse used in the United States. They have an interrupting capacity of 200 000 RMS symmetrical amperes. HRCI-T fuses are physically smaller than HRCI-J fuses. The configuration of this type of fuse limits its use to fuseholders and switches that will reject all other types of fuses.

HRCI-T fuses rated 600 volts have electrical characteristics similar to those of HRCI-J fuses and are tested in a similar manner by CSA and UL. HRCI-T fuses rated at 300 volts have electrical characteristics superior to those of Class J fuses in that their peak let-through currents, as well as the I^2t values, are much lower than those of Class J fuses. The *CSA Standard C22.2 No. 106–M1985* is similar to the UL standard for HRCI-T (Class T) fuses.

CSA currently lists the 600-volt HRCI-T fuses in sizes from 0 to 800 amperes. CSA lists the 300-volt HRCI-T fuses in sizes from 1 to 800 amperes. The 300-volt HRCI-T fuses may be installed on single-phase line to neutral circuits derived from a three-phase, four-wire solidly grounded system when the line-to-neutral voltage does not exceed 300 volts. On a 277/480-volt system, the line-to-line voltage is 480 volts, and the line-to-neutral voltage is 277 volts. On a 347/600-volt system, the 300-volt HRCI-T fuse would not be permitted; a 600-volt HRCI-T-rated fuse would be required.

HRCI-T fuses may be used to protect mains, feeders, branch circuits, and circuit breakers when the equipment is designed to accept this type of fuse.

Supplemental Fuses. These fuses are referred to as midget fuses and are cartridge fuses with small physical dimensions of 10 mm × 38 mm. They are used on circuits up to 600 volts. These fuses are similar to UL-listed Class G fuses, Figure 17-17, and are available in sizes ranging from 0 to 60 amperes. They have an interrupting capacity of 100 000 RMS symmetrical amperes. All Class G fuses have a diameter of 10 mm.

Courtesy of Eaton

FIGURE 17-16 Class T current-limiting, fast-acting fuse; 200 000-ampere interrupting rating. Links are made of silver. Has very little time delay. Good for the protection of circuit breakers and on circuits that *do not* have high inrush loads. Size at 300% for motors and other high inrush loads.

Courtesy of Eaton

Ampere rating	Dimensions
0–15	13/32 in. × 1 5/16 in.
16–20	13/32 in. × 1 32/32 in.
21–30	13/32 in. × 1 5/8 in.
31–60	13/32 in. × 2 1/4 in.

FIGURE 17-17 Class G fuses.

These fuses are current-limiting. They may be used for the protection of ballasts, electric heaters, small motors, and motor control transformers. They are also available in a fast-acting version used for semiconductor protection.

HRCI-MISC (Class CC). HRCI-MISC fuses are primarily used for the protection of motor control circuits, ballasts, and small transformers in industrial applications. HRCI-MISC fuses are also referred to as Class CC fuses and are rated at 600 volts or less and have a 200 000-ampere interrupting rating in sizes from $\frac{1}{10}$ ampere through 30 amperes. These fuses measure 10 mm × 38 mm and can be recognized by a "button" on one end of the fuse, Figure 17-18. This button is unique to HRCI-MISC Class CC fuses. When a fuseblock or fuseholder that has the matching Class CC rejection feature installed, it is impossible to insert any other fuse type. Only a Class CC fuse will fit into these special fuseblocks and fuseholders. A Class CC fuse can be installed in a standard fuse-holder but would have to be derated to an interrupting capacity of 10 000 amperes.

FIGURE 17-18 Class CC fuse with rejection feature.

Plug Fuses (Rules 14–202 to 14–208)

Rules 14–202 through *14–208* cover the requirements for plug fuses. The electrician will be most concerned with the following requirements for plug fuses, fuseholders, and adapters:

- They shall not be used in circuits exceeding 125 volts between conductors, except on systems having a grounded neutral with no conductor having more than 150 volts to ground. This situation is found in the 120/208-volt system in the commercial building covered in this text, or in the case of a 120/240-volt, single-phase system.

- They shall have ampere ratings of 0 to 30 amperes.

- The screw shell must be connected to the load side of the circuit.

- All plug fuses must be installed such that they cannot be interchanged with a fuse of a higher ampacity.

- Type S plug fuses are classified 0 to 15 amperes, 16 to 20 amperes, and 21 to 30 amperes.

Prior to the installation of Type S fuses, the electrician must determine the ampere rating of the various circuits. An adapter of the proper size is then inserted into the Edison-base fuseholder; only the proper size of Type S fuse can be inserted into the fuseholder. The adapter makes the fuseholder nontamperable and noninterchangeable. For example, after a 15-ampere adapter is inserted into a fuseholder for a 15-ampere circuit, it is impossible to insert a Type S or an Edison-base fuse having a larger ampere rating into this fuse-holder without removing the 15-ampere adapter. The adapters are designed so that they are extremely difficult to remove.

Type S fuses and suitable adapters, Figure 17-19, are available in a large selection of ampere ratings ranging from 3/10 ampere to 30 amperes. When a Type S fuse has dual elements and a time-delay feature, it does not blow unnecessarily under momentary overloads, such as the current surge caused by the startup of an electric motor. On heavy overloads or short circuits, this type of fuse opens very rapidly.

Courtesy of Eaton

A
Type S fuse

B
Adapter

Courtesy of Eaton

FIGURE 17-19 Type S fuses and adapter.

Standard Edison base plug fuses may also be used along with a fuse rejector. A fuse rejector is a plastic disc with a hole in the middle which corresponds to the diameter of the nipple on the bottom of the plug fuse. This disc is screwed into the bottom of the fuseholder. Because the nipple on the bottom of the plug fuse increases in diameter with an increase in ampacity, it is impossible to put a higher ampacity fuse into a fuseholder with a lower ampacity fuse rejector.

Tables 17-1, 17-2, and 17-3 provide ratings and application information for the fuses supplied by various manufacturers. The tables indicate the class, voltage rating, current range, interrupting ratings, and typical applications for the various fuses. Tables of this type may be used to select the proper fuse to meet the needs of a given situation.

Special Interrupting Ratings

Some HRCI-R, HRCI-J, and HRCI-L fuses have passed special tests that qualify them to be marked "Interrupting rating 300 000 amperes." This information is found on the label of the fuse. In addition, these fuses will have the letter designation "SP." However, CSA currently only recognizes fuseholders rated up to 200 000 amperes interrupting capacity.

▶ TESTING FUSES

When Power Is Turned On

On live circuits, extreme caution must be exercised when checking fuses. Many different voltage readings can be taken, such as line-to-line, line-to-ground, line-to-neutral, and so on.

Using a voltmeter, the first step is to be sure to set the scale to its highest voltage setting, then change to a lower scale after you are sure the voltage is within the range of the voltmeter. For example, when testing what you believe to be a 120-volt circuit, it is wise to first use the 600-volt scale, then try the 300-volt scale, and then use the 150-volt scale—**just to be sure**!

Taking a voltage reading across the bottom (load side) of fuses—either fuse-to-fuse, fuse-to-neutral, or fuse-to-ground—can show full voltage even though a fuse might have opened, because there can be "feedback" through the load. You could come to a wrong conclusion. Taking a voltage reading from the line side of a fuse to the load side of a fuse will show open-circuit voltage if the fuse has "blown" and the load is still connected. This also can result in a wrong conclusion.

Reading from the line side to the load side of a good fuse should show zero voltage or else an extremely small voltage.

Always carefully read the instructions furnished with electrical test equipment such as voltmeters, ohmmeters, and so on.

When Power Is Turned Off

This is the safest way to test fuses. Remove the fuse from the switch, then take a resistance reading across the fuse using an ohmmeter. A good fuse will show zero to very minimal resistance. An open (blown) fuse will generally show a very high resistance reading.

Cable Protectors

Cable protectors, sometimes referred to as cable limiters, are used quite often in commercial and industrial installations where parallel cables are used on service entrances and feeders.

Cable protectors differ in purpose from fuses, which are used for overload protection. Cable protectors are short-circuit devices that can *isolate* a faulted cable, rather than having the fault open the entire phase. They are selected on the basis of conductor size; for example, a 500-kcmil cable protector would be used on a 500-kcmil conductor.

Cable protectors are available for cable-to cable or cable-to-bus installation, for either aluminum or copper conductors. The type of cable protector

TABLE 17-1
Cross-Reference Fuse Guide.

Fuse (CSA Stnd. No.)		Volt	Ferraz Shawmut	GEC	Bussmann		Littelfuse	Fusetek
FERRAZ SHAWMUT BY MANUFACTURER'S TYPE REFERENCE OR SERIES NUMBER					CROSS REFERENCE GUIDE AMPERE RATINGS MUST BE ADDED FOR ORDERING PURPOSES			
					Bussmann	Edison/ Brush/Dorman		
HRC (Current Limiting)								
CLASS R(RK1) [248.12]	Fast Acting	250	**A2K-R**	C-HG/HB	KTN-R	NCLR	KLNR	RHN
		600	**A6K-R**	C-HR/HA	KTS-R	SCLR	KLSR	RHS
	Time Delay	250	**A2D-R**	LON-RK	LPN-RK-SP	LENRK	LLNRK	—
		600	**A6D-R**	LOS-RK	LPS-RK-SP	LESRK	LLSRK	—
CLASS R (RK5) [248.12]	Time Delay	250	**TRN-R/TR-R**	CRNR	FRN-R	ECNR/DLNR	FLN-R	RDN
		600	**TRS-R/ TRS-R**	CRSR	FRS-R	ECSR/DLSR	FLS-R/ IDSR	RDS
CLASS L [248.10]	Fast Acting	600	**A4BY**	CLL/CLU	KTU	LCU	KLLU	LFA
	Time Delay	600	**A4BQ/A4BT**	CLF	KRP-C/KLU	LCL	KLPC	—
CLASS J [248.8]	Fast Acting	600	**A4J**	C-J/JA	JKS	JFL/JCL/CJ	JLS	JFC
	Time Delay	600	**AJT**	—	LPJ	JDL	JTD	
CLASS T [248.15]	Fast Acting	300	**A3T**	—	JJN	TJN	JLLN	—
		600	**A6T**	—	JJS	TJS	JLLS	—
CLASS G [248.51]	Time Delay	480	**AG**	—	SC	SEC	SLC	—
CLASS CC [248.4]	Time Delay	600	**ATDR**	—	LP-CC	EDCC/HCTR	CCMR	—
	Time Delay	600	**ATQR**	—	FNQ-R	KLDR	KLDR	—
	Fast Acting	600	**ATMR**	CTK-R	KTK-R	HCLR	KLKR	FLKR
CLASS C [248.2]	Offset		**FES**	CIA/932	CGL	H07C	H07C/AA0	F11	2CO
	Blade	600	**FES**	CIS/933	CGL	K07C	K07C/BA0	F11	2CO
			FES	CCP/944	CGL	L14C	L14C/CEO	F11	2CO
	Centre		**FESC**	CF/965	CGL	M09C	M09C/DD	F11C	2CC
	Blade	600	**FESC**	CM/976	CGL	P11C	P11C/EF	F11C	2CC
			FESC	CLM/977	CGL	R11C	R11C	F11C	2CC
	Centre		**FESC**	CC/964	L09C	L09C	L09C	F11M	2CM
	Offset	600	**FES**	CFP/945	M14C	M14C	M14C/DE0/CD	F11M	2CM
	Centre		**FESF†**	CMF/966t	P09Cf	P09Ct	N09C/ED/FFt	F11Mt	2CMt
	Offset	600	**NIT***	NIT	NITD	CIF21	F21/NITD	—	N2B
			NSG/NS*	NS	NK	CIF06	F06/NSD/NSC	—	N2C
CLASS CA [248.3]	Fast Acting	600	**C-N**	C-N		CIF21		—	—
CLASS CB [248.3]	Fast Acting	600	**CNS**	CNS		CIF06B/NK		—	NIC
			CES	CES		EK		—	—

*Voltage ratings should be verified.   †SUPPLEMENTAL [248.14]

(*continues*)

TABLE 17-1 (*Continued*)

			FERRAZ SHAWMUT BY MANUFACTURER'S TYPE REFERENCE OR SERIES NUMBER		CROSS REFERENCE GUIDE AMPERE RATINGS MUST BE ADDED FOR ORDERING PURPOSES				

Fuse (CSA Stnd. No.)		Volt	Ferraz Shawmut	GEC	Bussmann			Littelfuse	Fusetek
					Bussmann	Edison/ Brush/Dorman			
Supplemental									
SUPPLEMEN-TAL (13/32″ × 11/2″) [248.14]	Fast Acting	600	**ATM**	CTK	KTK	MCL		KLK	FLK
		600	**SBS**	—	BBS	EBS		BLS	—
		250	**OTM**	—	BAF/BAN	MOL		OTM	FLF
	Time Delay	500	**ATQ**	—	FNQ	MEQ		FLQ	—
		250	**TRM**	—	FNM	MEN		FLM	FRM
		125/250	**G FN**	—	FNA	MID		FLA	—
	Offset Blade	600	**ESK**	C-K	CIH07 CIK07 CIL14			—	—
Standard (Code)									
CLASS K [248.9]	Fast Acting	250	**OTN**[*][†]**/OT**	50KOTN	NON[**]	KON[**]		NLN	OFN
		600	**OTS**[†]	50KOTS	NOS[**]	KOS[***]		NLS	OFS
CLASS H RENEWABLE [248.7]	Renewable	250	**RF**	—	REN	ERN/ELN		RLN	—
		600	**RFS**	—	RES	ERS/ELS		RLS	—
	Renewable Fuse Links	250	**RL**	—	LKN	ELNE		LKN	—
		600	**RLS**	—	LKS	—		LKS	—
CLASS H NON-RENEWABLE [248.6]	One Time	250	**NRN**	—	NON[**]	KON/PONC/ PON KOS		NLKP	OFN
		600	**NRS**	—	NOS[**]			—	OFS
	Time Delay	250	**CRN**	—	—	CDNC/CDN		FLN	ODN
		600	**CRS**	—	—	CDSC/CDS		FLS	ODS

[†](1-60) AMPS 50 000 AIR   [*]OTN 15-60 & NRN 15-60 ARE TYPE CSA-"P"   [**](1-60) AMPS 50 000 AIR, (70-600) AMPS 10 000 AIR   [***](15-60) AMPS 50 000 AIR

North American Semiconductor (Rectifier)

Ferraz Shawmut	Volt	Bussmann	Littelfuse	Brush Reliance	International Rectifier
A15QS	150	KAH/KAA	L15S	RFA	SF13X
A30QS	300	FWX-A/KAX/KAB	L25S(70-800)	REN/RFN	SF25X
A50P/A050FA	500	FWH(35-60)B/KBH	L50S	RFV(40-1200)	SF50P/XL50P, F
A60X/A060FA	600	KBC	L60S	RFS*	SF60X
A70P/A070FA	700	FWP/KBP/KAC*	L60N	—	SF70P/XL70P, F
A60X-4K	600	KAC*	—	—	SF60C*
A100P/-	1000	—	—	RFK	SF100P

*Mounting dimensions should be verified.

Note: Fuse characteristics can vary between manufacturers and should be checked for proper co-ordination as required.

Courtesy of Ferraz Shawmut

TABLE 17-2

WHICH FUSE? —A "Quick Guide" for Low-Voltage Distribution Applications 600 A and Less.

Fuse Clips or Contacts	Fuse Type		Feeders	Motors	Transformers	Breakers & Panelboards	Electric Heating — Industrial	Electric Heating — Residential	Capacitors	Welders
Class J	HRCI-J 600 V or less	Time Delay	AJT	AJT	AJT	AJT	AJT		AJT	
		Nontime Delay	A4J	*			A4J		A4J	
Class R	HRCI-R (UL Class RK1) 250 & 600V	Time Delay	A2D-R A6D-R	A2D-R A6D-R	A2D-R A6D-R		A2D-R A6D-R			
		Nontime Delay	A2KR A6KR	*			A2KR A6KR			
	HRCI-R (UL Class RK5) 250 & 600 V	Time Delay	TRR TRSR	TRR TRSR	TRR TRSR		TRR TRSR			TRR TRSR
Class C	HRCII-C 600V or less	Form II		FES FESC FESF					FES FESC FESF	
Standard (Code) *Note: Class RK1 and RK5 fuses may also be used in Standard Clips.*	10 kA Time Delay Type D 250 & 600 V		CRN CRS	CRN CRS	CRN CRS		CRN CRS	CRN —		CRN CRS
	50 kA Super One Time 250 & 600 V		OTN OTS†	*			OTN	OTN		
	10 kA One Time 250 & 600 V		NRN NRS†				NRN	NRN		
	10 kA Renewable 250 & 600 V		RFN RFS†							

Type P Ratings only 250 V, 15–60 A

Courtesy of Ferraz Shawmut

* Time Delay usually preferred for easier application and lower ratings.

† For Non-Cycling loads only.

Voltage Ratings—where two catalogue numbers are shown above, the upper is 250 Volts A.C. and the lower is 600 Volts A.C.

Note: When upgrading a Standard (Code) fuse installation where One Time or Renewable fuses have been rated at 300% of the motor FLC, it is sometimes more practical to substitute Nontime Delay HRCI-R fuses, rating for rating, rather than downsizing the clips to accommodate Time Delay fuses at 175% maximum of the motor FLC.

TABLE 17-3

Maximum Permitted Ratings for Overcurrent Protection (*CEC*, 2009).

Application Low-Voltage ⇒ Distribution	Feeders	Motors Full Voltage Start		Transformers and Secondary Protection				Electric Heating	Capacitors	Welders	
		Time Delay	Non T-Delay	Dry Type	Liquid Filled—Primary Rated at					Trans Arc	Resistance
					9A and over	2A up to 9A	less than 2A				
HRCI	100%	175%–225%	300%–400%	125%	*150%	*167%	*300%	125%	250%	200%	300%
Standard Code	100%	175%–225%	†300%–400%	125%	*150%	*167%	*300%	125% (P&D types only)	—	200% (D type only)	300% (D type only)
HRCII-C *Rule 14-212(c)*	—	—	255%–340%	—	†127%	†142%	†255%	†106%	212%	†170%	†255%
Fuse Rating as a percent of	Cable Amps	Motor Full Load Amps		Rated Primary Amps				Cable Amps	Capacitor Amps	Rated Primary Amps	
CEC, Rule ⇒	14-104	28-200		26-256	26-254			62-114	26-210	42-008	42-016

Total connected load must not exceed 80% of the fuse rating.

*Usually requires Time Delay fuse to prevent nuisance blowing on transformer switching current. Small transformers with very high inrush currents may need HRCI-R or Standard Code Time Delay fuses.

†Not recommended for these applications.

Courtesy of Ferraz Shawmut

illustrated in Figure 17-20 is for use with a 500-kcmil copper conductor, and is rated at 600 volts with an interrupting rating of 200 000 amperes.

Figure 17-21 is an example of how cable protectors may be installed on the service of the commercial building. A cable protector is installed at each end of each 500-kcmil conductor. Three conductors per phase terminate in the main switchboard where the service is then split into two 800-ampere bolted pressure switches.

Figure 17-22 explains how cable protectors are used where more than one customer is connected to the same transformer.

Courtesy of Eaton

FIGURE 17-20   Cable protector for 500-kcmil copper conductor.

© Cengage Learning®

FIGURE 17-21   Use of cable protectors.

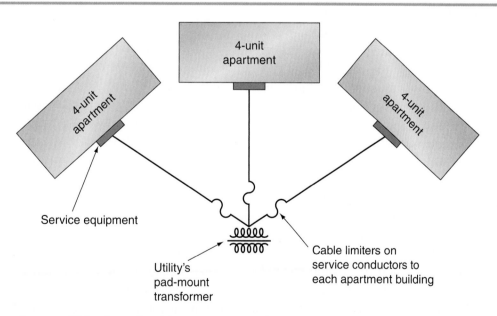

© Cengage Learning®

FIGURE 17-22   Some utilities install cable protectors where more than one customer is connected to one transformer. Thus, if one customer has problems in the main service equipment, the cable limiter on that service will open, isolating the problem from other customers on the same transformer. This is common for multiunit apartments, condominiums, and shopping centres.

Cable protectors are installed in the "hot" phase conductors only. They are not installed in the grounded neutral conductor.

TIME–CURRENT CHARACTERISTIC CURVES AND PEAK LET-THROUGH CHARTS

The electrician must have a basic amount of information concerning fuses and their application to be able to make the correct choices and decisions for everyday situations that arise on an installation.

The electrician must be able to use the following types of fuse data:

- time–current characteristic curves, including total clearing and minimum melting curves;

- peak let-through charts.

Fuse manufacturers furnish this information for each of the fuse types they produce.

The Use of Time–Current Characteristic Curves

The use of the time–current characteristic curves shown in Figures 17-23 and 17-24 can be demonstrated by considering a typical problem. Assume that an electrician must select a fuse to protect a motor that is governed by *Section 28*. This section indicates that the running overcurrent protection shall be based on not over 125% of the full-load running current of the motor. The motor in this example is assumed to have a full-load current of 24 amperes. Therefore, the size of the required protective fuse is determined as follows:

$$24 \times 1.25 = 30 \text{ amperes}$$

Now the question must be asked: Is this 30-ampere fuse capable of holding the inrush current of the motor (which is approximately four to five times the running current) for a sufficient length of time for the motor to reach its normal speed? For this problem, the inrush current of the motor is assumed to be 100 amperes.

Refer to Figure 17-23. At the 100-ampere line, draw a horizontal line until it intersects with the 30-ampere fuse line. Then draw a vertical line down to the base line of the chart. At this point, it can be

seen that the 30-ampere fuse will hold 100 amperes for approximately 40 seconds. In this amount of time, the motor can be started. In addition, the fuse will provide running overload protection as required.

If the time–current curve for thermal overload relays in a motor controller is checked, it will show that the overload element will open the same current in much less than 40 seconds. Therefore, in the event of an overload, the thermal overload elements will operate before the fuse opens. If the overload elements do not open for any reason, or if the contacts of the controller weld together, then the properly sized dual-element fuse will open, reducing the possibility of motor burnout. The above method is a simple way to obtain backup or double motor protection.

Referring again to Figure 17-23, assume that a short circuit of approximately 500 amperes occurs. Find the 500-ampere line at the left of the chart and then, as before, draw a horizontal line until it intersects with the 30-ampere fuse line. From this intersection point, drop a vertical line to the base line of the chart. This value indicates that the fuse will open the fault in slightly over $\frac{2}{100}$ (0.02) seconds. On a 60-hertz system (60 cycles per second), one cycle equals 0.016 seconds. Therefore, a 30-ampere fuse will clear a 500-ampere fault in just over one cycle.

The Use of Peak Let-Through Charts

The withstand ratings of electrical equipment (such as bus duct, switchboard bracing, controllers, and conductors) and the interrupting ratings of circuit breakers are given in the published standards of the Canadian Standards Association, Underwriters Laboratories, NEMA, CEMA, and the Insulated Cable Engineering Association (ICEA). The withstand ratings may be based either on the peak current (I_p) or on the root mean square (RMS) current.

As an example of the use of the let-through chart in Figure 17-25, assume that it is necessary to protect moulded-case circuit breakers. These breakers have an interrupting rating of 10 000 RMS symmetrical amperes. The available short-circuit current at the panel is 40 000 RMS symmetrical amperes. The fuse protecting the circuit-breaker panel is rated at 100 amperes. This 100-ampere fuse may be located either in the panel or at some distance away from the panel in a separate

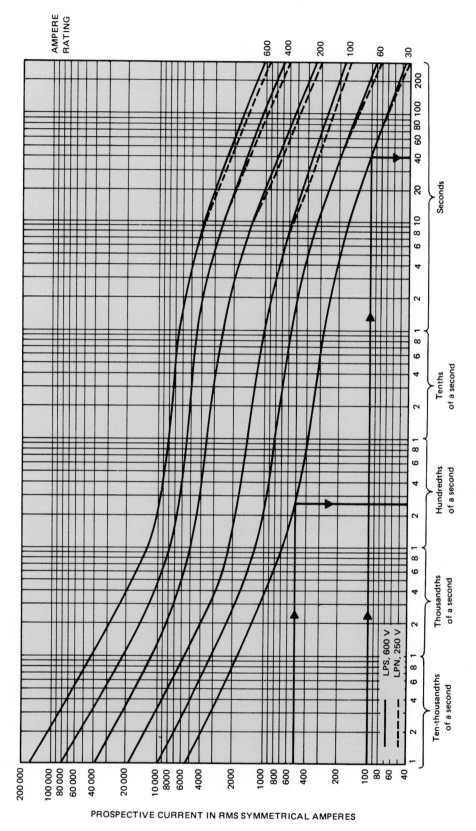

FIGURE 17-23   Average melting time-current curves for one family of current-limiting, dual-element, time-delay fuses rated 15 through 600 amperes.

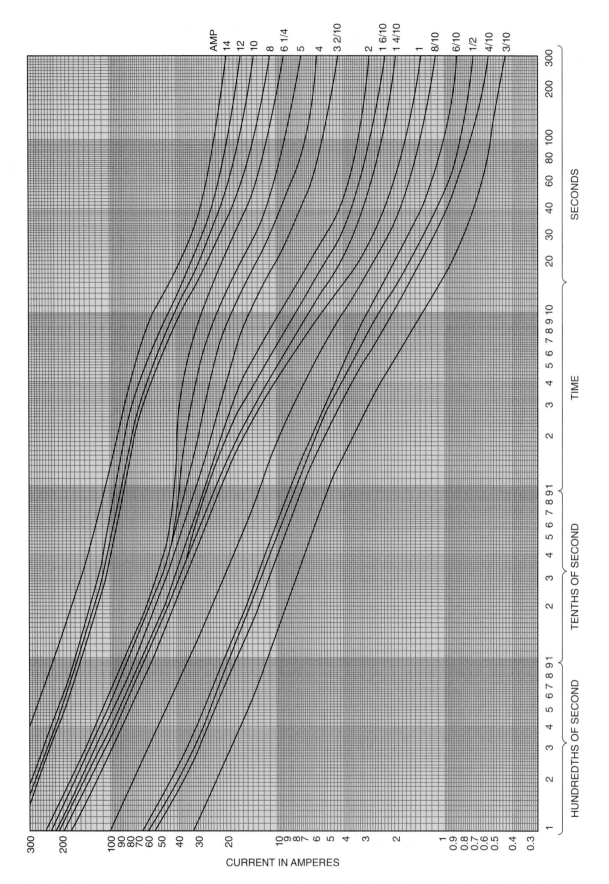

FIGURE 17-24 Average melting time-current curves for one family of current-limiting, dual-element, time-delay fuses rated 0.1 through 12 amperes.

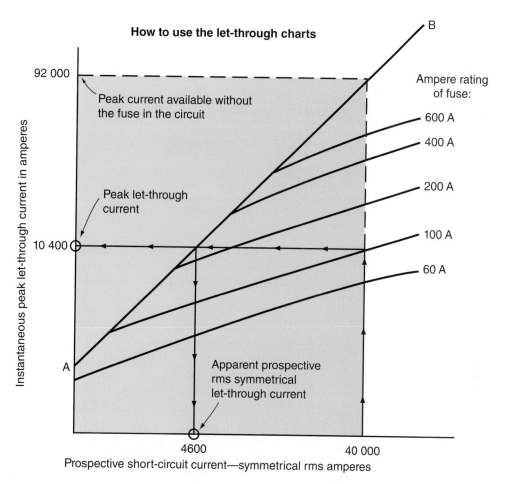

How to use the let-through charts

92 000

← Peak current available without
the fuse in the circuit

Peak let-through
current

10 400

A

Apparent prospective
rms symmetrical
let-through current

Instantaneous peak let-through current in amperes

Ampere rating
of fuse:

600 A

400 A

200 A

100 A

60 A

B

4600   40 000

Prospective short-circuit current—symmetrical rms amperes

Courtesy of Eaton

FIGURE 17-25   Using the let-through charts to determine peak let-through current and apparent prospective RMS symmetrical let-through current.

disconnect switch placed at the main service or distribution panel, Figure 17-26.

Now refer to Figure 17-25. Find the 40 000-ampere point on the base line of the chart and then draw a vertical line upward until it intersects the 100-ampere fuse curve. Move horizontally to the left until Line A-B is reached. Then move vertically downward to the base line. The value of this point on the base line is 4600 amperes and is the apparent RMS amperes. This means that the current-limiting effect of the fuse, when subjected to a 40 000 RMS symmetrical ampere fault, permits an apparent let-through current of only 4600 RMS symmetrical amperes. Therefore, the 100-ampere fuse selected readily protects the circuit breaker (with a 10 000-ampere interrupting rating) against the 40 000 RMS symmetrical ampere fault. Actual tests conducted by manufacturers have shown that the same fuse can be used to protect a circuit

breaker with a 10 000-ampere interrupting rating when 100 000 amperes are available.

In Figure 17-25, note that at a current of 40 000 RMS symmetrical amperes, the peak current, I_p, available at the first half-cycle is 92 000 amperes if there is no fuse in the circuit to limit the let-through current. When the 100-ampere fuse is installed, the instantaneous peak let-through current, I_p, is limited to just 10 400 amperes.

The magnetic stresses that occur in the example given can be compared. When the current-limiting fuse is installed, the magnetic stresses are only 1/80 of the stresses present when a noncurrent-limiting overcurrent device is used to protect the circuit.

Figure 17-27 shows a peak let-through chart for a family of dual-element fuses. Fuse manufacturers provide peak let-through charts for the various sizes and types of fuses they produce.

Interrupting rating	Test circuit power factor	Peak current first half-cycle
10 000 and less	0.45 – 0.50	1.7 × symmetrical rating
10 001 to 20 000	0.25 – 0.30	2.0 × symmetrical rating
above 20 000	0.15 – 0.20	2.2 × symmetrical rating

Available short-circuit current
40 000 amperes rms symmetrical

100 ampere fuse, current-limiting 200 000 AIR

120/240 volt moulded case circuit breakers;
10 000 AIR

© Cengage Learning®

FIGURE 17-26   Example of a fuse used to protect circuit breakers. This installation meets the requirements outlined in *Rule 14–014.*

FIGURE 17-27   Current-limiting effect of low-peak, dual-element, time-delay Class RK1 fuses.

Adapted with permission from Eaton

CSA Standard C22.2 No. 5.1-M91 describes the testing of moulded-case circuit breakers. One of the tests performed on these circuit breakers is a short-circuit test. The test circuit is calibrated for a given amount of fault current at a specific power factor. The lower the power factor, the higher will be the first half-cycle peak current.

At present, there are three basic levels of short-circuit tests for moulded-case circuit breakers.

Knowing how to use peak let-through charts to determine both instantaneous peak let-through current and apparent RMS let-through current of a current-limiting fuse enables the user to select current-limiting fuses that will protect the downstream circuit breakers. Manufacturers of circuit breakers also provide technical data that show fuse and breaker combinations for use at various available fault-current levels.

Some inspection authorities will no longer accept the use of the peak let-through charts. Only the manufacturers' technical data for series-tested combinations are accepted in some jurisdictions.

Peak let-through charts are also available for current-limiting circuit breakers. The current-limiting effect of a circuit breaker is different from

that of a current-limiting fuse. The current-limiting effect of a fuse is much greater than that of a breaker having the same ampere rating. Therefore, refer to fuse peak let-through charts when applying current-limiting fuses. Use breaker peak let-through charts when applying current-limiting breakers.

Figure 17-28 shows how a manufacturer might indicate the maximum size and type of fuse that will protect circuit breakers against fault currents that exceed the breakers' interrupting rating. Circuit breakers are tested according to the requirements of

INTEGRATED EQUIPMENT RATINGS   E·T·N Cutler-Hammer

POW-R-LINE 2A PANELBOARD: INTEGRATED EQUIPMENT RATINGS

VOLTAGE	MAX. FAULT LEVEL AVAIL. AT MAIN OR UPSTREAM DEVICE (kA SYM.)	BRANCH BREAKERS	MAIN OR UPSTREAM BREAKER TYPE			
			100A	225A	250A	400A
208Y/120	100	GHB, GBH	EDH, HFD	EDH, HFD	HJD	HKD, CHKD
and 240	200		FDC	FDC	JDC	KDC, LCL
	25		FD, HFD	FD, HFD	JD, HJD	KD, HKD, CKD, CHKD
480Y/277	65	GHB, GBH	HFD	HFD	HJD (1)	HKD/CHKD
(2)	100		FDC	FDC	JDC, LCL	KDC (1), LCL
	200		(FD/FDB)+ LFD	(FD/FDB)+ LFD (3)	----	----
	14		FDB, FD, HFD	FDB, FD, HFD	----	KD, HKD, CKD, CHKD, KDC
600Y/347	18	GBH	FD, HFD	FD, HFD	----	KD, HKD, CKD, CHKD, KDC
(2)	25		HFD	HFD	----	----
	35		FDC	FDC	----	----
	50		(FD/FDB)+LFD	(FD/FDB)+LFD (3)	----	----

Notes:
(1) Branch breakers: 1 pole 15-100A; 2,3 pole 15-50A.
(2) GHB/GBH breakers are suitable for use only on 3 phase 4 wire grounded wye systems.
(3) 150A maximum

POW-R-LINE 2A PANELBOARD: INTEGRATED EQUIPMENT RATINGS

VOLTAGE	MAX. FAULT LEVEL AVAIL. AT MAIN OR UPSTREAM DEVICE (KA SYM.)	BRANCH BREAKERS	MAIN OR UPSTREAM FUSE TYPE		
			100A	200A	400A
208Y/120	100	GHB, GBH	J,T,R	J,T,R	J,T
and 240	200		J,T,R	J,T	J,T
480Y/277	100	GHB, GBH	J,T,R	J,T,R	J,T
(1)	200		J,T,R	J,T	J,T
600Y/347	100	GBH	J, T	J, T	----
(1)					

Note:
1. GHB/GBH breakers are suitable for use only on 3 phase 4 wire grounded wye systems.

FIGURE 17-28 Series-tested equipment ratings.

Courtesy of Eaton

CSA Standard C22.2 No. 5.1-M91. Panelboards are tested according to the requirements of *CSA Standard C22.2 No. 29-M89*. Motor controllers are tested to *CSA Standard C22.2 No. 14-M91*. Switchboards are tested to *CSA Standard C22.2 No. 31-M89*.

Therefore, to protect circuit breakers for fault levels that exceed their interrupting rating, refer to the manufacturers' data, or use acceptable engineering practices for the selection of current-limiting overcurrent devices as discussed in Units 18 and 19 of this text.

CIRCUIT BREAKERS

The *CEC* defines a circuit breaker as a device that is designed to open and close a circuit by nonautomatic means and to open the circuit automatically on a predetermined overcurrent without being damaged itself when properly applied within its rating, *Section 0, Definitions*.

Moulded-case circuit breakers are the most common type in use today, Figure 17-29. The tripping mechanism of this type of breaker is enclosed in a moulded plastic case. The *thermal-magnetic* type of circuit breaker is covered in this unit. Another type is the power circuit breaker, which is larger, of heavier construction, and more costly. Power circuit breakers are used in large industrial applications. However, these breakers perform the same electrical functions as the smaller moulded-case breakers.

Circuit breakers are also available with solid-state adjustable trip settings. When compared to the standard moulded-case circuit breaker, these breakers are very costly.

Moulded-case circuit breakers are approved by CSA testing laboratories and are covered by NEMA and UL standards. *Rules 14–300* through *14–308* state the basic requirements for circuit breakers and are summarized as follows:

- Breakers shall be trip-free so that the internal mechanism trips the breaker to the *Off* position even if the handle is held in the *On* position.

- The breaker shall clearly indicate if it is in the *On* or *Off* position.

- The breaker shall be nontamperable; that is, it cannot be readjusted (to change its trip point or time required for operation) without dismantling the breaker or breaking the seal.

- The rating shall be durably marked on the breaker. For the smaller breakers with ratings of 100 amperes or less and 600 volts or less, this marking must be moulded, stamped, or etched on the handle or other portion of the breaker that will be visible after the cover of the panel is installed.

- Every breaker having an interrupting rating other than 5000 amperes shall have its interrupting rating shown on the breaker.

- *Rule 30–710(3)* and the CSA require that when a circuit breaker is to be used as a toggle switch for fluorescent lighting, the breaker must be CSA-approved for switching duty. Circuit breakers are used as toggle switches when it is desired to control 120-or 347-volt lighting circuits (turn them on and off) from the panel instead of installing separate toggle switches. Breakers approved for use as toggle switches will bear the letters "SWD" on the label. Breakers not marked in this manner must not be used as switches.

- A circuit breaker should not be loaded to more than 80% of its current rating for loads over 225 amperes that could be on for more than three hours in any six-hour period, unless the breaker is marked otherwise, such as those breakers approved by the CSA for 100% loading (100% rated), *Rule 8–104(3,4)*.

Robert Asento/Shutterstock.com

A   **B**

FIGURE 17-29   Moulded-case circuit breakers.

- If the voltage rating on a circuit breaker is marked with a single voltage (example: 240 volts), the breaker may be used in grounded or ungrounded systems where the voltage between any two conductors does not exceed the breaker's marked voltage rating.

- If the voltage rating on a circuit breaker is marked with a slash voltage (example: 347/600 volts or 120/240 volts), the breaker may be used where the voltage to ground does not exceed the breaker's lower voltage marking, and the voltage between any two conductors does not exceed the breaker's higher voltage marking. This type of breaker is thus restricted to systems having a grounded neutral.

Thermal-Magnetic Circuit Breakers

A thermal-magnetic circuit breaker contains a bimetallic element. On a continuous overload, the bimetallic element moves until it unlatches the inner tripping mechanism of the breaker. Harmless momentary overloads do not move the bimetallic element far enough to trip the breaker. If the overload is heavy, or if a short circuit occurs, then the magnetic mechanism within the circuit breaker trips instantly. The time required for the breaker to open the circuit completely depends upon the magnitude of the fault current and the mechanical condition of the circuit breaker. This time may range from approximately one-half cycle to several cycles.

Circuit-breaker manufacturers calibrate and set the tripping characteristic for most moulded-case breakers. Breakers are designed so that it is difficult to alter the set tripping point, *Rule 14–304*. For certain types of breakers, however, the trip coil can be changed physically to a different rating. Adjustment provisions are made on some breakers to permit the magnetic trip range to be changed. For example, a breaker rated at 100 amperes may have an external adjustment screw with positions marked HI-MED-LO. The manufacturer's application data for this breaker indicates that the magnetic tripping occurs at 1000 amperes, 750 amperes, or 500 amperes respectively for the indicated positions. These settings usually have a tolerance of ±10%.

The *ambient-compensated* type of circuit breaker is designed so that its tripping point is not affected by an increase in the surrounding temperature. An ambient-compensated breaker has two elements: The first element heats due to the current passing through it and because of the heat of the surrounding air; the second element is affected only by the ambient temperature. These elements act in opposition to each other. In other words, as the tripping element tends to lower its tripping point because of a high ambient temperature, the second element exerts an opposing force that stabilizes the tripping point. Therefore, current through the tripping element is the only factor that causes the element to open the circuit.

Factors that can affect the proper operation of a circuit breaker include moisture, dust, vibration, corrosive fumes and vapours, and excessive tripping and switching. As a result, care must be taken when locating and installing circuit breakers and all other electrical equipment.

The interrupting rating of a circuit breaker is marked on the breaker label. The electrician should check the breaker carefully for the interrupting rating since the breaker may have several voltage ratings with a different interrupting rating for each. For example, assume that it is necessary to select a breaker having an interrupting rating at 240 volts of at least 50 000 amperes. A close inspection of the breaker may reveal the following data:

VOLTAGE	INTERRUPTING RATING
240 volts	65 000 amperes
480 volts	25 000 amperes
600 volts	18 000 amperes

Recall that for a fuse, the interrupting rating is marked on the fuse label. This rating is the same for any voltage up to and including the maximum voltage rating of the fuse.

The standard full-load ampere ratings of non-adjustable circuit breakers are the same as those for fuses. As previously mentioned, additional standard ratings of fuses are 1, 3, 6, and 10 amperes.

The time–current characteristics curves for circuit breakers are similar to those for fuses. A typical circuit breaker time–current curve is shown in Figure 17-30. Note that the current is indicated

Current in percent of breaker trip unit rating

Average unlatching times
Breaker tripping magnetically

Current rms amps	Time seconds
1 000	0.0080
3 000	0.0042
5 000	0.0032
10 000	0.0023

Minimum →   ← Maximum

Maximum interrupting time

Current in percent of breaker trip unit rating

Courtesy of Eaton

FIGURE 17-30   Typical moulded-case circuit-breaker, time–current curve.

in percentage values of the breaker trip unit rating. Therefore, according to this graph, the 100-ampere breaker being considered will:

1. carry its 100-ampere (100%) rating indefinitely;

2. carry 300 amperes (300%) for a minimum of 25 seconds and a maximum of 70 seconds;

3. unlatch its tripping mechanism in 0.0032 seconds (approximately one-quarter cycle) with a current of 5000 amperes;

4. interrupt the circuit in a maximum time of 0.016 seconds (one cycle) with a current of 5000 amperes (5000%).

The same time–current curve can be used to determine that a 200-ampere circuit breaker will:

1. carry its 200-ampere (100%) rating indefinitely;

2. carry 600 amperes (300%) for a minimum of 25 seconds and a maximum of 70 seconds;

3. unlatch its tripping mechanism in 0.0032 seconds (approximately one-quarter cycle) with a current of 5000 amperes;

4. interrupt the circuit in a maximum time of 0.016 seconds (one cycle) with a current of 5000 amperes (2500%).

This example shows that if a short circuit in the magnitude of 5000 amperes occurs on a circuit with both the 100-ampere breaker and the 200-ampere breaker installed, they will open together because they have the same unlatching times. In many instances, this action is the reason for otherwise unexplainable power outages (see Unit 19). This is rather common when heavy (high value) fault currents occur on circuits protected with moulded-case circuit breakers.

Common Misapplication

A possible violation of *Rules 14–012* and *14–014* is the installation of a main circuit breaker (such as a 100-ampere breaker) that has a high interrupting rating (such as 50 000 amperes) while making the assumption that the branch-circuit breakers (with interrupting ratings of 10 000 amperes) are protected adequately against the 40 000-ampere short circuit, Figure 17-31.

Short-circuit current 40k at main breaker

*100 ampere main breaker with interrupting rating of 50 000 amperes

Moulded case circuit breakers interrupting rating of 5000 amperes

*Note that the main breaker cannot protect the branch breakers against a short circuit of 40 000 amperes; branch breakers are capable of interrupting 5000 amperes.

Cengage Learning®

FIGURE 17-31   Common *CEC* violation, *Rules 14–012* and *14–014*, when installing standard-type moulded-case circuit breakers.

Standard circuit breakers with high interrupting ratings cannot protect a standard circuit breaker with a lower interrupting rating.

Since the opening time of a 100-ampere circuit breaker generally exceeds one-half cycle, almost the full amount of current (40 000 amperes) will enter the branch-circuit breaker section of the panel where the breaker interrupting rating is 10 000 amperes. **The probable outcome of such a situation is an electrical explosion with risk of serious personal injury or property loss resulting from a fire.**

A similar situation exists when a fuse with an interrupting rating of 50 000 amperes is installed ahead of a plug or cartridge fuse having an interrupting rating of 10 000 amperes. As in the case of circuit breakers, a fuse having a high interrupting rating cannot protect a fuse having a low interrupting rating against fault currents in excess of the interrupting rating of the lower-rated fuse. Therefore, *all* fuses in a given panel must be capable of interrupting the maximum fault current available at that panel.

SERIES-TESTED COMBINATIONS

Series-rated circuit breakers are also called *series-connected* circuit breakers.

The CSA has tested circuit-breaker panels and load centres under the test conditions of two breakers (and, in a few cases, three) installed in series. Series combinations with current-limiting fuses ahead of circuit breakers are also tested. In the example shown in Figure 17-32, both the 20-ampere and the 100-ampere breakers will trip if they are properly matched. The series arrangement of breakers results in a cushioning effect when the two breakers trip simultaneously. Both circuit breakers **must** trip in order for the series combination to obtain the CSA certification. A panel tested under these conditions may be rated as an entire unit. For the situation in Figure 17-32, the panel is rated as suitable for connection to a system that can deliver no more than 22 000 amperes of short-circuit current.

The disadvantage of series-rated circuit breakers is that for heavy short circuits on any of the branch circuits, the main breaker also trips and causes the complete loss of power to the entire panel. Figure 17-28, page 299, is an example of a manufacturer's series-tested combination datasheet.

On normal overload conditions on a branch circuit, the branch breaker will trip off without causing the main breaker to trip off. Refer to Unit 18 for more data on nonselectivity.

The advantage of installing series-rated breakers is that their cost is somewhat lower than installing breakers "fully rated" for the available fault current at the panelboard or load centre.

SAFETY ALERT

Be extremely careful when installing series-rated breakers. Do *not* "mix" different manufacturers' breakers. Do *not* use any breakers that have not been CSA-approved as being suitable for use in combination with one another.

Series-connected breakers have been described as incomplete in certain constructional features or restricted in performance capabilities and are intended for use as components of complete equipment submitted for investigation rather than for direct separate installation in the field. The final acceptance of the component is dependent upon its installation and use in complete equipment submitted to the CSA.

Rule 14–014(d) requires that the manufacturer of a series-rated panel must legibly mark the equipment. The label will show the many ampere ratings and catalogue numbers of those breakers that are suitable for use with that particular piece of equipment.

Rule 14–014(e) requires that there be "field marking" of series-related equipment. Most electrical inspectors expect the electrical contractor to attach this labelling to the equipment. For example, the main distribution equipment of the commercial building is located away from the individual tenants' panelboards. The electrical contractor will affix legible and durable labels to each disconnect switch in the main distribution equipment, identifying the

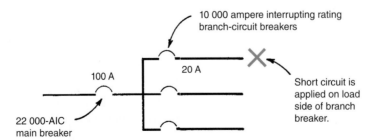

FIGURE 17-32 Series-rated circuit breakers. In this example, both the 20-ampere breaker and the 100-ampere main breaker trip off under high-level fault conditions.

© Cengage Learning®

CAUTION
SERIES-RATED SYSTEM
_____AMPERES AVAILABLE
IDENTIFIED REPLACEMENT COMPONENT REQUIRED

FIGURE 17-33 Example of label on series-rated breaker.

size and type of overcurrent device that has been installed for the proper protection of the downstream panelboards. See Figure 17-33.

This "Caution—Series-Rated System" labelling will alert future users of the equipment that they should not indiscriminately replace existing overcurrent devices or install additional overcurrent devices with sizes and types of fuses or circuit breakers not compatible with the series-rated system as originally designed and installed. Otherwise, the series-rated system might not operate properly, creating a hazard to life and property.

CURRENT-LIMITING BREAKERS

A current-limiting circuit breaker will limit the let-through energy (I^2t) to something less than the I^2t of a one-half cycle wave.

The label on this type of breaker will show the words "Current-Limiting." The label will also indicate the breaker's let-through characteristics or will indicate where to obtain the let-through characteristic data. These let-through data are necessary to ensure adequate protection of downstream components (wire, breakers, controllers, bus bar bracing, etc). Therefore, it is important when installing circuit breakers to ensure not only that the circuit breakers have the proper interrupting rating (capacity), but also that all of the components connected to the circuit downstream from the breakers are capable of withstanding the let-through current of the breakers. This means that the branch-circuit conductors must be capable of withstanding

40 000 amperes for approximately one cycle (the opening time of the breaker).

COST CONSIDERATIONS

As previously discussed, there are a number of different types of circuit breakers to choose from. The selection of the type to use depends upon a number of factors, including the interrupting rating, selectivity, space, and cost. When an installation requires something other than the standard moulded case–type circuit breaker, the use of high-interrupting or current-limiting types may be necessary. The electrician and/or design engineer must complete a short-circuit study to ensure that the overcurrent protective devices provide proper protection against short-circuit conditions as required by the *CEC*.

The following list-price cost comparisons are for various 100-ampere, 240-volt, panelboard-type circuit breakers.

TYPE	INTERRUPTING RATING	LIST PRICE
Standard, plug-in	10 000	$ 165.00
Standard, bolt-on	22 000	322.00
Standard, bolt-on	100 000	390.00
Current-limiting	200 000	1 163.00
Electronic solid-state having adjustable trips and ground fault protection, 225 frame size	65 000	3 323.00

Moulded-case circuit breakers with an integral current-limiting fuse are available. The thermal element in this type of breaker is used for low overloads, the magnetic element is used for low-level short circuits, and the integral fuse is used for short circuits of high magnitude. The interrupting capacity of this breaker/fuse combination (sometimes called a *limiter*) is higher than that of the same breaker without the fuse.

Table 29 shows that on motor branch circuits, the maximum setting of an inverse-time circuit

SAFETY ALERT

Carefully read the label on the equipment to be connected. When the label states "fuses," fuses must be used for the overcurrent protection. Circuit breakers are not permitted in this case. Refer to Figure 17-34 and *Rule 28–202*.

FIGURE 17-34   This installation conforms to *Rules 28–202* and *28–604(5)*. The disconnect switch is within sight of and within 3 m of the unit and contains the 40-ampere overcurrent protection in panel called for on the air-conditioner **nameplate**.

breaker should be sized at not over 250% of the full-load current of the motor. *Rule 28–210* indicates that instantaneous trip breakers should be set at not over 1300% of the motor full-load current or 215% LRC.

Some labels might read "maximum size overcurrent device," in which case fuses or circuit breakers could be used for branch-circuit overcurrent protection.

REVIEW

Note: Refer to the *CEC* or the blueprint package as necessary.

1. What is the purpose of overcurrent protection? _____

    _____

    _____

    _____

2. List the four factors that must be considered when selecting overcurrent protective devices.

    (1) _____

    (2) _____

    (3) _____

    (4) _____

3. What sections of the *CEC* require that overcurrent protective devices have adequate interrupting ratings? _____

    _____

4. Indicate the ampere rating of the fuse that is selected to protect the following copper conductors. (Refer to *Table 13.*) No motors are connected.

   a. No. 12 TW75 _____-ampere overcurrent device

   b. No. 8 T90 Nylon _____-ampere overcurrent device

   c. No. 3 R90 _____-ampere overcurrent device

   d. No. 3/0 RW90 XLPE (600 V) _____-ampere overcurrent device

5. A single-phase motor draws a full-load current of 40 amperes. Complete parts (a), (b), and (c) for this motor.

   a. Install _____-ampere dual-element fuses in a _____-ampere switch.

   b. Install _____-ampere current-limiting fuses in a _____-ampere switch.

   c. Install a _____-ampere circuit breaker, standard inverse-time type.

6. The best current-limiting fuses generally contain _____ links surrounded by_____ arc-quenching filler. Current-limiting fuses can also contain _____ links.

7. In your own words, explain the meaning of the phrase "interrupting rating."

   _____

   _____

   _____

   _____

   _____

8. In your own words, explain the meaning of the term "current-limiting."

   _____

   _____

   _____

9. Define the following terms:

   a. I^2t _____

   b. I_P _____

   c. RMS_____

10. Provide the correct information for the fuse classes indicated:

	AMPERE RANGE	VOLTAGE RATING	INTERRUPTING RATING
Class G fuses	_____A	_____V	_____A
Class H fuses	_____A	_____V	_____A
Class K fuses	_____A	_____V	_____A
Class J fuses	_____A	_____V	_____A
Class L fuses	_____A	_____V	_____A
Plug fuses	_____A	_____V	_____A
Class R fuses	_____A	_____V	_____A
Class T fuses	_____A	_____V	_____A
Class CC fuses	_____A	_____V	_____A

11. Class T and Class J fuses (will) (will not) fit into standard fuse clips. (Circle the correct answer.)

12. Using Figure 17-23, determine the opening time for a 60-ampere fuse that is loaded to 300 amperes. _____ seconds.

13. A short-circuit current of 800 amperes will open a 60-ampere fuse of the type shown in Figure 17-23 in _____ seconds. A short-circuit current of 5000 amperes will open the same fuse in _____ seconds.

14. Using the charts in Figures 17-25 and 17-27, determine the approximate instantaneous peak let-through values and the apparent (equivalent) values of current for the following short-circuit currents:

PROSPECTIVE SHORT-CIRCUIT CURRENT	FUSE	INSTANTANEOUS PEAK LET-THROUGH	APPARENT RMS CURRENT AMPERES
a. 100 000 A	60 A, 250 V	_____ A	_____ A
b. 40 000 A	60 A, 250 V	_____ A	_____ A
c. 200 000 A	200 A, 600 V	_____ A	_____ A
d. 30 000 A	100 A, 250 V	_____ A	_____ A

15. A section of plug-in bus duct is braced for 14 000 RMS amperes. The available short-circuit current is 30 000 RMS symmetrical amperes. Using Figure 17-27, determine if a 200-ampere, 250-volt, low-peak, dual-element fuse will limit the current sufficiently to protect the bus duct against a short circuit. Explain. _____

_____

_____

_____

16. If the same section of plug-in bus duct (Question 15) is connected to a 200-ampere, standard moulded-case circuit breaker having an interrupting rating of 42 000 amperes, will the duct be protected properly? Explain. _____

_____

_____

_____

17. In a thermal-magnetic circuit breaker, overloads are sensed by the _____ element; short circuits are sensed by the _____ element.

18. Is it possible to install

a. a 20-ampere Type D plug fuse in a 15-ampere fuseholder? _____

b. a 30-ampere Class T fuse in a standard (Class H) 20-ampere fuseholder? _____

c. a 100-ampere Class H fuse in a 100-ampere fuseholder for Class J fuses? _____

d. a Class L fuse in a 400-ampere standard switch? _____

19. A cable limiter

    a. is a short-circuit device only.

    b. is not to be used for overload protection.

    c. is generally connected to both ends of large paralleled conductors so that if a fault occurs on one of the conductors, that faulted cable is isolated from the system.

    d. is rated at 600-volt, 200 000-ampere interrupting capacity.

20. When the label on equipment states "Maximum size fuse 50 amperes," is it permitted to connect the equipment to a 50-ampere circuit breaker? _____

21. Class CC fuses are control circuit fuses easily recognized by the _____ on one end. These fuses are rated at (125 V) (250 V) (480 V) (600 V). (Fill in and circle the correct answers.)

22. Current-limiting circuit breakers will limit the energy to something (less than) (equal to) (more than) the available fault current. (Circle the correct answer.)

23. The interrupting rating of a circuit breaker (automatically qualifies) (does not qualify) the equipment in which the breaker is installed to the same interrupting rating. (Circle the correct answer.)

24. It has been learned from the electric utility that the available fault current at the main service equipment will be approximately 55 000 RMS symmetrical amperes. (Label the following statements as TRUE or FALSE.)

    It is permissible to select:

    a. a standard main breaker rated at 65 000 AIC and branch breakers rated at 10 000 AIC. _____

    b. Class L and Class RK5 fuses in equipment CSA approved at 20 000 amperes. _____

    c. main and branch-circuit breakers rated at 42 000 AIC. _____

    d. main fuses of the current-limiting type, then determine if, under fault conditions, the fuses will limit the let-through current to some value less than 10 000 amperes. Branch breakers can then be 10 000 amperes interrupting rating. _____

    e. "series-connected" breakers properly matched by the manufacturer in their CSA-recognized combination panel. The panel is to be rated at 65 000 RMS symmetrical amperes. It is understood that should a fault occur on a branch breaker, the branch breaker and the main breaker will trip off, thereby shutting off all power in the building. _____

    f. all breakers having a 65 000 AIC rating (be sure other components, such as the conductors, motor controllers, and wire can withstand 65 000 amperes). _____

25. What two sections of the *CEC* cover the marking of series-rated systems with the words "Caution—Series-rated system?"

26. You are asked to remove an existing fusible panel and replace it with a panel containing circuit breakers. List the information from the fusible panel that would be required to place an order for the new breaker panel.

_____   _____

_____   _____

_____   _____

_____   _____

27. Magnetic forces vary as the square of the peak current. This is a very significant factor under short-circuit conditions. Recall that the peak for a normal sine wave is 1.4 times the RMS value. When a fault occurs, most of the resistance of the circuit has been bypassed (short circuited) so the major opposition to the flow of current is the reactance of the circuit. Thus, under fault conditions, the power factor of the faulted circuit becomes very low. This causes the peak current for the first half-cycle to become quite high.

The peak current for the first half-cycle for circuits having a power factor of 50% (0.50) is _____ times the RMS current.

The peak current for the first half-cycle for circuits having a power factor of 20% (0.20) is _____ times the RMS current.

28. Draw a line between each of the following statements and the correct percentage of FLA for dual-element time-delay fuses used for short-circuit and ground-fault protection of a motor branch circuit.

    a. Maximum percentage per *Table 29* is   125%

    b. Absolute maximum per *Rule 28–200* is   175%

    c. Fuse manufacturer's recommendation is   225%

29. In general, motor overload protection is based upon a percentage of the motor's full-load current. Draw a line between each motor type and the correct choice.

    a. Motors having a service factor of not less than 1.15   115%

    b. Motors having a service factor of less than 1.15   125%

    c. All other motors   115%

30. Referring to Figure 17-28, it is necessary to install the feeder supplying a panelboard in which Type GB branch circuit breakers are installed. The available fault-current at the panelboard has been calculated to be approximately 75 000 amperes RMS symmetrical. The breakers in the panel have a marked interrupting rating of 10 000 amperes. What is the maximum size and type of fuse that must be used to protect this panelboard properly? _____

_____

31. For Question 30, would the installation comply with the *CEC* if the main distribution board consisted of circuit breakers? The feeder to the panelboard is a 225-ampere circuit breaker having an interrupting rating of 75 000 amperes. Explain your answer.

_____

_____

Short-Circuit Calculations and Coordination of Overcurrent Protective Devices

OBJECTIVES

After completing the study of this unit, the student will be able to

- perform short-circuit calculations using the point-to-point method
- calculate short-circuit currents using the appropriate tables and charts
- define the terms "coordination," "selective systems," and "nonselective systems"
- define the term "interrupting rating" and explain its significance
- use time–current curves

The student must understand the intent of *Rules 14–012* and *14–014,* and *Appendix B.* These rules ensure that the fuses and/or circuit breakers selected for an installation are capable of interrupting—at the rated voltage—the current that may flow under any condition (overload, short circuit, or ground fault) with complete safety to personnel and without damage to the panel, load centre, switch, or electrical equipment in which the protective devices are installed.

An overloaded condition resulting from a miscalculation of load currents will cause a fuse to blow or a circuit breaker to trip in a normal manner. However, a miscalculation, a guess, or ignorance of the magnitude of the available short-circuit currents may result in the installation of breakers or fuses having inadequate interrupting ratings. Such a situation can occur even though the load currents in the circuit are checked carefully. Breakers or fuses having inadequate interrupting ratings need only be subjected to a short circuit to cause them to explode, which could injure personnel and seriously damage the electrical equipment. The interrupting rating of an overcurrent device is its maximum rating and must not be exceeded.

In any electrical installation, individual branch circuits are calculated as has been discussed previously in this text. After the quantity, size, and type of branch circuits are determined, these branch-circuit loads are then combined to determine the size of the feeder conductors to the respective panelboards. Most consulting engineers will specify that a certain number of spare branch-circuit breakers be installed in the panelboard, plus a quantity of spaces that can be used in the future.

For example, a certain computation may require a minimum of sixteen 20-ampere branch circuits. The specification might call for a 24-circuit panelboard with 16 active circuits, four spares, and four spaces. *Rule 8–108(2)* requires that two additional spaces be left in the panel in dwelling units, but this rule does not apply to commercial buildings.

The next step is to determine the interrupting rating requirements of the fuses or circuit breakers to be installed in the panel. *Rule 14–012* is an all-encompassing rule that covers the interrupting rating requirements for services, mains, feeders, subfeeders, and branch-circuit overcurrent devices. For various types of equipment, normal currents can be determined by checking the equipment nameplate current, voltage, and wattage ratings. In addition, an ammeter can be used to check for normal and overloaded circuit conditions.

A number of formulas can be used to calculate short-circuit currents. Manufacturers of fuses, circuit breakers, and transformers publish numerous tables and charts showing approximate values of short-circuit current. A standard ammeter must *not* be used to read short-circuit current because this practice will result in damage to the ammeter and possible injury to personnel.

▶ SHORT-CIRCUIT CALCULATIONS

The following sections will cover several of the basic methods of determining available short-circuit currents. As the short-circuit values given in the various tables are compared to the calculations, it will be noted that there are slight variances in the results. These differences are due largely to (1) the rounding off of the numbers in the calculations, and (2) the variations in the resistance and reactance data used to prepare the tables and charts. For example, the value of the square root of three (1.732) is used frequently in three-phase calculations. Depending on the accuracy required, values of 1.7, 1.73, or 1.732 can be used in the calculations.

In actual practice, the available short-circuit current at the load side of a transformer is less than the values shown in Problem 1 (below). However, this simplified method of finding the available short-circuit currents will result in values that are conservative.

The actual impedance value on a Canadian Standards Association (CSA)-approved 25-kVA or larger transformer can vary ±10% from the transformer's marked impedance. The actual impedance could be as low as 1.8% or as high as 2.2%. This will affect the available fault-current calculations.

For example, in Problem 1, the *marked* impedance is reduced by 10% to reflect the transformer's possible *actual* impedance. The calculations show this "worst case" scenario. All short-circuit examples in this text have the marked transformer impedance values reduced by 10%.

Another factor that affects fault-current calculations is voltage. Utility companies are allowed to vary voltage to their customers within a certain range. This might be ±10% for power services, and ±5% for lighting services. The higher voltage will result in a larger magnitude of fault current.

Another source of short-circuit current comes from electric motors that are running at the time the fault occurs. This is covered later on in this unit.

Thus, it can be seen that no matter how much data we plug into our fault-current calculations, many variables are out of our control. What we hope for is to arrive at a result that is reasonably accurate so that our electrical equipment is reasonably safe insofar as interrupting ratings and withstand ratings are concerned.

In addition to the methods of determining available short-circuit currents that are provided below, there are computer programs that do the calculations. These programs are fast, particularly where there are many points in a system to be calculated.

Determining the Short-Circuit Current at the Terminals of a Transformer Using the Impedance Formula

PROBLEM 1 Assume that the three-phase pad-mounted transformer installed by the utility company for the commercial building has a rating of 300 kVA at 120/208 volts with an impedance of 4% (from the transformer nameplate). The available short-circuit current at the secondary terminals of the transformer must be determined. To simplify the calculation, it is also assumed that the utility can deliver unlimited short-circuit current to the primary of the transformer. In this case, the transformer primary is known as an *infinite bus* or an *infinite primary*.

The first step is to determine the normal full-load current rating of the transformer

$$I \text{ (at the secondary)} = \frac{\text{kVA} \times 1000}{E \times 1.73} = \frac{300 \times 1000}{208 \times 1.73}$$
$$= 834 \text{ amperes normal full load}$$

Using the impedance value given on the plate of the transformer, the next step is to find a multiplier that can be used to determine the short-circuit

current available at the secondary terminals of the transformer.

The factor of 0.9 shown in the calculations below reflects the fact that the transformer's actual impedance might be 10% less than that marked on the nameplate and would be a worst case condition. In electrical circuits, the lower the impedance, the higher the current.

If the transformer is marked 4% impedance, then

$$\text{multiplier} = \frac{100}{4 \times 0.9} = 27.778$$

and short-circuit current $= 834 \times 27.778$
$$= 23\ 167 \text{ amperes.}$$

If the transformer is marked 2% impedance, then

$$\text{multiplier} = \frac{100}{2 \times 0.9} = 55.556$$

and short-circuit current $= 834 \times 55.556$
$$= 46\ 334 \text{ amperes.}$$

If the transformer is marked 1% impedance, then

$$\text{multiplier} = \frac{100}{1 \times 0.9} = 111.111$$

and short-circuit current $= 834 \times 111.111$
$$= 92\ 667 \text{ amperes}$$

Determining the Short-Circuit Current at the Terminals of a Transformer Using Tables

Table 18-1 shows the short-circuit currents for a typical transformer. Transformer manufacturers publish short-circuit tables for many sizes of transformers having various impedance values. Table 18-1 provides data for a 300-kVA, three-phase transformer with an impedance of 2%.

According to the table, the symmetrical short-circuit current is 42 090 amperes at the secondary terminals of a 120/208-volt three-phase transformer (refer to the zero-foot row of the table). This value is on the low side because the manufacturer that developed the table did not allow for the ± impedance variation allowed by the CSA standard. Problem 1 indicates that the available short-circuit current at the secondary of the transformer is 46 334 amperes at 2% impedance.

TABLE 18-1

Symmetrical Short-Circuit Currents at Various Distances from a Liquid Filled Transformer (300 kVA transformer, 2% impedance).

WIRE-SIZE (COPPER)

	DIST. (FT.)	#14	#12	#10	#8	#6	#4	#1	0	00	000	2-000	0000
	0	42090	42090	42090	42090	42090	42090	42090	42090	42090	42090	42090	42090
	5	6910	10290	14730	19970	25240	29840	34690	35770	36640	37340	39610	37930
	10	3640	5610	8460	12350	17090	22230	29030	30760	32210	33410	37340	34420
	25	1500	2360	3670	5650	8430	12150	18930	21170	23240	25090	31710	26750
208	50	760	1200	1890	2950	4530	6810	11740	13670	15610	17510	25090	19320
VOLTS	100	380	600	960	1510	2350	3610	6610	7920	9320	10810	17510	12320
	200	190	300	480	760	1190	1860	3510	4280	5140	6090	10810	7110
	500	80	120	190	310	480	760	1460	1800	2180	2630	4990	3130
	1000	40	60	100	150	240	380	740	910	1110	1350	2630	1620
	5000	10	10	20	30	50	80	150	180	230	280	550	330
	0	37820	37820	37820	37820	37820	37820	37820	37820	37820	37820	37820	37820
	5	7750	11330	15810	20720	25260	28940	32560	33340	33960	34460	36080	34870
	10	4140	6320	9400	13430	18040	22670	28230	29560	30660	31550	34460	32290
	25	1720	2700	4180	6360	9360	13190	19640	21620	23380	24920	30240	26260
240	50	870	1380	2160	3360	5130	7620	12730	14630	16480	18230	24920	19850
VOLTS	100	440	700	1100	1730	2680	4100	7380	8770	10220	11720	18230	13220
	200	220	350	550	880	1370	2130	3990	4830	5770	6790	11720	7880
	500	90	140	220	350	560	870	1670	2050	2490	2990	5610	3540
	1000	40	70	110	180	280	440	850	1050	1280	1550	2990	1850
	5000	10	10	20	40	60	90	170	210	260	320	630	380
	0	18910	18910	18910	18910	18910	18910	18910	18910	18910	18910	18910	18910
	5	10450	12820	14750	16150	17080	17690	18200	18310	18400	18470	18690	18520
	10	6750	9170	11630	13780	15400	16530	17540	17740	17910	18040	18470	18150
	25	3180	4740	6770	9150	11520	13570	15690	16160	16540	16840	17830	17100
480	50	1680	2590	3900	5680	7840	10170	13190	13960	14600	15120	16840	15560
VOLTS	100	860	1350	2090	3180	4680	6600	9820	10810	11690	12460	15120	13130
	200	440	690	1080	1680	2560	3810	6370	7320	8240	9110	12460	9930
	500	180	280	440	700	1080	1670	3040	3640	4290	4960	8010	5560
	1000	90	140	220	350	550	860	1620	1970	2370	2800	4960	3270
	5000	20	30	40	70	110	180	340	420	510	620	1210	750
	0	15130	15130	15130	15130	15130	15130	15130	15130	15130	15130	15130	15130
	5	10210	11790	12920	13690	14180	14500	14770	14820	14870	14900	15010	14930
	10	7270	9270	11010	12350	13280	13890	14410	14520	14610	14680	14900	14730
	25	3740	5370	7280	9230	10920	12200	13410	13670	13870	14040	14570	14170
600	50	2040	3080	4500	6270	8170	9950	11940	12400	12770	13060	14040	13310
VOLTS	100	1060	1650	2510	3730	5290	7080	9650	10350	10930	11420	13060	11840
	200	540	850	1330	2040	3040	4390	6840	7640	8390	9050	11420	9640
	500	220	350	550	860	1330	2010	3550	4180	4830	5480	8180	6110
	1000	110	170	280	440	680	1050	1950	2360	2800	3270	5480	3760
	5000	20	40	60	90	140	220	420	520	640	770	1470	910

250 kcmil	2-250 kcmil	3-300 kcmil	350 kcmil	2-350 kcmil	3-350 kcmil	3-400 kcmil	500 kcmil	2-500 kcmil	750 kcmil	4-750 kcmil
42090	42090	42090	42090	42090	42090	42090	42090	42090	42090	(42090)
38270	40100	40870	38840	40410	40960	41030	39300	40650	39650	41460
35030	38270	39710	36040	38840	39870	40010	36850	39300	37480	40840
27780	33590	36560	29550	34780	36930	37230	31020	35730	32190	(39090)
20520	27780	32250	22660	29550	32850	33340	24520	31020	26050	36480
13380	20520	26010	15400	22660	26850	27530	17250	24520	18860	32190
7860	13380	18660	9360	15400	19590	20370	10820	17250	12150	26050
3500	6510	10030	4290	7820	10770	11400	5100	9120	5870	16570
1820	3500	5650	2250	4290	6140	6560	2710	5100	3160	10310
380	740	1260	470	930	1380	1490	570	1130	670	2560
37820	37820	37820	37820	37820	37820	37820	37820	37820	37820	37820
35120	36420	36960	35520	36640	37020	37070	35840	36800	36090	37370
32730	35120	36140	33470	35520	36260	36350	34060	35840	34510	36930
27090	31650	33860	28480	32530	34130	34340	29610	33230	30510	35680
20900	27090	30610	22740	28480	31060	31430	24300	29610	25570	33780
14240	20900	25600	16150	22740	26280	26830	17860	24300	19310	30510
8650	14240	19200	10190	16150	20030	20710	11650	17860	12960	25570
3960	7230	10890	4820	8590	11620	12250	5700	9920	6520	17200
2080	3960	6300	2560	4820	6820	7270	3070	5700	3570	11130
430	860	1440	540	1070	1580	1710	660	1290	770	2910
18910	18910	18910	18910	18910	18910	18910	18910	18910	18910	18910
18550	18730	18800	18610	18760	18810	18810	18650	18780	18690	18850
18210	18550	18690	18320	18610	18710	18720	18400	18650	18470	18800
17250	18040	18380	17490	18180	18410	18440	17690	18280	17840	18630
15820	17250	17870	16260	17490	17940	18000	16610	17690	16890	18360
13540	15820	16930	14240	16260	17060	17170	14810	16610	15260	17840
10450	13540	15300	11370	14240	15530	15710	12150	14810	12780	16890
6140	9360	11820	7050	10320	12190	12500	7880	11150	8600	14550
3610	6140	8520	4300	7050	8940	9290	4960	7880	5560	11830
840	1610	2600	1040	1980	2820	3020	1250	2350	1450	4730
15130	15130	15130	15130	15130	15130	15130	15130	15130	15130	15130
14940	15040	15070	14970	15050	15080	15080	15000	15060	15010	15100
14770	14940	15020	14820	14970	15020	15030	14870	15000	14900	15070
14250	14680	14850	14380	14750	14870	14890	14490	14800	14570	14980
13460	14250	14590	13700	14380	14620	14650	13900	14490	14050	14840
12090	13460	14080	12510	13700	14150	[14210	12850	13900	13120	14570
10010	12090	13160	10640	12510	13290	13390	11160	12850	11580	14050
6530	9210	10960	7310	9900	11210	11400	7990	10470	8560	12700
4110	6530	8540	4780	7310	8860	9120	5410	7990	5970	10940
1030	1930	3030	1260	2340	3270	3480	1510	2750	1740	5180

Determining the Short-Circuit Current at Various Distances From a Transformer Using Table 18-1

The amount of available short-circuit current decreases as the distance from the transformer increases, as indicated in Table 18-1. See Problem 2.

Determining Short-Circuit Currents at Various Distances From Transformers, Switchboards, Panelboards, and Load Centres Using the Point-to-Point Method

A simple method for determining the available short-circuit currents (also referred to as fault current) at various distances from a given location is the *point-to-point method*. Reasonable accuracy is obtained when this method is used with three-phase and single-phase systems.

The following procedure demonstrates the use of the point-to-point method:

Step 1: Determine the full-load rating of the transformer in amperes from the transformer nameplate, Table 18-2, or the following formulas:

a. For three-phase transformers:

$$I_{FLA} = \frac{kVA \times 1000}{E_{L-L} \times 1.73}$$

where E_{L-L} = line-to-line voltage b.

b. For single-phase transformers:

$$I_{FLA} = \frac{kVA \times 1000}{E_{L-L}}$$

Step 2: Find the percent impedance (Z) on the nameplate of the transformer.

Step 3: Find the transformer multiplier M_1:

$$M_1 = \frac{100}{\text{transformer \% impedance} \times 0.9}$$

Note: Because the marked transformer impedance can vary ±10% per the CSA standard, the 0.9 factor above takes this into consideration to show "worst case" conditions.

TABLE 18-2

Short-Circuit Currents Available From Various Size Transformers.

VOLTAGE+ AND PHASE	KVA	FULL LOAD AMPS	% IMPEDANCE++ (NAME PLATE)	+SHORT-CIRCUIT AMPS
	25	104	1.6	11 431
	37.5	156	1.6	16 961
120/240	50	209	1.7	21 065
1 ph.*	75	313	1.6	32 789
	100	417	1.6	42 779
	167	695	1.8	60 038
	150	417	2.0	23 166
	225	625	2.0	34 722
	300	834	2.0	46 333
	300	834	4.0	23 166
120/208	500	1388	2.0	77 111
3 ph.	750	2080	5.0	66 036
	1000	2776	5.0	88 127
	1500	4164	5.0	132 180
	2000	5552	5.0	123 377
	2500	6950	5.0	154 444
	112.5	135	1.0	15 000
	150	181	1.2	16 759
	225	271	1.2	25 062
	300	361	1.2	33 426
277/480	500	601	1.3	51 368
3 ph.	750	902	5.0	28 410
	1000	1203	5.0	36 180
	1500	1804	5.0	57 261
	2000	2406	5.0	53 461
	2500	3007	5.0	66 822

† Three-phase short-circuit currents based on "infinite" primary.

* Single-phase values are L–N values at transformer terminals. These figures are based on change in turns ratio between primary and secondary, 100 000 KVA primary, zero feet from terminals of transformer, 1.2 (%X) and 1.5 (%R) for L–N vs. L–L reactance and resistance values, and transformer X/R ratio v 3.

†† UL listed transformers 25 KVA or greater have a ±10% impedance tolerance. "Short Circuit Amps" reflect a worst case scenario.

+ Fluctuations in system voltage will affect the available short-circuit current. For example, a 10% increase in system voltage will result in a 10% increase in the available short-circuit currents shown in the table.

Notes: The interrupting rating for an overcurrent protective device is the device's *maximum* rating under standard test conditions. This interrupting rating must not be exceeded.

Three-phase line-to-ground (neutral) can vary from 25% to 125% of the L–L–L bolted fault-current value. Use 100% as typical.

For single-phase, centre-tapped transformers, it is common practice to multiply the L–L bolted fault current value by 1.5 to determine the approximate L–G (L–N) fault current value. Maximum short-circuit current will occur across the 120-volt transformer terminals.

Step 4:   Determine the transformer let-through short-circuit current at the secondary terminals of the transformer. Use tables or the following formula:

a. For three-phase transformers (L–L–L):

$$I_{SCA} = \text{transformer FLA} \times M_1$$

b. For single-phase transformers (L–L):

$$I_{SCA} = \text{transformer FLA} \times M_1$$

c. For single-phase transformers (L–N):

$$I_{SCA} = \text{transformer FLA} \times M_1 \times 1.5$$

At the secondary terminals of a single-phase centre-tapped transformer, the L–N fault current is higher than the L–L fault current. At some distance from the terminals, depending on the wire size and type, the L–N fault current is lower than the L–L fault current. The L–N fault current can vary from 1.33 to 1.67 times the L–L fault current. These figures are based on the different turns ratios between the primary and the secondary, infinite source current, zero distance from the terminals of the transformer, and $1.2 \times \%$ reactance (X) and $1.5 \times \%$ resistance (R) for the L–N vs L–L resistance and reactance values. For simplicity, in Step 4c we used an approximate multiplier of 1.5. See Figure 18-2 (page 322), for example.

Step 5:   Determine the f factor:

a. For three-phase faults:

$$f = \frac{1.73 \times L \times I_{SCA}}{N \times C \times E_{L-L}}$$

b. For single-phase, line-to-line (L–L) faults on single-phase, centre-tapped transformers:

$$f = \frac{2 \times L \times I_{SCA}}{N \times C \times E_{L-L}}$$

c. For single-phase, line-to-neutral (L–N) faults on single-phase, centre-tapped transformers:

$$f = \frac{2 \times L \times I_{SCA}}{N \times C \times E_{L-N}}$$

where

   L = the length of the circuit to the fault, in feet
   C = the constant derived from Table 18-3 for the specific type of conductors and wiring method
   E_{L-L} = the line-to-line voltage

E_{L-N} = the line-to-neutral voltage
N = the number of conductors in parallel

Step 6:   After finding the f factor, refer to Table 18-4 and locate the corresponding value of the multiplier M_2, or calculate it as follows:

$$M_2 = \frac{1}{1 + f}$$

Step 7:   Multiply the available fault current at the beginning of the circuit by the multiplier M_2 to determine the available symmetrical fault current at the fault.

$$I_{SCA} \text{ at fault} = I_{SCA} \text{ at beginning of circuit} \times M_2$$

Motor Contribution.   All motors running at the instant a short circuit occurs contribute to the short-circuit current. The amount of current from the motors is approximately equal to the starting (locked rotor) current for each motor. This current value depends upon the type of motor and its characteristics. It is a common practice to multiply the full-load ampere rating of the motor by four or five to obtain a close approximation of the locked rotor current and provide a margin of safety. For energy-efficient motors, multiply the motor's full-load current rating by six to eight times for a reasonable approximation of the fault-current contribution. The current contributed by running motors at the instant a short circuit occurs is added to the value of the short-circuit current at the main switchboard prior to the start of the point-to-point calculations for the rest of the system. To simplify the following problems, motor contributions have not been added to the short-circuit currents.

PROBLEM 2   For a 300-kVA transformer with a secondary voltage of 208 volts and 2% impedance, find the available short-circuit current at a main switch that is located 7.6 m (25 ft) from the transformer. The main switch is supplied by four 750-kcmil copper conductors per phase in steel conduit.

Refer to Table 18-1 and read the value of 39 090 amperes in the column on the right-hand side of the table for a distance of 25 feet.

TABLE 18–3

Table of C Values

AWG OR KCMIL	COPPER CONDUCTORS THREE SINGLE CONDUCTORS						COPPER CONDUCTORS THREE CONDUCTOR CABLE					
	Steel Conduit			Nonmagnetic Conduit			Steel Conduit			Nonmagnetic Conduit		
	600 V	5 KV	15 KV	600 V	5 KV	15 KV	600 V	5 KV	15 KV	600 V	5 KV	15 KV
14	389	—	—	389	—	—	389	—	—	389	—	—
12	617	—	—	617	—	—	617	—	—	617	—	—
10	981	—	—	981	—	—	981	—	—	981	—	—
8	1557	1551	1557	1556	1555	1558	1559	1557	1559	1559	1558	1559
6	2425	2406	2389	2430	2417	2406	2431	2424	2414	2433	2428	2420
4	3806	3750	3695	3825	3789	3752	3830	3811	3778	3837	3823	3798
3	4760	4760	4760	4802	4802	4802	4760	4790	4760	4802	4802	4802
2	5906	5736	5574	6044	5926	5809	5989	5929	5827	6087	6022	5957
1	7292	7029	6758	7493	7306	7108	7454	7364	7188	7579	7507	7364
1/0	8924	8543	7973	9317	9033	8590	9209	9086	8707	9472	9372	9052
2/0	10755	10061	9389	11423	10877	10318	11244	11045	10500	11703	11528	11052
3/0	12843	11804	11021	13923	13048	12360	13656	13333	12613	14410	14118	13461
4/0	15082	13605	12542	16673	15351	14347	16391	15890	14813	17482	17019	16012
250	16483	14924	13643	18593	17120	15865	18310	17850	16465	19779	19352	18001
300	18176	16292	14768	20867	18975	17408	20617	20051	18318	22524	21938	20163
350	19703	17385	15678	22736	20526	18672	19557	21914	19821	22736	24126	21982
400	20565	18235	16365	24296	21786	19731	24253	23371	21042	26915	26044	23517
500	22185	19172	17492	26706	23277	21329	26980	25449	23125	30028	28712	25916
600	22965	20567	17962	28033	25203	22097	28752	27974	24896	32236	31258	27766
750	24136	21386	18888	28303	25430	22690	31050	30024	26932	32404	31338	28303
1000	25278	22539	19923	31490	28083	24887	33864	32688	29320	37197	35748	31959

AWG OR KCMIL	ALUMINUM CONDUCTORS THREE SINGLE CONDUCTORS						ALUMINUM CONDUCTORS THREE CONDUCTOR CABLE					
	Steel Conduit			Nonmagnetic Conduit			Steel Conduit			Nonmagnetic Conduit		
	600 V	5 KV	15 KV	600 V	5 KV	15 KV	600 V	5 KV	15 KV	600 V	5 KV	15 KV
14	236	—	—	236	—	—	236	—	—	236	—	—
12	375	—	—	375	—	—	375	—	—	375	—	—
10	598	—	—	598	—	—	598	—	—	598	—	—
8	951	950	951	951	950	951	951	951	951	951	951	951
6	1480	1476	1472	1481	1478	1476	1481	1480	1478	1482	1481	1479
4	2345	2332	2319	2350	2341	2333	2351	2347	2339	2353	2349	2344
3	2948	2948	2948	2958	2958	2958	2948	2956	2948	2958	2958	2958
2	3713	3669	3626	3729	3701	3672	3733	3719	3693	3739	3724	3709
1	4645	4574	4497	4678	4631	4580	4686	4663	4617	4699	4681	4646
1/0	5777	5669	5493	5838	5766	5645	5852	5820	5717	5875	5851	5771

(continues)

TABLE 18-3 *(Continued)*

AWG OR KCMIL	ALUMINUM CONDUCTORS THREE SINGLE CONDUCTORS						ALUMINUM CONDUCTORS THREE CONDUCTOR CABLE					
	Steel Conduit			Nonmagnetic Conduit			Steel Conduit			Nonmagnetic Conduit		
	600 V	5 KV	15 KV	600 V	5 KV	15 KV	600 V	5 KV	15 KV	600 V	5 KV	15 KV
2/0	7186	6968	6733	7301	7152	6986	7327	7271	7109	7372	7328	7201
3/0	8826	8466	8163	9110	8851	8627	9077	8980	8750	9242	9164	8977
4/0	10740	10167	9700	11174	10749	10386	11184	11021	10642	11408	11277	10968
250	12122	11460	10848	12862	12343	11847	12796	12636	12115	13236	13105	12661
300	13909	13009	12192	14922	14182	13491	14916	14698	13973	15494	15299	14658
350	15484	14280	13288	16812	15857	14954	15413	15490	15540	16812	17351	16500
400	16670	15355	14188	18505	17321	16233	18461	18063	16921	19587	19243	18154
500	18755	16827	15657	21390	19503	18314	21394	20606	19314	22987	22381	20978
600	20093	18427	16484	23451	21718	19635	23633	23195	21348	25750	25243	23294
750	21766	19685	17686	23491	21769	19976	26431	25789	23750	25682	25141	23491
1000	23477	21235	19005	28778	26109	23482	29864	29049	26608	32938	31919	29135

AMPACITY	PLUG-IN BUSW RAY		FEEDER BUSWAY		HIGH IMPED. BUSWAY
	Copper	Aluminum	Copper	Aluminum	Copper
225	28700	23000	18700	12000	—
400	38900	34700	23900	21300	—
600	41000	38300	36500	31300	—
800	46100	57500	49300	44100	—
1000	69400	89300	62900	56200	15600
1200	94300	97100	76900	69900	16100
1350	119000	104200	90100	84000	17500
1600	129900	120500	101000	90900	19200
2000	142900	135100	134200	125000	20400
2500	143800	156300	180500	166700	21700
3000	144900	175400	204100	188700	23800
4000	—	—	277800	256400	—

SHORT-CIRCUIT CURRENT VARIABLES

Phase-to-Phase-to-Phase Fault

The three-phase fault current determined in Step 7 is the *approximate* current that will flow if the three "hot" phase conductors of a three-phase system are shorted together, in what is commonly referred to as a "bolted fault." This is the worst case condition.

Phase-to-Phase Fault

To obtain the *approximate* short-circuit current values when two "hot" conductors of a three-phase system are shorted together, use 87% of

TABLE 18-4

Simple Chart to Convert f Values to M_2 Multiplier When Using the Point-to-Point Method.

F	M_2	F	M_2
0.01	0.99	1.50	0.40
0.02	0.98	2.00	0.33
0.03	0.97	3.00	0.25
0.04	0.96	4.00	0.20
0.05	0.95	5.00	0.17
0.06	0.94	6.00	0.14
0.07	0.93	7.00	0.13
0.08	0.93	8.00	0.11
0.09	0.92	9.00	0.10
0.10	0.91	10.00	0.09
0.15	0.87	15.00	0.06
0.20	0.83	20.00	0.05
0.30	0.77	30.00	0.03
0.40	0.71	40.00	0.02
0.50	0.67	50.00	0.02
0.60	0.63	60.00	0.02
0.70	0.59	70.00	0.01
0.80	0.55	80.00	0.01
0.90	0.53	90.00	0.01
1.00	0.50	100.00	0.01
1.20	0.45		

$$M_2 = \frac{1}{1 + f}$$

the three-phase current value. In other words, if the three-phase current value is 20 000 amperes when the three "hot" lines are shorted together (L–L–L value), then the short-circuit current to two "hot" lines shorted together (L–L) is approximately

$$20\ 000 \times 0.87 = 17\ 400 \text{ amperes}$$

PROBLEM 3 It is desired to find the available short-circuit current at the main switchboard of the commercial building. Once this value is known, the

electrician can provide overcurrent devices with adequate interrupting ratings and the proper bus bar bracing within the switchboard (*Rules 14–012, 14–014,* and *Appendix* B). Figure 18-1 shows the actual electrical system for the commercial building. As each of the following steps in the point-to-point method is examined, refer to the entries given in Tables 18-3 and 18-4 to determine the necessary values of C, f, and M_2.

Step 1: $I_{FLA} = \dfrac{\text{kVA} \times 1000}{E_{L-L} \times 1.73}$

$= \dfrac{300 \times 1000}{208 \times 1.73} = 834$ amperes

Step 2: Multiplier,

$M_1 = \dfrac{100}{\text{Transformer \% impedance} \times 0.9}$

$= \dfrac{100}{4 \times 0.9}$

$= 27.78$

Step 3: $I_{SCA} = 834 \times 27.78 = 23\ 168.5$

Step 4: $f = \dfrac{1.73 \times L \times I}{3 \times C \times E_{L-L}}$

$= \dfrac{1.73 \times 80 \times 23\ 168.5}{3 \times 22\ 185 \times 208} = 0.2316$

Step 5: $M_2 = \dfrac{1}{1 + f} = \dfrac{1}{1 + 0.2316} = 0.812$

Step 6: The short-circuit current available at the line-side lugs on the main switchboard is: $23\ 168.5 \times 0.812 = 18\ 812$ RMS symmetrical amperes.

Phase-to-Neutral (Ground)

For solidly grounded three-phase systems, such as the 120/208-volt system that supplies the commercial building, the phase-to-neutral (ground) bolted short-circuit current can vary from 25% to 125% of the L–L–L bolted short-circuit current. Therefore, it is common practice to consider the L–N or L–G short-circuit current value to be the same as the L–L–L short-circuit current value.

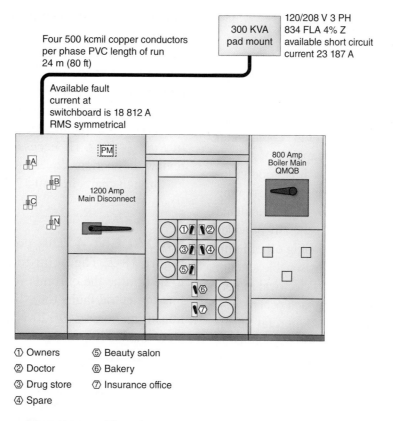

Four 500 kcmil copper conductors per phase PVC length of run 24 m (80 ft)

300 KVA pad mount

120/208 V 3 PH
834 FLA 4% Z
available short circuit current 23 187 A

Available fault current at switchboard is 18 812 A RMS symmetrical

PM

800 Amp Boiler Main QMQB

1200 Amp Main Disconnect

A
B
C
N

① Owners   ⑤ Beauty salon
② Doctor   ⑥ Bakery
③ Drug store   ⑦ Insurance office
④ Spare

FIGURE 18-1 Electrical system for the commercial building.

EXAMPLE

If the three-phase L–L–L fault current has been calculated to be 20 000 amperes, then the L–N fault current is approximately:

20 000 × 1.00 = 20 000 amperes

In summary:

L–L bolted short-circuit current 5 87% L–L–L

L–N bolted short-circuit current 5 100% L–L–L

The main concern is to provide the proper interrupting rating for the overcurrent protective devices, and the adequate withstand rating for the equipment. Therefore, for most three-phase electrical systems, the line-line-line bolted fault-current value will provide the desired level of safety.

Arcing Fault Values (Approximate)

There are times when one wishes to know the currents in arcing faults. These currents vary considerably. The following multipliers give acceptable approximations.

TYPE OF FAULT	480 VOLTS	208 VOLTS
L–L–L	0.89	0.12
L–L	0.74	0.02
L–G	0.38[*]	0

[*]Some reference books indicate this value to be 0.19.

EXAMPLE

What is the approximate line-to-line arcing fault value on a 208/120-volt system where the line–line–line fault current has been calculated to be 30 000 amperes?

ANSWER

$$30\ 000 \times 0.02 = 600\ \text{amperes}$$

As shown in Problem 3, the fuses or circuit breakers located in the main switchboard of the commercial building must have an interrupting capacity of at least 39 847 RMS symmetrical amperes. It is good practice to install protective devices having an interrupting rating at least 25% greater than the actual calculated available fault current. This practice generally provides a margin of safety to permit the rounding off of numbers, as well as compensating for a reasonable amount of short-circuit contribution from any electrical motors that may be running at the instant the fault occurs.

The fuses specified for the commercial building have an interrupting rating of 200 000 amperes (see the Specifications in the Appendix of this text).

In addition, the switchboard bracing is specified to be 50 000 amperes.

If current-limiting fuses are installed in the main switchboard feeders protecting the various panelboards, breakers having an interrupting rating of 10 000 amperes may be installed in the panelboards. This installation meets the requirements of *Rule 14–012*. (If necessary, the student should review the sections covering the application of peak let-through charts in Unit 17.)

PROBLEM 4 This problem is illustrated in Figure 18-2. The point-to-point method is used to determine the currents for both line-to-line and line-to-neutral faults for a 167-kVA, 2% impedance transformer on a 240/120-volt, single-phase system. Note that the impedance has been reduced by 10%.

Infinite primary

Transformer:
167 kVA, 120/240 volts, single-phase
2% impedance, 696 FLA

(A) Fault A_{L-L}

$XFMR.FLA = 696$ amperes

$$M_1 = \frac{100}{2 \times 0.9} = 55.556$$

$$XFMR_{SCA} = 696 \times 55.556 = 38\ 667\ \text{amperes}$$

(C) Fault C_{L-L}

$$f = \frac{2 \times L \times I}{N \times C \times E_{L-L}}$$

$$= \frac{2 \times 25 \times 38\ 667}{2 \times 22\ 185 \times 240} = 0.18$$

$$M_2 = \frac{1}{1+f} = \frac{1}{1+0.18} = 0.85$$

$$I_{SCA} = 38\ 667 \times 0.85 = 32\ 867\ \text{amperes}$$

(E) Fault E_{L-L}

$$f = \frac{2 \times L \times I}{N \times C \times E_{L-L}}$$

$$= \frac{2 \times 20 \times 32\ 867}{1 \times 5906 \times 240} = 0.93$$

$$M_2 = \frac{1}{1+f} = \frac{1}{1+0.93} = 0.52$$

$$I_{SCA} = 32\ 867 \times 0.52 = 17\ 091\ \text{amperes}$$

L₁   N   L₂

7.6 m (25 feet)
Two 500-kcmil copper conductors per phase in each of two steel raceways

Main disconnect

6.1 m (20 feet)
One 2 AWG copper conductor per phase in steel raceway

Panelboard

(B) Fault B_{L-N}

$XFMR.FLA = 696$ amperes

$$\frac{100}{2 \times 0.9} = 55.556$$

$$XFMR_{SCA} = 696 \times 55.556 \times 1.5 = 58\ 000\ \text{amperes}$$

(D) Fault D_{L-N}

$$f = \frac{2 \times L \times I}{N \times C \times E_{L-N}}$$

$$M_1 = \frac{2 \times 25 \times 58\ 000}{2 \times 22\ 185 \times 120} = 0.54$$

$$M_2 = \frac{1}{1+f} = \frac{1}{1+0.54} = 0.65$$

$$I_{SCA} = 58\ 000 \times 0.65 = 37\ 700\ \text{amperes}$$

(F) Fault F_{L-N}

$$f = \frac{2 \times L \times I}{N \times C \times E_{L-N}}$$

$$= \frac{2 \times 20 \times 37\ 700}{1 \times 5906 \times 120} = 2.13$$

$$M_2 = \frac{1}{1+f} = \frac{1}{1+2.13} = 0.32$$

$$I_{SCA} = 37\ 700 \times 0.32 = 12\ 064\ \text{amperes}$$

© Cengage Learning®

FIGURE 18–2 Point-to-point calculation for single-phase, centre-tapped transformer. Calculations show L–L and L–N values.

If noncurrent-limiting overcurrent devices (standard moulded-case circuit breakers) are to be installed in the main switchboard, breakers having adequate interrupting ratings must be installed in the panelboards. A short-circuit study must be made for each panelboard location to determine the value of the available short-circuit current.

The cost of circuit breakers increases as the interrupting rating of the breakers increases. The most economical protection system generally results when current-limiting fuses are installed in the main switchboard to protect the breakers in the panelboards or series-rated breaker combinations are used. In this case, the breakers in the panelboards will have the standard 10 000-ampere interrupting rating.

Summary

1. To meet the requirements of *Rules 14–012* and *14–014*, it is absolutely necessary to determine the available fault currents at various points on the electrical system. If a short-circuit study is not done, the selection of overcurrent devices may be in error, resulting in a hazard to life and property.

2. For an installation using fuses only, the fuses must have an interrupting rating *not less* than the available fault current. The electrical equipment to be protected by the fuses must be capable of withstanding the let-through current of the fuse.

3. For an installation using fuses to protect panels that contain circuit breakers having inadequate interrupting ratings (less than the available fault current), use the peak let-through charts or fuse-breaker application tables from manufacturers to ensure the proper selection of fuses that will adequately protect the downstream circuit breakers.

4. For an installation using standard circuit breakers only, the breakers must have an interrupting rating *not less* than the available fault current at the point of application. It is recommended that the breakers used have interrupting ratings *at least 25% greater* than the available fault current.

5. To use current-limiting circuit breakers, refer to the manufacturer's application data.

6. To use series-connected (rated) circuit breakers, refer to the manufacturer's application data. (An example of series-rated breakers may be found in Figure 17-32, on page 304.) Remember that, under heavy fault conditions, the above installations will be nonselective. This situation is discussed in detail on the following pages.

COORDINATION OF OVERCURRENT PROTECTIVE DEVICES

While this text cannot cover the topic of electrical system coordination (selectivity) in detail, the following material will provide the student with a working knowledge of this important topic.

What Is Coordination? (Rule 14–012, Appendix B)

A situation known as *nonselective coordination* occurs when a fault on a branch circuit opens not only the branch-circuit overcurrent device, but also the feeder overcurrent device, Figure 18-3. Nonselective systems are installed unknowingly and cause needless power outages in portions of an electrical system that should not be affected by a fault.

A *selectively coordinated system,* Figure 18-4, is one in which *only* the overcurrent device immediately upstream from the fault opens. Obviously, the installation of a selective system is much more desirable than a nonselective system, in that all the other circuits are unaffected by a fault on an adjacent circuit.

The fault on the branch circuit trips both the branch-circuit breaker and the feeder-circuit breaker. As a result, power to the panel is cut off, and circuits that should not be affected are now off.

© Cengage Learning®

FIGURE 18–3   Nonselective system.

Only the branch-circuit fuse opens. All other circuits and the main feeder remain on.

FIGURE 18–4   Selective system.

The importance of selectivity in an electrical system is of primary concern in health-care facilities, where maintaining essential electrical systems is paramount. The unexpected loss of power in certain areas of hospitals, nursing care centres, and similar health-care facilities can be catastrophic.

The importance of selectivity (system coordination) is also critical to industrial installations where additional hazards would be introduced should a disorderly shutdown occur. Coordination may be defined as the proper localizing of a fault condition to restrict outages to the equipment affected.

Some local electrical codes require that all circuits, feeders, and mains in buildings such as schools, shopping centres, assembly halls, nursing homes, retirement homes, churches, restaurants, and any other places of public occupancy be selectively coordinated so as to minimize the dangers associated with total power outages. Nonselectivity of an electrical system is not considered good design practice, and is generally accepted only as a trade-off for a low-cost installation.

It is advisable to check with the authority enforcing the *Canadian Electrical Code (CEC)* before proceeding too far in the selection of overcurrent protective devices for a specific installation.

By knowing how to determine the available short-circuit current and ground-fault current, the electrician can make effective use of the time–current curves and peak let-through charts (Unit 17) to find the length of time required for a fuse to open or a circuit breaker to trip.

What Causes Nonselectivity?

In Figure 18-5, a short circuit in the range of 3000 amperes occurs on the load side of a 20-ampere breaker. The magnetic trip of the breaker is adjusted permanently by the manufacturer to unlatch at a current value equal to ten times its rating, or 200 amperes. The feeder breaker is rated at 100 amperes; the magnetic trip of this breaker is set by the manufacturer to unlatch at a current equal to ten times its rating, or 1000 amperes. This type of breaker generally cannot be adjusted in the field. Therefore, a current of 200 amperes or more will cause the 20-ampere breaker to trip instantly. In addition, any current of 1000 amperes or more will cause the 100-ampere breaker to trip instantly.

For the breakers shown in Figure 18-5, a momentary fault of 3000 amperes will trip (unlatch) both breakers. Since the flow of current in a series circuit is the same in all parts of the circuit, the 3000-ampere fault will trigger both magnetic trip mechanisms. The time–current curve shown in Figure 17–30 (page 302) indicates that for a 3000-ampere fault, the unlatching time for both breakers is 0.0042 second and the interrupting time for both breakers is 0.016 second.

FIGURE 18–5   Nonselective system.

FIGURE 18–6   Selective system.

The term "interrupting time" refers to the time it takes for the circuit breakers' contacts to open, thereby stopping the flow of current in the circuit. Refer to Figures 18–5 and 18–6.

This example of a nonselective system should make apparent to the student the need for a thorough study and complete understanding of time–current curves, fuse selectivity ratios, and unlatching time data for circuit breakers. Otherwise, a blackout may occur, such as the loss of exit and emergency lighting. The student must be able to determine available short-circuit currents (1) to ensure the proper selection of protective devices with adequate interrupting ratings, and (2) to provide the proper coordination as well.

Figure 18-6 shows an example of a selective circuit in which the available fault current is only 500 amperes. This current trips the 20-ampere breaker instantly (the unlatching time of the breaker is approximately 0.0080 seconds and its interrupting time is 0.016 seconds). The graph in Figure 17-30 (page 302) indicates that the 100-ampere breaker interrupts the 500-ampere current in a range from 7 to 20 seconds. This relatively lengthy trip time range is due to the fact that the 500-ampere fault acts upon the current thermal trip element only, and does not affect the magnetic trip element, which operates on a current of 1000 amperes or more.

Selective System Using Fuses

The proper choice of the various classes and types of fuses is necessary if selectivity is to be achieved, Figure 18-7. Indiscriminate mixing of fuses of different classes, time–current characteristics, and even manufacturers may cause a system to become nonselective.

To ensure selective operation under low-overload conditions, it is necessary only to check and compare the time–current characteristic curves of fuses. Selectivity occurs when the curves do not cross one another.

Fuse manufacturers publish *selectivity guides,* similar to the one shown in Figure 18-8, to be used for short-circuit conditions. When using these guides, selectivity is achieved by maintaining a specific amperage ratio between the various classes and types of fuses. A selectivity chart is based on any fault current up to the maximum interrupting ratings of the fuses listed in the chart.

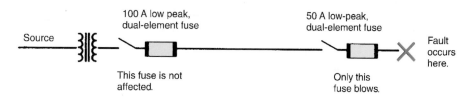

FIGURE 18–7   Selective system using fuses.

RATIOS FOR SELECTIVITY										
LINE-SIDE FUSE	LOAD-SIDE FUSE									
	KRP-C HI-CAP time-delay Fuse 601-6000A Class L	KTU LIMITRON fast-acting Fuse 601-6000A Class L	KLU LIMITRON time-delay Fuse 601-4000A Class L	KTN-R, KTS-R LIMITRON fast-acting Fuse Class K 0-600A	JJS, JJN TRON fast-acting Fuse Class T 0-600A	JKS LIMITRON quick-acting Fuse Class J 0-600A	FRN-R, FRS-R FUSETRON dual-element Fuse Class K 0-600A	LPN-R, LPS-R LOW-PEAK dual-element Fuse Class K 0-600A	LPJ LOW-PEAK time-delay Fuse Class J 0-600A	SC Type Fuse (Class G) 0-60A
KRP-C HI-CAP time-delay Fuse 601-6000A Class L	2:1	2:1	2.5:1	2:1	2:1	2:1	4:1	2:1	2:1	..
KTU LIMITRON fast-acting Fuse 601-6000A Class L	2:1	2:1	2.5:1	2:1	2:1	2:1	6:1	2:1	2:1	..
KLU LIMITRON time-delay Fuse 601-4000A Class L	2:1	2:1	2:1	2:1	2:1	2:1	4:1	2:1	2:1	N/A
KTN-R, KTS-R LIMITRON fast-acting Fuse 0-600A Class RK1	N/A	N/A	N/A	3:1	3:1	3:1	8:1	3:1	3:1	4:1
JJN, JJS TRON fast-acting Fuse 0-600A Class T	N/A	N/A	N/A	3:1	3:1	3:1	8:1	3:1	3:1	4:1
JKS LIMITRON quick-acting Fuse 0-600A Class J	N/A	N/A	N/A	3:1	3:1	3:1	8:1	3:1	2:1	4:1
FRN-R, FRS-R, FUSETRON dual-element Fuse 0-600A Class K	N/A	N/A	N/A	1.5:1	1.5:1	1.5:1	2:1	1.5:1	1.5:1	1.5:1
LPN-R, LPS-R LOW-PEAK dual-element Fuse 0-600A Class RK1	N/A	N/A	N/A	3:1	3:1	3:1	8:1	2:1	2:1	4:1
LPJ LOW-PEAK Fuse 15-600A Class J	N/A	N/A	N/A	3:1	1.5:1	1.5:1	8:1	1.5:1	2:1	2:1
SC Type Fuse 0-60A Class G	N/A	N/A	N/A	2:1	2:1	2:1	4:1	3:1	3:1	2:1

N/A = NOT APPLICABLE

FIGURE 18–8 Selectivity guide.

Selective System Using Circuit Breakers

Circuit-breaker manufacturers publish time–current characteristic curves and unlatching information, see Figure 17-30 (page 302).

For normal overload situations, a circuit breaker having an ampere rating lower than the ampere rating of an upstream circuit breaker will trip. The upstream breaker will not trip. The system is selective for normal overload currents.

For low-level faults (short circuits) less than the instantaneous trip setting of an upstream circuit breaker, a circuit breaker having an ampere rating lower than the ampere rating of the upstream breaker will trip, and the upstream breaker will not trip. The system is selective, Figure 18-6.

For fault-current levels above the instantaneous trip setting of the upstream circuit breaker, both the branch-circuit breaker and the upstream circuit breaker will trip off. The system is nonselective, Figure 18-5.

There are no ratio selectivity charts for breakers as there are for fuses.

▶ SINGLE PHASING

The *CEC* requires that all three-phase motors be provided with running overcurrent protection to each phase, *Rule 28–200*. A line-to-ground fault generally will blow one fuse. There will be a resulting increase of 173% to 200% in the line current in the remaining two connected phases. This increased current will be sensed by the motor fuses and overload relays when such fuses and relays are sized at 125% or less of the full-load current rating of the motor. Thus, the fuses and/or the overload relays will open before the motor windings are damaged. When properly matched, the overload relays open before the fuses. The fuses offer backup protection if the overload relays fail to open for any reason.

A line-to-line fault in a three-phase motor will blow two fuses. In general, the operating coil of the motor controller will drop out, thus providing protection to the motor winding.

To reduce *single-phasing* problems, each three-phase motor must be provided with individual overload protection through the proper sizing of the overload relays and fuses. Phase failure relays are also available. However, can other equipment be affected by a single-phasing condition?

In general, loads that are connected line-to-neutral or line-to-line—such as lighting, receptacles, and electric heating units—will not burn out under a single-phasing condition. In other words, if one main fuse blows, then two-thirds of the lighting, receptacles, and electric heat will remain on. If two main fuses blow, then one-third of the lighting remains on, and a portion of the electric heat connected line-to-neutral will stay on.

Nothing can prevent the occurrence of single-phasing. What must be detected is the increase in current that occurs under single-phase conditions. This is the purpose of motor overload protection as required in *Rule 28–300*.

It is essential to maintain some degree of lighting in occupancies such as stores, schools, offices, and health-care facilities (such as nursing homes). A total blackout in these public structures has the potential for causing panic and extensive personal injury. A loss of one or two phases of the system supplying a building should not cause a complete power outage in the building.

PROBLEM 5 It is desired to install 100-ampere, low-peak time-delay Class J fuses in a main switch, and a 50-ampere time-delay Class J fuse in the feeder switch. Is this combination of fuses selective?

Refer to the selectivity guide in Figure 18-8. The chart indicates that a 2:1 minimum ratio must be maintained for this type of fuse. Since 100:50 is a 2:1 ratio, the installation is selective. In addition, any fuse of the same type having a rating of less than 50 amperes will also be selective with the 100-ampere main fuse. That is, if a fault occurs on the load side of the 50-ampere fuse, only the 50-ampere fuse will open.

REVIEW

Note: Refer to the *CEC* or the blueprint package as necessary.

1. Using Table 18-1, determine the available short-circuit currents on a 208-volt system for the following:

   a. 15.24 m (50 ft) of No. 1 conductor _____ A

   b. 7.62 m (25 ft) of No. 3/0 AWG conductor _____ A

   c. 15.24 m (50 ft) of 500-kcmil conductor _____ A

   d. 15.24 m (50 ft) of two No. 3/0 conductors per phase _____ A

   e. 30.48 m (100 ft) of No. 4/0 conductor _____ A

2. a. Define *selectivity.* _____

   _____

   _____

   _____

   b. Define *nonselectivity.* _____

   _____

   _____

   _____

3. Indicate whether the following systems are selective or nonselective:

   a. a KRP-C 800-ampere fuse installed upstream of an LPS-R 200-ampere fuse

   _____

   b. an FRS-R 600-ampere Fusetron fuse installed ahead of an LPS-R 400-ampere low-peak fuse _____

   c. a KTS-R 400-ampere Limitron fuse installed ahead of an FRS-R 200-ampere Fusetron fuse _____

   d. a 225-ampere main circuit breaker in a panel that contains 20-ampere branch-circuit breakers, where the breakers have adequate interrupting capacity for the 20 000 amperes of available fault current _____

4. A series-connected panel contains a 1200-ampere main breaker with many branch-circuit breakers rated at 100 amperes and 225 amperes. The branch-circuit breakers have an individual interrupting rating of 10 000 amperes. The main circuit breaker has an individual interrupting rating of 42 000 amperes. In combination, the panel is CSA-listed as suitable for use at fault currents not to exceed 65 000 amperes.

   a. For a fault current above 10 000 amperes on the load side of one of the 225-ampere breakers, the main breakers (will trip) (will not trip). (Circle the correct answer.)

   b. For a fault current of 4000 amperes on the load side of one of the 100-ampere breakers, the main breaker (will trip) (will not trip). (Circle the correct answer.)

5. Refer to Figure 18-9 and calculate the available short-circuit current at Panel "A." This panel is supplied by a 7.6 m (25 ft) run of No. 3/0 copper conductors in steel conduit. Use the point-to-point method and show all calculations.

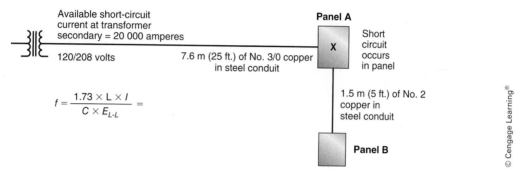

FIGURE 18-9   Diagram for Questions 5 and 6.

6. Calculate the available fault current if a short circuit occurs in Panel "B," Figure 18-9. Use the point-to-point method and show all calculations.

7. Calculate the available fault current if a short circuit occurs in the bakery panel of the commercial building shown in the blueprint package. Use the power riser diagram and building drawings to determine the size and length of conductor from the main switchboard to the bakery panel. Show all calculations.

8. A three-phase, L–L–L bolted short-circuit current has been calculated to be 40 000 RMS symmetrical amperes. What is the approximate value of:

   a. a line-to-line fault?   _____

   b. a line-to-ground fault? _____

Equipment and Conductor Short-Circuit Protection

OBJECTIVES

After completing the study of this unit, the student will be able to

- understand that all electrical equipment has a withstand rating

- discuss the withstand rating of conductors

- understand important *CEC* sections that pertain to interrupting rating, available short-circuit current, current-limitation, effective grounding, bonding, and temperature limitation of conductors

- understand what the term "ampere-squared seconds" means

- perform calculations to determine how much current a copper conductor can safely carry for a specified period of time before being damaged or destroyed

- refer to charts to determine conductor withstand ratings

- understand that the two forces present when short circuits or overloads occur are *thermal* and *magnetic*

- discuss the 3 m "tap" conductor installation, *Rule 14–100*

All electrical equipment—including switches, motor controllers, conductors, bus duct, panelboards, load centres, switchboards, and so on—has an ability to withstand a certain amount of electrical energy for a given amount of time before damage to the equipment occurs. This gives rise to the term "withstand rating."

Canadian Standards Association (CSA) standards specify certain test criteria for the above equipment. For example, switchboards must be capable of withstanding a given amount of fault current for at least three cycles. Note that both the amount of current and the length of time for the test are specified.

Simply stated, withstand rating is the ability of the equipment to "hold together" for the time it takes the overcurrent protective device (fuse or circuit breaker) to respond to the fault condition. Where the equipment is intended to break current, such as a fuse or circuit breaker, the equipment is marked with its interrupting rating. (See *Rule 14–012* for the definition of "interrupting rating.") Electrical equipment manufacturers conduct exhaustive tests to determine the withstand or interrupting ratings of their products.

Equipment tested in this manner is usually marked with the size and type of overcurrent protection required. For example, the label on a motor controller might indicate a maximum size fuse for different sizes of thermal overloads used in that controller. If the marking indicates that fuses must be used for the overcurrent protection, a circuit breaker would not be permitted.

The ability of a current-limiting overcurrent device to limit the let-through energy to a value less than the amount of energy that the electrical system is capable of delivering means that the equipment can be protected against fault-current values of high magnitude. Equipment manufacturers specify current-limiting fuses to minimize the potential damage that might occur in the event of a high-level short circuit or ground fault.

The electrical engineer and/or electrical contractor must perform short-circuit studies, and then determine the proper size and type of current-limiting overcurrent protective device that can be used ahead of the electrical equipment that does not have an adequate withstand or interrupting rating for the available fault current to which it could be

Study the normal-, overloaded-, and short-circuit diagrams (Figure 19-1, A–C) and observe how Ohm's law is applied to these circuits. The calculations for the ground-fault circuit (Figure 19-1, D) are not shown because the impedance of the return ground path can vary considerably.

■ CONDUCTOR WITHSTAND RATING

Up to this point in the text, we have covered in detail how to compute conductor ampacities for branch circuits, feeders, and service-entrance conductors, with the basis being the connected load and/or volt-amperes per square metre. Then the demand factors and other diversity factors are applied. If we have followed the *Canadian Electrical Code (CEC)* rules for calculating conductor size, the conductors will not be damaged under overload conditions because the conductors will have the proper ampacity and will be protected by the proper size and type of overcurrent protective device.

Tables in the CEC, such as *Table 2,* cover the ampacity ratings for various sizes and insulation types of conductors. These tables consider normal loading conditions. But what happens to these conductors under high-level short-circuit and ground-fault conditions? Small conductors often are the weak link in an electrical system.

When short circuits and/or ground faults occur, the connected loads are bypassed, thus the term "short circuit." All that remains in the circuit are the conductors and other electrical devices such as breakers, switches, and motor controllers. The impedance (AC resistance) of these devices is extremely low, so for all practical purposes, the most significant opposition to the flow of current when a fault occurs is the conductor impedance. Fault-current calculations are covered in Unit 18.

The following discussion covers the actual ability of a conductor to maintain its integrity for the time it is called upon to carry fault current, instead of burning away and causing additional electrical problems. It is important that the electrician, electrical contractor, electrical inspector, and consulting engineer have

A **Normal circuit**

The conductors can safely carry the current; they do not get hot. The 15 ampere fuses do not open.

Conductor rated 15 amperes
Conductor resistance = 0.001 ohm

15 A fuse   9.99875 amperes

240 volt source

Source resistance
= 0.001 ohm

15 A fuse

24 Ω   Load resistance
       = 24 ohms

Conductor rated 15 amperes
Conductor resistance = 0.001 ohm

$$I \text{ (through circuit)} = \frac{E}{R} = \frac{240}{24 + 0.001 + 0.001 + 0.001} = \frac{240}{24.003} = 9.99875 \text{ amperes}$$

B **Overloaded circuit**

The conductors begin to get "hot," but the 15 ampere fuses will open in some period of time before the conductors are damaged.

Conductor rated 15 amperes
Conductor resistance = 0.001 ohm

15 A fuse   19.995 amperes

240 volt source

Source resistance
= 0.001 ohm

15 A fuse

Load resistance
$\frac{R1}{N} = \frac{24}{2} = 12$ ohms

24 Ω   24 Ω

Conductor rated 15 amperes
Conductor resistance = 0.001 ohm

$$I \text{ (through circuit)} = \frac{E}{R} = \frac{240}{12 + 0.001 + 0.001 + 0.001} = \frac{240}{12.003} = 19.995 \text{ amperes}$$

FIGURE 19-1   These simple diagrams clearly illustrate the current flow in (A) a normal circuit, (B) an overloaded circuit, (C) a short circuit, and (D) a line-to-ground fault.

The withstand rating of a conductor, such as an equipment grounding conductor, main bonding jumper, or any other current-carrying conductor, indicates how much current the conductor can

Let us review some *CEC* sections that focus on the importance of equipment and conductor withstand rating, available fault currents, and circuit impedance.

- *Rule 14–012, Interrupting Rating: Equipment*

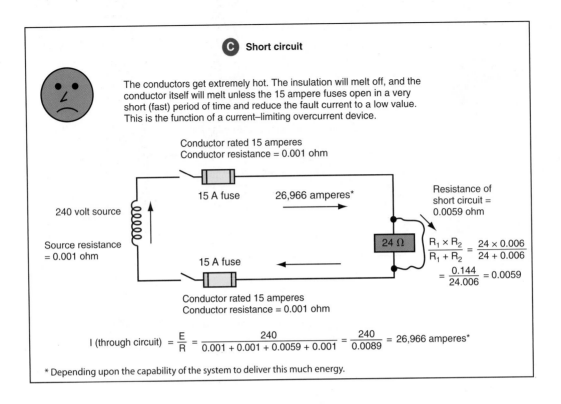

C Short circuit

The conductors get extremely hot. The insulation will melt off, and the conductor itself will melt unless the 15 ampere fuses open in a very short (fast) period of time and reduce the fault current to a low value. This is the function of a current–limiting overcurrent device.

Conductor rated 15 amperes
Conductor resistance = 0.001 ohm

15 A fuse   26,966 amperes*

240 volt source

Source resistance = 0.001 ohm

15 A fuse

Conductor rated 15 amperes
Conductor resistance = 0.001 ohm

Resistance of short circuit = 0.0059 ohm

24 Ω

$$\frac{R_1 \times R_2}{R_1 + R_2} = \frac{24 \times 0.006}{24 + 0.006}$$
$$= \frac{0.144}{24.006} = 0.0059$$

$$I \text{ (through circuit)} = \frac{E}{R} = \frac{240}{0.001 + 0.001 + 0.0059 + 0.001} = \frac{240}{0.0089} = 26{,}966 \text{ amperes*}$$

* Depending upon the capability of the system to deliver this much energy.

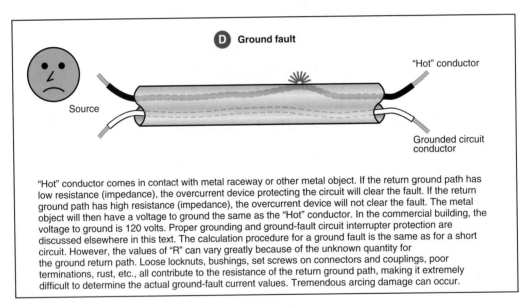

D Ground fault

"Hot" conductor

Source

Grounded circuit conductor

"Hot" conductor comes in contact with metal raceway or other metal object. If the return ground path has low resistance (impedance), the overcurrent device protecting the circuit will clear the fault. If the return ground path has high resistance (impedance), the overcurrent device will not clear the fault. The metal object will then have a voltage to ground the same as the "Hot" conductor. In the commercial building, the voltage to ground is 120 volts. Proper grounding and ground-fault circuit interrupter protection are discussed elsewhere in this text. The calculation procedure for a ground fault is the same as for a short circuit. However, the values of "R" can vary greatly because of the unknown quantity for the ground return path. Loose locknuts, bushings, set screws on connectors and couplings, poor terminations, rust, etc., all contribute to the resistance of the return ground path, making it extremely difficult to determine the actual ground-fault current values. Tremendous arcing damage can occur.

© Cengage Learning®

FIGURE 19-1 (Continued)

system voltage and the current that is available at the line terminals of the equipment.

- *Section 0, Definition of Overcurrent Device:* An overcurrent device is any device capable of automatically opening an electric circuit, under

- *Rule 10–002:* System and circuit conductors are grounded to limit voltages due to lightning, line surges, or unintentional contact with higher voltage lines, and to stabilize the voltage to ground during normal operation. Equipment grounding conductors are bonded to the system operation of overcurrent devices under ground fault conditions. Grounding and bonding limits alternating-current (AC) voltages to ground to 150 volts or less on circuits supplying interior wiring systems. It also protects life from the danger of electric shock and property damage by bonding noncurrent-carrying metal systems.

- *Rules 10–400* and *10–402:* Metal raceways, cable trays, cable armour, cable sheath, enclosures, frames, fittings, and other metal noncurrent-carrying parts shall be effectively bonded where necessary to assure electrical continuity and the capacity to conduct safely any fault current likely to be imposed on them. Any nonconductive paint, enamel, or similar coating shall be removed at threads, contact points, and contact surfaces or the equipment shall be connected by means of fittings so designed as to make such removal unnecessary.

- *Rule 10–812, Grounding Conductor Size for AC Systems:* The size of the grounding conductor for a grounded system shall not be less than No. 6 AWG.

- *Rule 10–814, Bonding Conductor Size:* The size of copper or aluminum equipment bonding conductors shall not be less than that given in *Table 16A.*

- *Rule 4–004(8), Ampacity of Wires and Cables:* Where the ambient temperature exceeds or is anticipated to exceed 30°C (86°F), the temperature correction factors from *Table 5A* shall apply.

CONDUCTOR HEATING

The amount of energy (heat) generated during a fault varies as the square of the RMS (root-mean-square) current multiplied by the time (duration of fault) in seconds. This value is expressed as I^2t. The term I^2t is called "ampere-squared seconds."

Since watts = I^2R, we could say that the damage that might be expected under severe fault conditions can be related to (1) the amount of current flowing during the fault, (2) the time in seconds that the fault current flows, and (3) the resistance

conditions is generally extremely low, however, we can simply think in terms of how much current (I) is flowing and for how long (t) it will flow.

It is safe to say that whenever an electrical system is subjected to a high-level short circuit or ground fault, less damage will occur if the fault current can be limited to a low value, and if the faulted circuit is cleared as quickly as possible.

It is important to understand the time-current characteristics of fuses and circuit breakers in order to minimize equipment damage. Time-current characteristic curves were discussed in Unit 17.

CALCULATING AN INSULATED (75°C THERMOPLASTIC) CONDUCTOR'S SHORT-TIME WITHSTAND RATING

Copper conductors can withstand

- one ampere (RMS current),
- for five seconds,
- for every 42.25 cmil (0.0214 mm^2) of cross-sectional area.

Note in the above statement that both current (how much) and time (how long) are included.

Let's take a real-world situation. If we wish to provide an equipment bonding conductor for a circuit fed with 60-ampere conductors, we find in *Table 16A* that a No. 10 copper conductor is the *minimum* size permitted. Referring to Figure 19-2,

AWG	CM AREA
14	4 110
12	6 530
10	10 380
8	16 510
6	26 240
4	41 740
3	52 630
2	66 360
1	83 690
0	105 600
00	133 100
000	167 800
0000	211 600

we find that the cross-sectional area of a No. 10 conductor is 10 380 cmil. This No. 10 copper conductor has a 5-second withstand rating of:

$$\frac{10\ 380\ \text{circular cmil}}{42.25\ \text{cmil}} = 246\ \text{amperes.}$$

This means that 246 amperes is the maximum amount of current that a No. 10 copper insulated conductor can carry for 5 seconds without being damaged. This is the No. 10 conductor's short-time withstand rating.

Stating this information using the *thermal stress (heat)* formula, we have:

$$\text{Thermal stress (heat)} = I^2 t$$

where

I = RMS current in amperes
t = time in seconds

Thus, the No. 10 copper conductor's 5-second withstand rating is:

$$I^2 t = 246 \times 246 \times 5$$

$$= 302\ 580\ \text{ampere-squared seconds}$$

With this basic information, we can easily determine the short-time withstand rating of this No. 10 copper-insulated conductor for other values of time and/or current. The No. 10 copper conductor's 1-second withstand rating is:

$$I^2 t = \text{ampere-squared seconds}$$

$$I^2 = \frac{\text{ampere-squared seconds}}{t}$$

$$I = \sqrt{\frac{\text{ampere-squared seconds}}{t}}$$

$$= \sqrt{\frac{302\ 580}{1}}$$

$$= 550\ \text{amperes}$$

The approximate opening time of a typical moulded-case circuit breaker is one cycle, that is, $\frac{1}{60}$ of a second or 0.0167 second. The No. 10 copper conductor's one-cycle withstand rating is:

$$I = \sqrt{\frac{302\ 580}{}}$$

The typical clearing time for a current-limiting fuse is approximately one-fourth of a cycle or 0.004 second. The No. 10 copper conductor's one-quarter-cycle withstand rating is:

$$I = \sqrt{\frac{302\ 580}{0.004}}$$

$$= 8697\ \text{amperes}$$

Therefore, a conductor can be subjected to large values of fault current if the clearing time is kept very short.

When applying current-limiting overcurrent devices, it is important to use peak let-through charts to determine the apparent RMS let-through current before applying the thermal stress formula. For example, in the case of a 60-ampere Class RK1 current-limiting fuse, the apparent RMS let-through current with an available fault current of 40 000 amperes is approximately 3000 amperes. See Figure 17-27 (page 298), and Figures 19-3 and 19-4.

$$\begin{aligned} I^2 t\ \text{let-through of current-limiting fuse} \\ = 3000 \times 3000 \times 0.004 \\ = 36\ 000\ \text{ampere-squared seconds} \end{aligned}$$

Therefore, the No. 10 AWG copper equipment grounding conductor could be used where the available fault current is 40 000 amperes, if the circuit is protected by current-limiting fuses.

With an available fault current of 100 000 amperes, the apparent RMS let-through current of the 60-ampere Class RK1 current-limiting fuse is approximately 4000 amperes. In this case

$$\begin{aligned} I^2 t = 4000 \times 4000 \times 0.004 \\ = 64\ 000\ \text{ampere-squared seconds} \end{aligned}$$

Remember that, as previously discussed, the withstand $I^2 t$ rating of a No. 10 copper conductor is 302 580 ampere-squared seconds.

EXAMPLE

A 75°C thermoplastic-insulated Type TW75 copper conductor can withstand 4200 amperes for one cycle.

Current Limitation Curves

FIGURE 19-3 Current-limiting effect of Class RK1 fuses. The technique required to use these charts is covered in Unit 17.

b. What is the I^2t let-through value for a non-current-limiting circuit breaker that takes one cycle to open? The available fault current is 40 000 amperes.

c. What is the I^2t let-through value for a current-limiting fuse that opens in 0.004 second when subjected to a fault of 40 000 amperes? The apparent RMS let-through current is approximately 4600 amperes. Refer to Figure 17-27 (page 298).

d. Which overcurrent device (b or c) will properly protect the conductor under the 40 000-ampere available fault current?

ANSWERS

a. I^2t = 4200 × 4200 × 0.016

   = 282 240 ampere-squared seconds

b. I^2t = 40 000 × 40 000 × 0.016

   = 25 600 000 ampere-squared seconds

c. I^2t = 4600 × 4600 × 0.004

   = 84 640 ampere-squared seconds

d. Comparing the I^2t withstand rating of the conductor to the I^2t let-through values of the breaker (b) and the fuse (c), the proper choice of protection for the conductor is (c).

The use of peak let-through, current-limiting charts is discussed in Unit 17. Peak let-through charts are available from all manufacturers of current-limiting fuses and current-limiting circuit breakers.

Conductors are particularly vulnerable to damage under fault conditions. Circuit conductors can be heated to a point where the insulation is damaged or completely destroyed. The conductor itself can actually burn away. In the case of equipment grounding conductors, if the conductor burns away under fault conditions, the equipment can become "hot," creating an electrical shock hazard.

Even if the equipment grounding conductor does not burn, it can become so hot that it melts the insulation on adjacent conductors, shorting out the other circuit conductors in the raceway or cable.

Of extreme importance are bonding jumpers, particularly the main bonding jumpers in service equipment, which must be capable of handling extremely high fault currents.

CALCULATING WITHSTAND RATINGS AND SIZING FOR BARE COPPER CONDUCTORS

A bare conductor can withstand higher levels of current than an insulated conductor of the same cross-sectional area. Do not exceed

- one ampere (RMS current),
- for five seconds,
- for every 30 cmil (0.0152 mm²) of cross-sectional area of the conductor.

Because bare equipment grounding conductors are often installed in the same raceway or cable as the insulated circuit conductors, the "weakest link" of the system would be the insulation on the circuit conductors. Therefore, the conservative approach when considering conductor safe withstand ratings is to apply the one ampere—for five seconds—for every 42.25 cmil formula, or simply refer to a conductor short-circuit withstand rating chart, Figure 19-5, and use the values for insulated conductors.

USING CHARTS TO DETERMINE A CONDUCTOR'S SHORT-TIME WITHSTAND RATING

The Insulated Cable Engineers Association, Inc. publishes much data on this subject. The graph in Figure 19-5 shows the withstand rating of 75°C thermoplastic-insulated copper conductors. Many engineers will use these tables for bare grounding conductors because, in most cases, the bare equipment grounding conductor is in the same raceway as the phase conductors. An extremely hot equipment grounding conductor in contact with

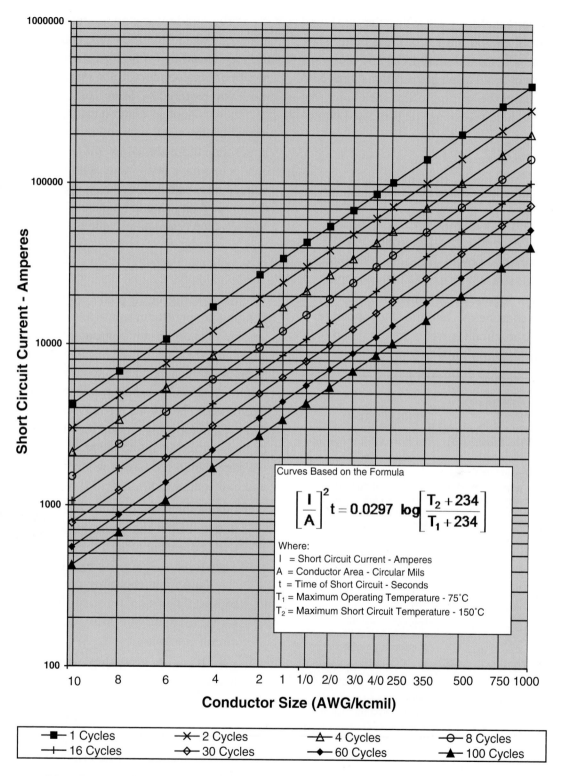

FIGURE 19-5 Allowable short-circuit currents for insulated copper conductors.

Table from ICEA P-32-382-2007 (R2013). Courtesy of Insulated Cable Engineers Association, Inc.

nonmetallic conduit is used, the equipment grounding conductor and the phase conductors are in the same raceway.

to the right, to the time of opening of the overcurrent protective device. Draw a line downward to the bottom of the chart to determine the conductor size short-circuit current chart is larger than the size found in *Table 16A*, then you must install the larger size. Installing a conductor too small to handle the available fault current could result in insulation damage or, in the worst case, the burning off of the equipment grounding conductor, leaving the protected equipment "hot."

EXAMPLE

A circuit is fed with No. 6 AWG Copper Type TW conductors. Determine the minimum-size equipment grounding conductor for available fault currents of 14 000 amperes and 40 000 amperes. Refer to the fuse peak let-through chart below, and the chart showing allowable short-circuit currents for insulated copper conductors, Figure 19-5.

AVAILABLE FAULT CURRENT	OVERCURRENT DEVICE	CONDUCTOR SIZE
40 000 amperes	Typical one-cycle breaker	No. 1/0 copper conductor
40 000 amperes	Typical RK1 fuse Clearing time 1/4 cycle or less	Using a 60-ampere RK1 current-limiting fuse, the apparent RMS let-through current is approximately 3000 amperes. *Table 16* shows a minimum No. 10 equipment bonding conductor. The allowable short-circuit current chart shows a conductor smaller than a No. 10. Therefore, No. 10 is the minimum-size equipment bonding conductor permitted.
14 000 amperes	Typical one-cycle breaker	No. 4 copper conductor
14 000 amperes	Typical RK1 fuse Clearing time 1/4 cycle or less	Using a 60-ampere RK1 current-limiting fuse, the apparent RMS let-through current is approximately 2200 amperes. *Table 16* shows a minimum No. 10 equipment bonding conductor. The allowable short-circuit current charts shows a conductor smaller than a No. 10. Therefore, No. 10 is the minimum-size equipment bonding conductor permitted.

If the conductor size as determined from the allowable short-circuit current chart is smaller than the size found in *Table 16A*, then install an equipment bonding conductor *not smaller* than the minimum size required by *Table 16A*.

There is another common situation in which it is necessary to install equipment bonding conductors larger than those shown by *Table 16A*. When circuit conductors have been increased in size because of a voltage-drop calculation, the equipment bonding conductor ampacity should be increased proportionately. Voltage-drop calculations were covered in Unit 3.

Summary

When selecting equipment bonding conductors, bonding jumpers, or current-carrying conductors, the amount of short-circuit or ground-fault current available and the clearing time of the overcurrent protective device must be taken into consideration so as to minimize damage to the conductor and associated equipment. The conductor must not become a fuse. The conductor must remain intact under any values of fault current it is subjected to. As stated earlier, the important issues are how much current will flow and how long the current will flow. The choices are:

1. Install current-limiting overcurrent devices that will limit the let-through fault current and will reduce the time it takes to clear the fault. Then refer to *Table 16A* to select the minimum-size equipment bonding conductor permitted by the *CEC*, and refer to the other conductor tables such as *Table 2* for the selection of circuit conductors.

2. Install conductors that are large enough to handle the full amount of available fault current for the time it takes a noncurrent-limiting overcurrent device to clear the short circuit or ground fault. Withstand ratings of conductors can be calculated or determined by referring to conductor withstand rating charts.

▶ MAGNETIC FORCES

Magnetic forces acting upon electrical equipment

expressed as $I_p{}^2$. Refer to Figure 19-4 for the case in which there is no fuse in the circuit. The peak current, I_p (available short-circuit current), is indicated as 30 000 amperes. The peak let-through current, I_p, resulting from the current-limiting effect of a fuse, is indicated as 10 000 amperes.

EXAMPLE

Because magnetic forces are proportional to the square of the peak current, the magnetic forces (stresses) on the electrical equipment subjected to the full 30 000-ampere peak current are *nine times* that of the 10 000-ampere peak current let-through by the fuse. Stated another way, a current-limiting fuse that can reduce the available short-circuit peak current from 30 000 amperes to only 10 000 amperes will subject the electrical equipment to *only one-ninth* the magnetic forces.

Visual signs indicating that too much current was permitted to flow for too long a time include conductor insulation burning, melting and bending of bus bars, arcing damage, exploded overload elements in motor controllers, and welded contacts in controllers.

A current-limiting fuse (either a dual-element fuse or a straight current-limiting fuse) must be selected carefully. The fuse must not only have an adequate interrupting rating to clear a fault safely without damage to the fuseholder, but it also must be capable of limiting the let-through current (I_p) and the value of I^2t to the withstand rating of the equipment it is to protect.

The graph shown in Figure 19-4 illustrates the current-limiting effect of fuses. The square of the area under the dashed line is energy (I^2t). I_p is the available peak short-circuit current that flows if there is no fuse in the circuit, or if the overcurrent device is not a current-limiting type. For a current-limiting fuse, the square of the shaded area of the graph represents the energy (I^2t), and the peak let-through current is I_p'. The melting time of the fuse element is t_m and the square of the shaded area corresponding to this time is the melting energy. The total clearing time, t_c, is the sum of the melting time and the arcing time. The square of the shaded area for time, t_c, is the total energy to which the circuit is subjected while the fuse clears. For the graph in Figure 19-4, the area under the dashed line is six times greater than the shaded area. Since energy is equal to the area squared, the circuit is subjected to 36 times as much energy when it is protected by noncurrent-limiting overcurrent devices.

TAP CONDUCTORS

Branch-circuit conductors, feeder conductors, or service-entrance conductors must have an ampacity not less than the maximum load served, or, put another way, the load must not exceed the ampere rating of the circuit, *Rule 8–104(1,2)*.

The basic overcurrent protection requirement in *Rule 14–100* is that conductors shall be protected at the point where they receive their supply of current and at each point where the size of conductor is reduced. There are seven exceptions allowed by *Rule 14–100*, with each pertaining to specific applications of conductor taps. These exceptions must be read and understood.

Of particular interest are feeder taps that are not over 3 m, Figure 19-6, and feeder taps that are not over 7.5 m, Figure 19-7. A number of requirements must be met when taps of this sort are made.

There are instances where a smaller conductor must be tapped from a larger conductor. For example, Figure 19-6 shows a large feeder protected by a 400-ampere fuse. The tap conductor shall have an ampacity not less than either the load supplied by the tap or the rating of the switch or panel it feeds.

The above example of tapping a small conductor from a larger conductor that is protected by an overcurrent device much larger than the small tap conductor is an indication that the *CEC* is concerned with the protection of the smaller conductor. Note that it is the length of the conductor—not the length of the conduit run between the splitter and the switch—that is used for the calculation.

If the length of the tap from the splitter block to the terminal in the switch or panel is over 3 m

FIGURE 19-7 *Rule 14–100(c),* the 1/3 tap rule. This rule allows the tap to be over 3 m but not over 7.5 m long. It must have an ampacity not less than 1/3 that of the larger conductor.

FIGURE 19-6 *Rule 14–100(b)* requires that the tap be not more than 3 m.

rule is often referred to as the "1/3 tap rule" because the ampacity of the tap conductor must be at least 1/3 the ampacity of the larger conductor feeding it. It also must terminate in an overcurrent device that has an ampacity equal to or less than the ampacity of the tap conductor. In Figure 19-7, the tap must

have an ampacity 1/3 that of the 430-ampere wire feeding the splitter. *Table 2* lists No. 1 Type R90 as wire suitable for 143 amperes.

Here again, the electrician must properly size the conductors for the load to be served, take into consideration the possibility of voltage drop, and then check the short-circuit withstand rating of the conductor to be sure that a severe fault will not cause damage to the conductor's insulation or, in the worst case, vaporize the conductor.

REVIEW

Note: Refer to the *CEC* or the blueprint package as necessary.

1. Define *withstand rating.* _____

   _____

2. The label of a motor controller is marked "Maximum size fuse." To conform to *Rule 2–024*, which of the following is correct? (Circle the letter of the best answer.)

   a. Install fuses as stated on the label.

   b. Install a circuit breaker ahead of the controller.

   c. Either (a) or (b) is correct.

3. According to *Rule 10–500*, an effective grounding path must have the capacity to
   _____

4. What does the label "current-limiting" on an overcurrent device refer to?
   _____
   _____
   _____

5. What table of the *CEC* shows the minimum-size equipment bonding conductor?
   _____

6. The heating effect caused by current flowing through a conductor can be expressed by the term I^2t. What does this term mean? _____
   _____
   _____
   _____

7. Copper 75°C insulated conductors can safely carry _____ amperes for _____ seconds for every _____ cmil of its cross-sectional area.

8. Calculate the maximum amount of current that a No. 12 copper 75°C insulated conductor can safely carry for five seconds.

9. What is the I^2t withstand rating of the No. 12 copper conductor in Question 8?

10. Calculate the maximum amount of fault current that the No. 12 copper conductor in Question 8 can withstand for:

   a. one cycle (0.016 seconds)

   b. one-quarter cycle (0.004 seconds)

11. Referring to Figure 19-3, what are the instantaneous peak let-through current and apparent RMS current for the following 600-volt Class RK1 fuses?

FUSE AMPERE RATING	AVAILABLE FAULT CURRENT	INSTANTANEOUS PEAK LET-THROUGH CURRENT	APPARENT RMS LET-THROUGH CURRENT
600	200 000	_____	_____
400	100 000	_____	_____
200	40 000	_____	_____
100	15 000	_____	_____
60	10 000	_____	_____

12. What is the minimum-size 75°C thermoplastic-insulated copper conductor permitted for the following available short-circuit currents? Derive answers by using Figure 19-5.

FAULT CURRENT	CLEARING TIME	CONDUCTOR SIZE
10 000 amperes	1 cycle	_____
10 000 amperes	$\frac{1}{2}$ cycle	_____
10 000 amperes	$\frac{1}{4}$ cycle	_____
30 000 amperes	1 cycle	_____
30 000 amperes	$\frac{1}{2}$ cycle	_____
30 000 amperes	$\frac{1}{4}$ cycle	_____

13. The available fault current is 50 000 amperes. A No. 3/0 Type TY75 copper feeder is protected by a 200-ampere overcurrent device. The conductors are installed in a nonmetallic raceway (PVC) and therefore require that a separate equipment bonding conductor be installed. The inspector states that if current-limiting fuses are installed to protect the feeder, the proper equipment bonding conductor is found directly in *Table 16.*

    a. This would be a(n) _____ equipment bonding conductor.

    b. If a standard moulded-case circuit breaker that has a one-cycle opening time is used as the feeder device, the inspector requires that a(n) _____ equipment bonding conductor be installed, according to Figure 19-5.

14. According to *Rule 14–100,* for taps not longer than 3 m, the ampacity of the tap conductor shall be (not less than) (three times) the load and (greater than) (not less than) (equal to) the rating of the switch or panel it is feeding. (Circle the correct answers.)

UNIT **20**

Low-Voltage Remote-Control Lighting

OBJECTIVES

After completing the study of this unit, the student will be able to

- list the components of a low-voltage remote-control wiring system

- understand the appropriate *CEC* rules governing the installation of a low-voltage remote-control wiring system

- demonstrate the correct connections for wiring a low-voltage remote-control system

LOW-VOLTAGE REMOTE-CONTROL LIGHTING

Conventional general lighting control is used in the major portion of the commercial building shown in the blueprint package. For the drugstore, however, a method known as *low-voltage remote-control lighting* is selected because of the number of switches required and because it is desired to have extensive control of the lighting.

Relays

A low-voltage remote-control wiring system is relay operated. The relay is controlled by a low-voltage switching system and in turn controls the power circuit connected to it, Figure 20-1. The low-voltage, split coil relay is the heart of the low-voltage remote-control system, Figure 20-2. When the *on* coil of the relay is energized, the solenoid mechanism causes the contacts to move into the *on* position to complete the power circuit. The contacts stay in this position until the *off* coil is energized. When this occurs, the contacts are withdrawn and the power circuit is opened. The red wire is the *on* wire, the black wire is *off*, and the blue wire is common to the transformer.

The low-voltage relay is available in two mounting styles. One style of relay is designed to mount through a 16-mm trade size knockout opening, Figure 20-3. For a 21-mm knockout, a rubber grommet is inserted to isolate the relay from the metal. This practice should ensure quieter relay operation.

FIGURE 20-2   Low-voltage relay.

The second relay mounting style is the *plug-in relay*. This type of relay is used in an installation where several relays are mounted in one enclosure. The advantage of the plug-in relay is that it plugs directly into a bus bar. As a result, it is not necessary to splice the line voltage leads.

Single Switch

The switch used in the low-voltage remote-control system is a normally open, single-pole, double-throw, momentary contact switch, Figure 20-4. This switch is approximately one-third the size of a standard single-pole switch and, in general, this type of switch has short lead wires for easy connections, Figure 20-5. These switch leads are colour-coded.

FIGURE 20-3   Low-voltage relay.

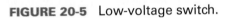

The white wire is common and is connected to the 24-volt transformer source, the red wire connects to the *on* circuit, and the black wire connects to the *off* circuit.

Master Control

It is often desirable to control several circuits from a single location. Up to eight low-voltage switches, Figure 20-6, will fit in the same space as that required for two conventional switches. This eight-switch master can be mounted on a $4\frac{11}{16}$-in (119-mm) box using an adapter provided with the master. Directory strips, identifying the switches' functions, can be prepared and inserted into the switch cover. An eight-switch master is used in the commercial building drugstore. The connection diagram is shown in Figure 20-11.

FIGURE 20-5   Low-voltage switch.

If the control requirements are complex or extensive, a master sequencer, Figure 20-7, can be installed. There is practically no limit to the number of relays that can be controlled, almost instantaneously, with this microprocessor-controlled electronic switch.

Master Control with Rectifiers

Relays can be wired such that they can be operated both individually by separate switches and as a group by a master control. The principle of

FIGURE 20-7 Master sequencer.

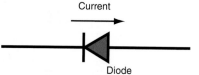

FIGURE 20-9 Rectifier connected to allow current in one direction and block current in opposite direction.

the switches, a master switching arrangement can be achieved, Figure 20-10. This method of master control is used in the drugstore.

Although a switching schedule is included in the specifications, the electrician may find it necessary to prepare a schematic diagram similar to that shown in Figure 20-11. The relays and rectifiers for the drugstore master control are located in the low-voltage control panel.

WIRING METHODS

Section 16 governs the installation of a low-voltage system. *Section 16* provisions apply to remote-control circuits, low-voltage relay switching, low-energy power circuits, and low-voltage circuits.

The drugstore low-voltage wiring is classified as a Class 2 circuit. By definition, the power source of a low-voltage circuit is limited. The circuit transformer, Figure 20-12, is designed so that in the event of an overload, the output voltage decreases as the current increases. Any overload can be counteracted by this energy-limiting characteristic of the specially designed transformer core. If the transformer is not self-protected, a thermal device may be used to open the primary side to protect the transformer from overheating. This thermal device resets automatically as soon as the transformer cools. Other transformers are protected with nonresetting fuse links or externally mounted fuses.

operation of a master control with *rectifiers* is based on the fact that a rectifier permits current in only one direction, Figure 20-8.

For example, if two rectifiers are connected as shown in Figure 20-9, current cannot flow from A to B, or from B to A; however, current can flow from A to C, or from B to C. Thus, if a rectifier is placed in one lead of the low-voltage side of the transformer, and additional rectifiers are used to isolate

Current

Diode

FIGURE 20-8 Rectifier showing direction of electron

Although the *CEC* does not require that the low-

FIGURE 20-10 Master control with rectifiers.

Courtesy of Sam Maltese

FIGURE 20-12   35-VA transformer.

this requirement. The advantage of using a raceway for the installation is that it provides a means for making future additions at a minimum cost. A disadvantage of this approach is the initial higher construction cost.

Conductors

Conductors smaller than No. 14 AWG may be used for low-voltage and remote-control systems. Conductors ranging from No. 16 AWG to No. 18 AWG are permitted for Class 1 circuits, *Rule 16–110*. Class 2 circuits may use conductors from No. 16 AWG to No. 26 AWG, *Rule 16–210*. No. 20 AWG is used as per the requirements found in Division 26 Electrical Specifications (Appendix). Larger conductors should be used for long runs to minimize the voltage drop. The

conductors may be either a double-cotton covering or plastic. *Rules 16–114* and *16–210* restrict the types of conductors allowed for use in low-voltage circuits. Regardless of the type of insulation used, the individual conductors in the cables can be identified readily. To simplify connections, low-voltage cables are available with various conductor combinations, including blue–white, red–black, and black–white–red. These colour-coded wires are connected like-colour to like-colour throughout the installation. A cable suitable for outdoor use, either overhead or underground, is available for low-voltage, remote-control systems.

Low-Voltage Panel

The low-voltage relays in the drugstore installation are to be mounted in an enclosure next to the power panel, Figure 20-13. A barrier in this low-voltage panel separates the 120-volt power lines from the low-voltage control circuits, in compliance with *Rule 16–212*.

◼ *SECTION 26 09 00* INSTRUMENTATION AND CONTROL FOR ELECTRICAL SYSTEMS

Low-Voltage, Remote-Control Switching

A low-voltage, remote-control switch system shall be installed in the drugstore as shown on the blueprints and detailed herein. All components shall be specification grade and constructed to operate on 24-volt control power. The transformer shall be a 120/24-volt, energy-limiting type for use on a Class 2 signal system.

Cabinet.   A metal cabinet matching the panelboard cabinets shall be installed for the installation of relays and other components. A barrier will separate the control section from the power wiring.

Relays.   Relays shall be 24-volt AC, split-coil

FIGURE 20-13   Typical low-voltage panel installation.

Switches.   The switches shall be complete with wall plate and mounting bracket. They shall be normally open, single-pole, double-throw, momentary contact switches with on-off identification.

Rectifiers.   Heavy-duty silicon rectifiers with 7.5-ampere continuous duty rating shall be provided.

Wire.   The wiring shall be in two- or three-conductor, colour-coded, No. 20 AWG wire.

Rubber Grommet.   An adapter will be installed on all relays to isolate the relay from the metal cabinet to reduce noise. The contractor is responsible to ensure the bonding of all metal parts.

Switching Schedule.   Connections will be made

SWITCH	RELAY	AREA SERVED	BRANCH CIRCUIT #
RCa	A	Main area lighting	7
RCb	B	" " "	9
RCc	C	" " "	9
RCd	D	Makeup area	11
RCe	E	Storage	11
RCf	F	Toilet	11
	G	Pharmacy	13
	H	"	13
	I	"	13
RCj	J	Stairway and basement	15
RCl, RCm	L,M	Show window	17, 19
	K	Sign	21
RCM-1	D		
RCM-2	E,F		
RCM-3	G,H,I		
RCM-4	I,J,K		
RCM-5	A,B,C		
RCM-6	J		
RCM-7	K		
RCM-8	L,M		

FIGURE 20-14   Drugstore low-voltage remote-

REVIEW

Note: Refer to the *CEC* or the blueprint package as necessary.

1. The low-voltage relay has a _____ coil design.

2. If the relay is in the *on* position and the *on* circuit is re-energized, the relay (will) (will not) change to the *off* position. (Circle the correct answer.)

3. Match the following items by placing the letter of the correct wire in the appropriate blank of the column on the left.

   a. _____ Relay *on* circuit   A. Red wire

   b. _____ Relay common circuit   B. Black wire

   c. _____ Relay *off* circuit   C. Blue wire

   d. _____ Switch *on* circuit   D. White wire

   e. _____ Switch *off* circuit   E. Green wire

   f. _____ Switch common circuit   F. Yellow wire

4. Indicate the current direction for the device shown by the symbol —|◁— (Diode)

   _____

5. What section of the *CEC* governs the installation of remote-control and signal circuits? _____

6. Low-voltage wiring (must) (need not) be installed in raceway to comply with the *CEC*. (Circle the correct answer.)

7. Complete the following connection diagram using rectifiers for the master control. Indicate wire colours.

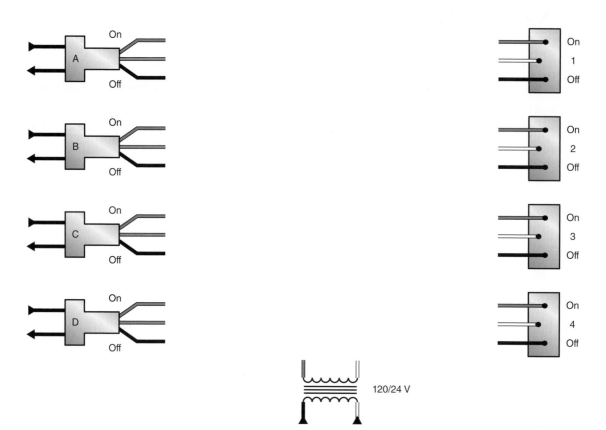

120/24 V

120 V source

Switching schedule

Switch	Control relay
1	A
2	B
3	C & D
4	A & B

Commercial Utility-Interactive Photovoltaic Systems

OBJECTIVES

After studying this chapter, you should be able to

- list the components of a utility-interactive solar photovoltaic system

- describe the function of utility-interactive solar photovoltaic system components

- apply the *Canadian Electrical Code (CEC)* to the design and installation of commercial utility-interactive solar photovoltaic system components

- interpret a typical utility-interactive solar photovoltaic system single-line diagram

INTRODUCTION

Utility-interactive solar photovoltaic (PV) systems have been around for many years, but the installation of these systems has recently accelerated due to energy conservation and environmental concerns. Because of some unique challenges involved with the electrical work for a photovoltaic system, *Sections 64* and *84* of the *CEC* will apply to the installation. *Section 84* of the *CEC* is involved because the PV system and the systems provided by the electric utility can operate in parallel. It is vital that commercial electricians be aware of these requirements, as commercial photovoltaic installations tend to be larger and more complex than residential systems. Engineering and installation of commercial photovoltaic systems will often involve the electrical, structural, and roofing trades. This unit will focus on electrical requirements of the *CEC* and related electrical installation techniques that are common in the industry.

Many of the *CEC* requirements for disconnecting means, grounding/bonding, labelling, and photovoltaic system interconnection to the building's electrical system exist because the photovoltaic system is a second supply of electricity for the building or structure. A utility-interactive solar photovoltaic system is not just another load for a building's electrical system: the photovoltaic system is another source, much like an emergency or standby generator. One big difference is that a photovoltaic supply cannot be turned off as easily as a generator; much of the photovoltaic system will be energized anytime the sun is shining. Proper installation is therefore critical for building occupants and first responders in the event of an emergency.

THE PHOTOVOLTAIC EFFECT

When particles of light (photons) are absorbed by specialized combinations of semiconductors (diodes), electrical current is produced; this is known as the photovoltaic effect. Although the effect was first observed in the mid-1800s, it was not until material used today for production of photovoltaic cells is crystalline silicon, which is fabricated into wafers and then doped (infused) with phosphorus gas to create the photovoltaic semiconductor.

Further processing produces a functioning photovoltaic cell. Completed cells are combined to produce the modules used for photovoltaic systems. Over the years, photovoltaic modules have become much more efficient and costs of manufacturing have dropped. The increased expense and environmental concerns associated with the use of fossil fuels have helped to make photovoltaic systems a cost-effective source of electricity.

THE BASIC UTILITY-INTERACTIVE PHOTOVOLTAIC SYSTEM

The photovoltaic effect can be put to practical use by supplying a commercial building with some or all of the electrical power needed to operate. System sizes range from 2 kW to 1 mW and larger. The photovoltaic modules in the array produce direct current (DC) electricity when exposed to sunlight. Direct current from the modules supplies an inverter that converts the DC into alternating current (AC) synchronized with utility electricity. The electrical energy produced by a photovoltaic system can be stored through the use of batteries that then supply the building's electrical needs at night or on overcast days. A large enough system like this would not even need a utility supply of electricity. Such stand-alone systems are not as common or as practical as utility-interactive photovoltaic systems, however, and the normal utility electrical supply is always present in a utility-interactive photovoltaic system. Electrical energy produced by the photovoltaic system reduces the amount of electricity supplied by the utility. Bidirectional electrical metering equipment installed by the utility accurately measures the power produced by the PV system, and if the photovoltaic system produces more electricity than is required for the building's electrical load, the surplus electricity is sold to the utility at predetermined rates. Many electrical utilities subsidize the programs that help to offset the cost of installation. A separate permit for installation of utility-interactive PV systems is usually required by local jurisdictions. Figures 21-1 and 21-2 identify the basic photovoltaic components and system configurations.

▶ UTILITY-INTERACTIVE PHOTOVOLTAIC SYSTEM COMPONENTS

Modules

Photovoltaic modules are the basic building block of a utility-interactive photovoltaic system. Individual photovoltaic semiconductor cells are electrically connected together, sealed in a protective material, and mounted to a frame at the factory (Figure 21-3). Modules are produced using different technologies, which result in modules of differing efficiencies, sizes, and production costs. The amount of electrical power or watts produced varies as well. All modules are tested and listed to UL Standard 1703. Manufacturers' specification sheets will list all pertinent information and allow comparison of modules.

Disconnects

Disconnecting means are required in various locations throughout a photovoltaic system, and

Notes:
1. These diagrams are intended to be a means of identification for photovoltaic system components, circuits, and connections.
2. Disconnecting means required by *CEC 64–060* are not shown.
3. System grounding and equipment grounding are not shown. See *CEC 64–064*.

FIGURE 21-1 Identification of solar photovoltaic system components.

Notes:
1. These diagrams are intended to be a means of identification for photovoltaic system components, circuits, and connections.
2. Disconnecting means and overcurrent protection required by *CEC 64–060* are not shown.
3. System grounding and equipment grounding are not shown. See *CEC 64–064*.
4. Custom designs occur in each configuration, and some components are optional.

FIGURE 21-2 Identification of solar photovoltaic system components in common system configurations.

FIGURE 21-3   Fused combiner box.

Rule 64–060 of the *CEC* provides requirements for their installation. *CEC 64–060(1) & (2)* require a disconnecting means for the source (array) ungrounded conductors, located either on the exterior of the building or nearest the point of entrance of photovoltaic system conductors. The DC grounded conductor is not generally permitted to be interrupted by disconnect contacts and must pass straight through the disconnect enclosure if operation of that switch, circuit breaker, or other device leaves the marked, grounded conductor in an ungrounded and energized state or condition, *CEC 64–060(5)*.

Any disconnect used for direct current must be suitable for use with DC and will be labelled as such by the manufacturer, *CEC 64–060(9) & (10)*. Specific instructions that manufacturers provide for using a disconnect with DC must be followed. For example, some manufacturers will require that two poles of a knife blade disconnect be used to interrupt a single DC circuit conductor. Direct current is more difficult to interrupt than alternating current, which is why this arrangement may be required. *CEC 64–060(4)* states that disconnecting means for inverters must be provided as well. Some inverters will have integral disconnects; those that don't will require installation of separate disconnects. Inverters will have connections

be energized even when the DC input is turned off. Why? Because the AC output of the inverter is connected to existing energized AC premises or service conductors. These conductors will still be energized even though the inverter is not producing electrical current. *CEC 64–060(8)* requires that equipment such as an inverter be able to be disconnected from ungrounded conductors of all sources. Where that equipment is energized from two sources (such as inverters), the disconnecting means is required to be grouped and identified. All equipment disconnecting means must be located within sight and 9 metres of the equipment, *CEC 64–060(2)(a)*. A disconnecting means is a function, not always a separate device.

Inverters

The equipment that converts the direct current produced by modules into an AC output is known as an inverter. A utility-interactive inverter will automatically synchronize the inverter output to the utility system frequency. Proper phase and voltage relationships must be maintained between an inverter output and the AC voltage and phase as supplied by the utility. Inverter output capacities range from 2 kW to 500 kW and more. See Figure 21-4 for an example of a medium size utility-interactive inverter. Commercial photovoltaic systems with multiple inverters producing a mega-watt or more of electrical energy are becoming more common.

PV system designers and installers should be aware of two key requirements of the standard.

1. In a process known as "anti-islanding," utility-interactive inverters are required to turn off automatically when utility power is lost. This prevents an inverter from feeding electricity back into the utility electrical grid, which would be extremely dangerous for utility workers. Remember that the 240 volts produced by an inverter could produce 12 000 volts or more on the primary side of utility transformers if the inverter were allowed to back-feed the utility system. In effect, the utility transformer would be operating as a step-up transformer.

2. There must be DC ground-fault protection (GFP). Ground faults in PV source circuits (array) do not result in the high fault currents produced by utility electrical services, as photovoltaic modules are a current-limited source. Short-circuit current from a module is slightly higher than normal module output to the inverter, but not enough to open an overcurrent device (fuse or circuit breaker). Ground-fault current will flow through bonded exposed metallic structures and raceways to the inverter and then back to the source (modules). Heating and arcing caused by fault current can cause fires and is a shock hazard as well. For these reasons, a ground-fault protection system is required to detect a ground fault in the photovoltaic array, interrupt the ground-fault current, indicate that a ground fault occurred, and disconnect the faulted circuit.

Ground-fault protection is very important for photovoltaic systems. A ground fault can continue as long as modules are producing current, which happens anytime the sun is shining. A single stage ground fault (positive or negative conductor to grounded surface) will be interrupted by

fault with both positive and negative conductors shorted together through common metallic structures/raceways on the roof will continue regardless of inverter action. Remember that modules in the array are the source of electrical current. If there is a complete path of current for the positive and negative conductors on the roof, current will flow. Either the positive or negative source (array) conductor will be grounded. Grounding of this current-carrying conductor will occur in the inverter and is determined by inverter design. *CEC 64–216* requires arc-fault protection when the DC source and/or output circuits on or in a building are operating at or above 80 volts. Arc-fault protection for the DC source/output circuits is required because damaging arc faults have occurred that the ground fault protection system cannot prevent. The *CEC* requires a listed system that will perform three functions:

1. Detect and interrupt arc faults in the DC PV source and output circuits, *CEC 26–216(2)(a)*.

2. Require a manual reset of the disabled or disconnected equipment, *CEC 26–216(2)(b)*.

3. Announce the arc fault, *CEC 26–216(2)(c)*.

RAPID SHUTDOWN OF PV SYSTEMS ON BUILDINGS

CEC 64–218 lists when a rapid shutdown device is required. Its primary function is to limit the photovoltaic source or output circuits to not more than 30 V and 240 VA within 10 seconds of initiation.

UTILITY-INTERACTIVE PHOTOVOLTAIC PLANS

A permit and associated inspection(s) by the authority having jurisdiction (AHJ) will be required for photovoltaic systems in most jurisdictions. Commercial PV projects will typically require engiof a permit. A utility-interactive solar photovoltaic plan will consist of a plot or site plan, an electrical single-line diagram, a structural plan, and roof penetration (flashing) details. Depending on the size and complexity of a project, there could be several pages of structural and electrical details. Manufacturers' specifications and installation instructions for key components should be provided because these form the basis for the design.

Site or Plot Plan

The site or plot plan will indicate the geographic locations of principal PV components and existing electrical service or affected panelboards. Photovoltaic module quantity and locations will be indicated along with locations of all disconnects, combiner boxes, inverters, and junction boxes. Conduit or raceway types and routing should also be indicated.

Single-Line Diagram

CEC Rule 84–030 requires a single-line diagram that should identify related components of the interconnected system, including switching arrangements, interlocks, isolation points, and their relative locations. See Figure 21-5.

Calculations

Calculations are needed to support the single-line diagram. The designer will use information from manufacturers' product specification sheets to make calculations for the sizing of module strings, arrays, inverters, disconnects, overcurrent devices, and the conductors/raceways that connect all of these components together. Required calculations can be broken down into two broad categories: AC and DC.

DC Calculations.   At a minimum, two DC calculations will be required.

1. Open circuit voltage (Voc). Quantity of modules in a series string will be limited by the open circuit voltage of the modules. The Voc (provided by the manufacturer's specification sheet) of each module in a series string is added together to provide total string voltage. For example, a string of 10 modules, each with a Voc of 50 volts, would have a total string Voc of 500 volts (10 modules × 50 volts = 500 volts). To comply with *CEC 64–202*, temperature must also be considered. *CEC 64–202(2)(a)* and *(b)* defines the maximum PV system voltage as the sum of the series-connected modules corrected to the lowest expected ambient temperature. Module Voc goes up as ambient temperature goes down—a colder module will have a higher Voc than a warmer module. The Voc provided by the manufacturer results from a standard test condition (STC) of 77°F (25°C). Actual Voc of the module/string will depend on ambient temperature where installed.

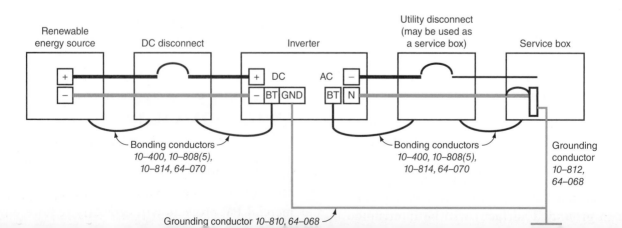

2. Current. The ampere ratings of photovoltaic source and output circuits are defined by *CEC 64–206*.

AC Calculations.   The AC side (inverter output) of a utility-interactive photovoltaic system will require calculations to size conductors and components as well. *CEC 64–100(1)* defines maximum inverter output current as the inverter continuous output rating. This information will be found in the manufacturer's specifications for the inverter. Dividing the inverter power rating (kW) by output voltage is another way to determine output current. A 10 kW (10 000 watt) inverter operating at 240 volts will have a rated output current of 42 amps (10 000 ÷ 240).

▶ UTILITY-INTERACTIVE PHOTOVOLTAIC SYSTEM INSTALLATION

The first order of business when starting a new commercial PV installation is to review the approved project plans and specifications. It is critical to verify that the site matches what is indicated on the approved plan, and any discrepancies should be reported to the engineer of record. A utility-interactive PV system will ultimately be integrated into the existing electrical system for the building. Does the existing electrical system match the design? Are locations of the existing service and panelboards accurate? Can the existing grounding electrode system be accessed? This is the time to ask these questions and inquire about any local requirements (AHJ and utility). Material, tool, and manpower needs based on the project schedule can be assessed at this point to keep the project running smoothly. Personal protective equipment (PPE) requirements should be assessed so that this safety equipment will be onsite when needed. Logistics such as site access, employee parking, and working hours should be coordinated with the owner or manager of the building.

Mounting of Array

determinant of whether the array is ground or roof mounted. A ground-mounted array may be constructed so that it can be used as covered vehicle parking. Where ground space is not available, it is common for a large commercial building to have most of its roof covered with modules. This could be a concern for local firefighters who could need roof access for firefighting. The *CEC* does not directly address this issue, but some jurisdictions have requirements that restrict the footprint of the array on a roof.

Array size and footprint must be reviewed for compliance with local requirements during the plan review process. Whether installed on the roof or in a parking lot, structural elements for support of the array will be the first priority. A ground-mounted array will have structural footings and framing designed by a structural engineer. A roof-mounted array may make use of existing supports, but many times additional structural framing is required. These additional elements will be detailed on the structural plan. The electrician is not normally directly responsible for the structural installation but may be involved in the layout of supports. Locations for conduit stub-ups for electrical components (whether underground or on the roof) must be coordinated with structural elements to access the final locations of combiner boxes, junction boxes, disconnects, inverters, panelboards, and services.

Once the structural layout and installation is started, the electrician can begin the electrical work. The electrical installation can be thought of as two separate systems: DC and AC. The DC system involves the conductors and components from the modules to the inverter. The AC system includes everything from the inverter output to the ultimate utility connection. Multiple AC panelboards, transformers, and disconnects may be required. The order of the electrical work will depend on the scheduling of the other elements of the project. For example, array source conductors may have to be installed as modules attached to the support racks. A large array of modules will be connected in a specific series-parallel combination in accordance with the single-line diagram.

Support of these conductors is of utmost importance. Array source conductors are permitted to be exposed, so good workmanship is the only protection against short circuits. Sharp edges must be avoided when routing the conductors. Only type *USE-2* or the newer type PV conductors are permitted to be installed as exposed source conductors. Cable ties used to secure the source conductors must be rated for ambient temperature and ultraviolet (UV) light exposure.

Module Grounding

Grounding of individual modules may have to be accomplished as the modules are installed. Module manufacturers will provide very specific grounding information in the installation instructions. The low currents associated with module ground faults make these connections extremely important. Ground-fault current must have a reliable low impedance path from the modules to the inverter for the life of the system. Module frames are anodized, clear-coated, or mil-finished aluminum. Aluminum oxidizes quickly, and copper conductors must be terminated properly to ensure a good connection. Using the grounding hardware and module frame grounding location specified in the module installation instructions is the only way to ensure connection integrity for the life of the system. The equipment grounding conductor attachment to module frames can be seen in Figure 21-6.

Sections 64 and *10* of the *CEC* list requirements for grounding and bonding; see Figure 21-6.

▶ SUMMARY

Although utility-interactive photovoltaic systems are constructed with many of the same electrical materials used in the industry, this does not mean that there are not differences. Significant portions of the array will always be energized anytime the sun is up. These always "hot" conductors will usually

FIGURE 21-6 Connection of equipment grounding conductor to module frames.

be operating close to the 600-volt insulation rating and exposed to weather. An array short circuit will not produce enough fault current for an overcurrent device to be effective. Contrast this with a typical roof-mounted HVAC unit. The 600-volt branch circuit for the HVAC unit is installed in conduit, originates at an overcurrent device (circuit breaker) in a panelboard, and usually terminates in a fused disconnect at the unit. Turning off the building main service disconnect will de-energize the HVAC branch circuit, but not the photovoltaic array conductors! Good workmanship in regard to module source string conductors and grounding/bonding of modules is very important. Inverter AC output to the building is a second supply of electrical power, and frequently a second electrical service connection is made for these systems. Location, disconnecting means, labelling, and grounding/bonding for this part of the system are vital. Utility-interactive photovoltaic systems are becoming more efficient by the day. Continued improvement of the systems along with a never-ending need for increased electrical power means that electricians who are able to install and maintain photovoltaic systems will be in demand for many years to come.

REVIEW

1. Is the electrical current produced by photovoltaic modules AC or DC?

   _____

   _____

2. A group of modules that form a complete photovoltaic power source is known as a(n)

   _____

   _____

3. As module temperature decreases, module open circuit voltage (Voc)

   _____.

4. Photovoltaic modules must be grounded per the _____ instructions.

5. Describe the function of a combiner box.

   _____

   _____

   _____

6. Describe the function of a utility-interactive inverter.

   _____

   _____

   _____

7. List three functions arc-fault protection must perform to provide protection for the DC source/output circuits.

   _____

   _____

   _____

8. Which *CEC* rule provides the requirements for grounding electrode systems in a photovoltaic installation?_____

9. Which *CEC* section provides the utility connection requirements for utility-interactive photovoltaic systems?_____

Appendix

Part A: Electrical Specifications

◼ SECTION 26 00 00 ELECTRICAL

General

Comply with *Division 01*—General Requirements. The specifications for this division are an integral part of the contract documents and shall be read accordingly.

Electrical drawings indicate only the general location of equipment and outlets. The contractor shall be responsible for the detail layout of equipment, outlets, raceways, and wiring.

Outlets shown on architectural room elevations take precedence over positions or mounting heights on electrical drawings.

The owner reserves the right to move the location of outlets 3 metres without additional charge providing the contractor is advised prior to installation.

Refer to architectural and structural drawings for accurate building dimensions.

Scope

The electrical contractor shall furnish all labour, tools, equipment, and material to install, test, and put into operation the electrical systems shown in the drawings or herein specified or both.

Submissions

Immediately after this contract is awarded the contractor shall submit a list of equipment, fixtures, and materials to be incorporated into the work including:

- Manufacturer's name
- Equipment numbers
- Catalogue numbers
- Equipment description
- Delivery date

The list must have the approval of the architect before work is commenced.

Shop Drawings

The contractor shall submit shop drawings, catalogue cuts, and descriptive literature of all equipment installed as part of this contract. Shop drawings are particularly required for the following: detail drawings of the engine-generator set and base, and complete wiring diagrams and assembly drawings including the autostart and safety controls.

Permits, Inspections, Certificates, and Fees

The contractor shall deliver to the owner final certificates of inspection and approval by the inspection authority having jurisdiction when all work has been completed, tested, and placed in operation.

Copies of electrical plans and specifications should be submitted as required by electrical inspection and supply authorities.

The contractor will notify inspection authorities in sufficient time to inspect work and pay all fees and costs.

Delivery, Handling, and Storage

The contractor shall assume full responsibility to receive, handle, store, and protect equipment, fixtures, and materials covered under this contract.

Record Drawings

The contractor will provide record drawings showing the locations of buried raceways and cables dimensioned to the centre line of building columns. Elevations should be shown with respect to finished grade or finished floor. The location of all access panels for junction boxes, pull boxes, and terminal cabinets should be shown, along with all changes in the location and sizes of equipment, outlets, and wiring, and such other changes as may occur during the work.

Satisfactory record drawings must be filed with the architect before a final certificate of acceptance

Relations With Other Trades

The work of the electrical trade shall be carried out in cooperation with other trades in such a manner as to avoid unnecessary delays and ensure the work of all trades is installed to the best advantage. The contractor shall assume full responsibility for laying out the work and for any damage to the owner or other trades caused by improper layout.

Workmanship

All work shall be in accordance with the *Canadian Electrical Code (CEC)* as well as provincial and local codes. Workmanship shall be of a uniformly high quality in regard to durability, efficiency, safety, and neatness of detail. Inferior work shall be replaced without cost when so ordered by the architect.

Cutting and Patching

The contractor shall bear all costs of cutting and patching resulting from the work of this division.

Cleaning

The contractor shall clean up all refuse caused by the work at the end of the day and remove waste on a weekly basis. On completion of the work, the contractor shall remove all surplus materials and waste.

Supports and Hangers

The contractor shall provide all supports, hangers, inserts, and sleeves necessary to properly execute the work.

Testing

The contractor will test and check electrical and communications systems for correct operation in the presence of the engineer.

An insulation resistance test will be conducted using a "megger" (500-volt instrument on circuits up to 350 volts) on all lighting and power circuits prior to connecting devices and equipment. All test results should be recorded and submitted to the or replace at no cost all circuits that do not meet the minimum requirements of the inspection authority.

In cooperation with mechanical trades, the contractor will take and record clamp-on ammeter readings of mechanical equipment motors while the equipment is operating under load.

Electrical system voltages should be checked after the facility has been in operation for 60 days and tap settings adjusted as required.

▶ BASIC MATERIALS AND METHODS

Products

The contractor will obtain from the engineer a list of acceptable equipment manufacturers. All material shall be new and bear the label of the Canadian Standards Association or other approval agency recognized by the inspection authority having jurisdiction. Unless otherwise indicated, factory finish all equipment with ANSI/ASA #61 grey paint.

Section 26 05 33 Raceway and Boxes for Electrical Systems

The contractor shall install exposed conduits symmetrical to the building construction. Conduits should be routed to avoid beams, columns, and other obstructions, and should cause minimum interference in spaces through which they pass.

Conduits should be grouped together whenever possible. Exposed conduits should be securely attached in place at intervals not exceeding the requirements of the *CEC*.

Three spare 27-mm trade size conduits shall be installed from each flush-mounted panel into the ceiling joist space above the panel.

Feeders shall be installed in either rigid metal conduit or rigid non-metallic conduit. The supply to the boiler shall be rigid metal conduit.

In the doctor's office, all branch-circuit conductors will be installed in electrical metallic tubing. In all other areas of the building, branch-circuit con-

Section 26 05 19 Low-Voltage Electrical Power Conductors and Cables

The conductor shall provide wires of the number and size shown in the drawings or described herein with spare conductors as indicated for the complete electrical system.

Pulling tensions and minimum bending radii should be limited to manufacturer's recommendations.

The contractor shall utilize cable lubricant when pulling cables through conduits and ducts.

Conductors shall have a 90°C rating, types RW90-XLPE (600 V unjacketed) or T90 Nylon unless specified otherwise. All conductors shall be copper.

The minimum-size conductors will be No. 12 AWG.

The neutral conductor of any feeder circuit shall not be smaller than the other circuit conductors.

Do not connect more than three lighting circuits from a three-phase panel or two circuits from a single-phase panel to a common neutral.

Do not exceed the *CEC* requirements for voltage drop on any branch circuit or feeder.

Colour code feeder and branch circuits maintaining phase and colour sequence throughout.

Underground Installation

Direct buried cables shall be installed in sifted sand free of rock, stone, and other sharp objects with a 75-mm layer above and below.

Where direct buried cables pass under a roadway or area subject to vehicular traffic, install in suitable concrete-encased ducts.

Section 26 05 33 Raceways and Boxes for Electrical Systems

Access panels shall be provided in ceilings where pull or junction boxes are not readily accessible.

The contractor shall neatly make cutouts for

Section 26 27 26 Wiring Devices

Switches shall be AC general-use, specification grade, 20 amperes, 120–277 volts, either single-pole, three-way, or four-way, as shown on the plans.

When two or more switches are required at the same location, they shall share a common wall plate unless otherwise noted.

The electrical contractor shall furnish and install, as indicated in blueprint E-1 (Electrical Symbol Schedule), receptacles meeting Canadian Standards Association (CSA) standards and approved by the CSA. No more than eight 5-15R receptacles are to be on one 2-wire circuit.

Wall plates for switches and receptacles shall be stainless steel unless otherwise noted on the drawings. Install wall plates after the painting of room surfaces has been completed.

Section 48 11 00 Fossil Fuel Plant Electrical Power Generation Equipment

The electrical contractor will furnish and install a motor-generator plan capable of delivering 12 KVA at 120/208 volts, wye-connected three-phase. Motor shall be for use with diesel fuel, liquid cooled, complete with 12-volt batteries and battery charger, mounted on anti-vibration mounts with all necessary accessories, including mufflers, exhaust piping, fuel tanks, fuel lines, and remote derangement annunciator.

■ SECTION 26 20 00 LOW-VOLTAGE ELECTRICAL TRANSMISSION

Service Entrance Board

The electrical contractor will furnish and install a service entrance as shown in the electrical power distribution diagram, which is found in the plans. The supply authority will install transformers and all primary service. The electrical contractor will provide a concrete pad and conduit rough-in as required by the supply authority.

500-kcmil conductors each set in rigid metallic or nonmetallic conduit. Conductors will be RW90-XLPE (600 V) copper.

The electrical contractor will furnish and install service entrance equipment, as shown in the plans and detailed herein. The equipment will consist of nine switches, and the metering equipment will be fabricated in three type NEMA 1 sections. A continuous neutral bus will be furnished for the length of the equipment and shall be isolated except for the main bonding jumper to the grounding bus, which shall also be connected to each section of the service-entrance equipment and to the water main. The switchboard shall be braced for 25 000 RMS symmetrical. The metering will be located in one section and shall consist of seven meters. Five of these meters shall be for the occupants of the building and two shall serve the owner's equipment.

The switches shall be as follows:

1. Bolted pressure switch, three pole, 1200 amperes with three 1000-amp fuses.

2. Bolted pressure switch, three pole, 800 amperes with three 700-amp fuses.

3. Quick make quick break switch, three pole, 100 amperes with three 90-amp fuses.

4. Quick make quick break switch, three pole, 200 amperes with three 175-amp fuses.

5. Quick make quick break switch, three pole, 200 amperes with three 125-amp fuses.

6. Quick make quick break switch, two pole, 100 amperes with two 100-amp fuses.

7. Quick make quick break switch, three pole, 100 amperes with three 90-amp fuses.

8. Quick make quick break switch, three pole, 60 amperes with three 45-amp fuses.

9. Quick make quick break switch, three pole, 100 amperes.

The bolted pressure switches shall be knife-type switches constructed with a mechanism that automatically applies a high pressure to the blade when the switch is closed. Switch shall be rated to interrupt 200 000 RMS symmetrical amperes when used in opening or closing the switch to minimize arcing. Switches shall be rated to interrupt 200 000 RMS symmetrical amperes when used with current-limiting fuses that have an equal rating. They shall have fuse clips for Class J fuses.

Fuses

Fuses 601 amps and larger shall have an interrupting rating of 200 000 RMS symmetrical amperes. They shall provide a time delay of not less than four seconds at 500% of their ampere rating. They shall be current-limiting Class L.

Fuses 600 amps and less shall have an interrupting rating of 200 000 RMS symmetrical amperes. They shall provide a time delay of not less than 10 seconds at 500% of their ampere rating. They shall be current-limiting Class J.

All fuses shall be selected to assure selective coordination.

Spare fuses shall be provided in the amount of 20% of each size and type installed, but in no case shall fewer than three spare fuses be supplied. These spare fuses shall be delivered to the owner at the time of acceptance of the project, and shall be placed in a spare fuse cabinet mounted on the wall adjacent to or located in the switchboard.

Fuse identification labels, showing the size and type of fuses installed, shall be placed inside the cover of each switch.

Bonding and Grounding

The electrical contractor shall install and test a complete bonding and grounding system as shown in the drawings or herein specified or both.

Transformers, panelboards, metal enclosures, exposed building steel, and metallic piping should be bonded to the main ground bus. The contractor shall provide ground connections to building steel with thermite welds or 9-mm silicon bronze alloy bolts, and peen ends of bolts after installation.

Grounding metal contact points should be cleaned of paint rust and other contaminates.

Exposed grounding conductors should be protected from mechanical damage with rigid metal

the grounding system efficiency for compliance with the *CEC*. Ohmic resistance shall be verified in the presence of the inspection authority. Two copies of the test report shall be supplied to the architect.

Panelboards

The electrical contractor shall furnish panelboards as shown on the plans and detailed in the panelboard schedules. Panelboards shall be approved by the CSA. All interiors will have 225-ampere bus with lugs, rated at 75°C, for incoming conductors. Boxes will be of galvanized sheet steel and will provide wiring gutters, as required by the *CEC*. Fronts will be suitable for either flush or surface installation and shall be equipped with a keyed lock and a directory card holder. All panelboards shall have an equipment grounding bus bonded to the cabinet. A typewritten directory identifying each breaker shall be provided in each panelboard. See Drawing E7 for a listing of the required poles and overcurrent devices.

Moulded-Case Circuit Breakers

Moulded-case circuit breakers shall be installed in branch-circuit panelboards as indicated on the panelboard schedules on Drawings E6 and E7. Each breaker shall provide inverse time delay under overload conditions and magnetic tripping for short circuits. The breaker operating mechanism shall be trip-free and multipole units will open all poles if one pole trips. Breakers shall have sufficient interrupting rating to interrupt 10 000 RMS symmetrical amperes and have a temperature termination rating of 75°C.

Transfer Switch

The electrical contractor will furnish and install a complete automatic load transfer switch capable of handling 12 KVA at 120/208 volts, three-phase. The switch control will sense a loss of power on any phase and signal the motor-generator to start. When emergency power reaches 80% of voltage and frequency, the switch will automatically transfer to the generator source. When the normal power has been restored for

▶ SECTION 26 50 00 LIGHTING

Luminaires

The electrical contractor will furnish and install luminaires, as described in the schedule shown in the plans. Luminaires shall be complete with all necessary supports, hangers, diffusers, and lamps. Ballast, when required, shall be as shown on the drawings or as specified. Magnetic ballasts shall have a sound rating of A. Alternative luminaires other than those specified shall only be used with the written approval of the architect.

Luminaires shall be adequately supported.

▶ SECTION 27 00 00 COMMUNICATIONS

Time Clock

A time clock shall be installed to control the lighting in the front and rear entries. The clock will be connected to Panel EM circuit No. 1. The clock will be 120 volts, one circuit, with astronomic control and a spring-wound carry-over mechanism.

Telephone System

The contractor shall supply and install a complete system of empty conduits, terminal cabinets, pull boxes, and outlets for the telephone wiring as shown on the drawings or described herein. Minimum size conduit shall be 21 mm. Conduit runs shall have not more than two quarter bends and bending radius shall not be less than 10 times the conduit diameter.

- The contractor shall supply and install a complete data network infrastructure in the insurance office.

- All data jacks, cable, terminations, patch panels, etc., shall be Category 5.
- Data jacks shall be installed in the surface raceway at intervals not to exceed 1.2 metres. All data jacks shall have off-white cover plates. Jacks shall be a contrasting colour to the telephone jacks. Data cabling shall be in the same compartment of the surface raceway as the telephone cabling.
- A 19-inch floor-mounted data rack shall be installed in the insurance office computer room. The rack shall be fastened to the floor with steel expansion type anchors. The location of the rack is to be determined on site by the occupant. All data cables are to terminate at the rack in patch panels supplied and installed by the contractor.
- All jacks and patch panel ports are to be labelled.
- All metal parts associated with the data network, such as boxes, raceways, racks, etc., shall be bonded to ground with a minimum #6 AWG copper conductor.
- Each cable shall be tested to guarantee performance to Category 5 standard. Two complete sets of test results shall be presented to the occupant.
- The contractor shall supply and install a 27-mm raceway from the ceiling space in the computer room to the location of the telephone equipment in the basement complete with a pullstring for the installation of the cable for Internet access. This cable shall be installed by others. There shall be no more than the equivalent of two quarter bends between pull points.

Part B: Division 23 Mechanical

▶ HEATING AND AIR-CONDITIONING SPECIFICATIONS

Boiler

The heating contractor will furnish and install a 200 kW electric hot water heating boiler complete

Hot Water Circulating Pumps

The heating contractor will furnish and install five circulating pumps. Each pump will serve a separate rental area and be controlled from a thermostat located in that area as indicated on the electrical plans. Pumps will be $\frac{1}{6}$ horsepower, 120 volts, single-phase. Overload protection will be provided by a manual motor starter.

Air-Conditioning Equipment

Air-conditioning equipment will be furnished and installed by the heating contractor in four of the rental areas indicated as follows.

Drugstore. A split-system packaged unit will be installed with a rooftop compressor-condenser and a remote evaporator located in the basement. The electrical characteristics of this system are:

- Voltage: 208 volts, three-phase, 3 wire, 60 hertz
- Hermetic refrigerant compressor-motor: Rated load current 20.2 amperes, 208 volts, three-phase
- Evaporator motor: Full load current 3.2 amperes, 208 volts, single-phase
- Condenser motor: Full load current 3.2 amperes, 208 volts, single-phase
- Minimum circuit ampacity: 31.65 amperes
- Overcurrent protection: 50 amperes, time-delay fuse

Insurance Office. A single package unit will be installed on the roof with electrical data identical to that of the unit specified for the drugstore.

Beauty Salon. A single package unit will be installed on the roof and will have the following electrical characteristics:

- Voltage: 208 volts, three-phase, 3 wire, 60 hertz
- Hermetic refrigerant compressor-motor: Rated load amperes 14.1, 208 volts, three-phase
- Condenser-evaporator motor: Full load amperes 3.3 at 208 volts, single-phase
- Minimum circuit ampacity: 20.92 amperes

- Overcurrent protection: 35 amperes, time-delay fuse

Doctor's Office. A single package will be installed on the roof and will have the following electrical characteristics:

- Voltage: 208 volts, single-phase, 2 wire, 60 hertz
- Hermetic refrigerant compressor-motor: Rated load amperes 16.8 at 208 volts, single-phase
- Condenser-evaporator motor: Full load amperes 3.7 at 208 volts, single-phase
- Minimum circuit ampacity: 24.7 amperes
- Overcurrent protection: 40 amperes, time-delay fuse

Heating Control

In cooled areas, the heating and cooling will be controlled by a combination thermostat located in the proper area as shown on the electrical plans.

▶ PLUMBING SPECIFICATIONS

Only those sections that pertain to the electrical work are listed here.

Motors

All motors will be installed by the contractor furnishing the motors. All electrical power wiring to the motors and the connection of the motors shall be made by the electrical contractor. All control wiring and control devices will be the responsibility of the contractor furnishing the equipment to be controlled.

Sump Pump

The plumbing contractor will furnish and install an electric motor-driven, fully automatic sump pump. Motor will be $\frac{1}{2}$ horsepower, 120 volts, single-phase. Overload protection will be provided by a manual motor starter.

Part C: Useful Formulas

Useful Formulas.

TO FIND	SINGLE PHASE	THREE-PHASE	DIRECT CURRENT
AMPERES when kilovolt-amperes are known	$\dfrac{kVA \times 1000}{E}$	$\dfrac{kVA \times 1000}{E \times 1.732}$	Not applicable
AMPERES when horsepower is known	$\dfrac{hp \times 1000}{E \times \% \text{ eff.} \times pf}$	$\dfrac{hp \times 746}{E \times 1.732 \times \% \text{ eff.} \times pf}$	$\dfrac{hp \times 746}{E \times \% \text{ eff.}}$
AMPERES when kilowatts are known	$\dfrac{kW \times 1000}{E \times pf}$	$\dfrac{kW \times 1000}{E \times 1.732 \times pf}$	$\dfrac{kW \times 1000}{E}$
WATTS	$E \times I \times pf$	$E \times I \times 1.732 \times pf$	$E \times I$
KILOWATTS	$\dfrac{E \times I \times pf}{1000}$	$\dfrac{E \times I \times 1.732 \times pf}{1000}$	$\dfrac{E \times I}{1000}$
VOLT-AMPERES	$E \times I$	$E \times I \times 1.732$	$E \times I$
KILOVOLT-AMPERES	$\dfrac{E \times I}{1000}$	$\dfrac{E \times I \times 1.732}{1000}$	Not applicable
HORSEPOWER	$\dfrac{E \times I \times \% \text{ eff.} \times pf}{746}$	$\dfrac{E \times I \times 1.732 \times \% \text{ eff.} \times pf}{746}$	$\dfrac{E \times I \times \% \text{ eff.}}{746}$

I = amperes E = volts kW = kilowatts kVA = kilovolt-amperes
hp = horsepower % eff. = percent efficiency pf = power factor

© Cengage Learning®

FIGURE A-1   Useful formulas.

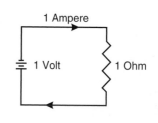

Ω = Ohm
1000 Ohms = 1 Kilohm (K)
1000 Kilohms (K) = 1 Megohm (Meg)
1 000 000 Ohms = 1 Megohm (Meg)

© Cengage Learning®

Code Index

Subject Index